Expansion in Finite Simple Groups of Lie Type

Expansion in Finite Simple Groups of Lie Type

Terence Tao

Graduate Studies
in Mathematics
Volume 164

American Mathematical Society
Providence, Rhode Island

EDITORIAL COMMITTEE
Dan Abramovich
Daniel S. Freed
Rafe Mazzeo (Chair)
Gigliola Staffilani

2010 *Mathematics Subject Classification.* Primary 05C81, 11B30, 20C33, 20D06, 20G40.

For additional information and updates on this book, visit
www.ams.org/bookpages/gsm-164

Library of Congress Cataloging-in-Publication Data
Tao, Terence, 1975
  Expansion in finite simple groups of Lie type / Terence Tao.
    pages cm. – (Graduate studies in mathematics ; volume 164)
  Includes bibliographical references and index.
  ISBN 978-1-4704-2196-0 (alk. paper)
  1. Finite simple groups. 2. Lie groups. I. Title.

QA387.T356   2015
512′.482–dc23
                                                                          2014049154

Copying and reprinting. Individual readers of this publication, and nonprofit libraries acting for them, are permitted to make fair use of the material, such as to copy select pages for use in teaching or research. Permission is granted to quote brief passages from this publication in reviews, provided the customary acknowledgment of the source is given.

Republication, systematic copying, or multiple reproduction of any material in this publication is permitted only under license from the American Mathematical Society. Permissions to reuse portions of AMS publication content are handled by Copyright Clearance Center's RightsLink® service. For more information, please visit: `http://www.ams.org/rightslink`.

Send requests for translation rights and licensed reprints to `reprint-permission@ams.org`.

Excluded from these provisions is material for which the author holds copyright. In such cases, requests for permission to reuse or reprint material should be addressed directly to the author(s). Copyright ownership is indicated on the copyright page, or on the lower right-hand corner of the first page of each article within proceedings volumes.

© 2015 by Terence Tao. All rights reserved.
Printed in the United States of America.

∞ The paper used in this book is acid-free and falls within the guidelines
established to ensure permanence and durability.
Visit the AMS home page at `http://www.ams.org/`
10 9 8 7 6 5 4 3 2 1   20 19 18 17 16 15

In memory of Garth Gaudry, who set me on the road

Contents

Preface	xi
Notation	xii
Acknowledgments	xiii

Part 1. Expansion in Cayley Graphs

Chapter 1. Expander graphs: Basic theory	3
§1.1. Expander graphs	4
§1.2. Connection with edge expansion	9
§1.3. Random walks on expanders	15
§1.4. Random graphs as expanders	17
Chapter 2. Expansion in Cayley graphs, and Kazhdan's property (T)	23
§2.1. Kazhdan's property (T)	27
§2.2. Induced representations and property (T)	37
§2.3. The special linear group and property (T)	47
§2.4. A more elementary approach	55
Chapter 3. Quasirandom groups	57
§3.1. Mixing in quasirandom groups	62
§3.2. An algebraic description of quasirandomness	67
§3.3. A weak form of Selberg's 3/16 theorem	67
Chapter 4. The Balog-Szemerédi-Gowers lemma, and the Bourgain-Gamburd expansion machine	85
§4.1. The Balog-Szemerédi-Gowers lemma	87

§4.2.	The Bourgain-Gamburd expansion machine	97
Chapter 5.	Product theorems, pivot arguments, and the Larsen-Pink nonconcentration inequality	101
§5.1.	The sum-product theorem	104
§5.2.	Finite subgroups of SL_2	110
§5.3.	The product theorem in $SL_2(k)$	120
§5.4.	The product theorem in $SL_d(k)$	125
§5.5.	Proof of the Larsen-Pink inequality	129
Chapter 6.	Nonconcentration in subgroups	135
§6.1.	Expansion in thin subgroups	137
§6.2.	Random generators expand	140
Chapter 7.	Sieving and expanders	143
§7.1.	Combinatorial sieving	146
§7.2.	The strong approximation property	156
§7.3.	Sieving in thin groups	160

Part 2. Related Articles

Chapter 8.	Cayley graphs and the algebra of groups	167
§8.1.	A Hall-Witt identity for 2-cocycles	177
Chapter 9.	The Lang-Weil bound	187
§9.1.	The Stepanov-Bombieri proof of the Hasse-Weil bound	194
§9.2.	The proof of the Lang-Weil bound	198
§9.3.	Lang-Weil with parameters	200
Chapter 10.	The spectral theorem and its converses for unbounded self-adjoint operators	203
§10.1.	Self-adjointness and resolvents	207
§10.2.	Self-adjointness and spectral measure	212
§10.3.	Self-adjointness and flows	218
§10.4.	Essential self-adjointness of the Laplace-Beltrami operator	224
Chapter 11.	Notes on Lie algebras	227
§11.1.	Abelian representations	233
§11.2.	Engel's theorem and Lie's theorem	235
§11.3.	Characterising semisimplicity	237
§11.4.	Cartan subalgebras	242

§11.5.	$\mathfrak{sl}_2$ representations	245
§11.6.	Root spaces	247
§11.7.	Classification of root systems	251
§11.8.	Chevalley bases	258
§11.9.	Casimirs and complete reducibility	263

Chapter 12.	Notes on groups of Lie type	267
§12.1.	Simple Lie groups over $\mathbf{C}$	268
§12.2.	Chevalley groups	278
§12.3.	Finite simple groups of Lie type	288

Bibliography	293
Index	301

Preface

Expander graphs are a remarkable type of graph (or more precisely, a family of graphs) on finite sets of vertices that manage to simultaneously be both sparse (low-degree) and "highly connected" at the same time. They enjoy very strong mixing properties: if one starts at a fixed vertex of an (two-sided) expander graph and randomly traverses its edges, then the distribution of one's location will converge exponentially fast to the uniform distribution. For this and many other reasons, expander graphs are useful in a wide variety of areas of both pure and applied mathematics.

There are now many ways to construct expander graphs, but one of the earliest constructions was based on the *Cayley graphs* of a finite group (or of a finitely generated group acting on a finite set). The expansion property for such graphs turns out to be related to a rich variety of topics in group theory and representation theory, including Kazhdan's property (T), Gowers' notion of a quasirandom group, the sum-product phenomenon in arithmetic combinatorics, and the Larsen-Pink classification of finite subgroups of a linear group. Expansion properties of Cayley graphs have also been applied in analytic number theory through what is now known as the *affine sieve* of Bourgain, Gamburd, and Sarnak, which can count almost prime points in thin groups.

This text is based on the lecture notes from a graduate course on these topics I gave at UCLA in the winter of 2012, as well as from some additional posts on my blog at `terrytao.wordpress.com` on further related topics. The first part of this text can thus serve as the basis for a one-quarter or one-semester advanced graduate course, depending on how much of the optional material one wishes to cover. While the material here is largely self-contained, some basic graduate real analysis (in particular, measure theory, Hilbert space theory, and the theory of L^p norms), graph theory, and linear algebra (e.g., the spectral theorem for unitary matrices) will be assumed. Some prior familiarity with the classical Lie groups (particularly the special linear group SL_n and the unitary group U_n) and representation theory will be helpful but not absolutely necessary. To follow Section 3.3 (which is optional) some prior exposure to Riemannian geometry would also be useful.

The core of the text is Part 1. After discussing the general theory of expander graphs in the first section, we then specialise to the case of Cayley graphs, starting with the remarkable observation[1] of Margulis linking Kazhdan's property (T) with expansion, and then turning to the more recent observations of Sarnak, Xue, Gamburd, and Bourgain linking the property of finite groups now known as *quasirandomness* with expansion, which is also related to the famous "3/16 theorem" of Selberg. As we will present in this text, this sets up a general "machine" introduced by Bourgain and Gamburd for verifying expansion in a Cayley graph, which in addition to quasirandomness requires two additional ingredients, namely a product theorem and a nonconcentration estimate. These two ingredients are then the focus of the next two sections of this part. The former ingredient uses techniques from arithmetic combinatorics related to the sum-product theorem, as well as estimates of Larsen and Pink on controlling the interaction between finite subgroups of a linear group and various algebraic varieties (such as conjugacy classes or maximal tori). The latter ingredient is perhaps the most delicate aspect of the theory, and often requires a detailed knowledge of the algebraic (and geometric) structure of the ambient group. Finally, we present an application of these ideas to number theory by introducing the basics of sieve theory, and showing how expansion results may be inserted into standard sieves to give new bounds on almost primes in thin groups.

Part 2 contains a variety of additional material that is related to one or more of the topics covered in Part 1, but which can be omitted for the purposes of teaching a graduate course on the subject.

Notation

For reasons of space, we will not be able to define every single mathematical term that we use in this book. If a term is italicised for reasons other than emphasis or for definition, then it denotes a standard mathematical object, result, or concept, which can be easily looked up in any number of references.

---

[1]This material in Section 2 is not absolutely required for subsequent sections of this part, although it does provide some helpful context for these later sections. Thus, this section may be abridged or even omitted altogether in a lecture course if desired.

(In the blog version of the book, many of these terms were linked to their Wikipedia pages, or other on-line reference pages.)

Given a subset E of a space X, the *indicator function* $1_E : X \to \mathbf{R}$ is defined by setting $1_E(x)$ equal to 1 for $x \in E$ and equal to 0 for $x \notin E$.

The cardinality of a finite set E will be denoted $|E|$. We will use[2] the asymptotic notation $X = O(Y)$, $X \ll Y$, or $Y \gg X$ denote the estimate $|X| \leq CY$ for some absolute constant $C > 0$. In some cases we will need this constant C to depend on a parameter (e.g., d), in which case we shall indicate this dependence by subscripts, e.g., $X = O_d(Y)$ or $X \ll_d Y$. We also sometimes use $X \sim Y$ as a synonym for $X \ll Y \ll X$. If n is a parameter going to infinity, we let $o_{n \to \infty}(1)$ denote a quantity depending on n and bounded in magnitude by $c(n)$ for some quantity $c(n)$ that goes to zero as $n \to \infty$. More generally, given an additional parameter such as k, we let $o_{n \to \infty; k}(1)$ denote a quantity that may depend on both k and n, which is bounded by $c_k(n)$ for some quantity $c_k(n)$ that goes to zero as $n \to \infty$ for each fixed k.

Acknowledgments

I am greatly indebted to my students of the course on which this text was based, as well as many further commenters on my blog, including Ian Agol, Abhishek Bhowmick, Nick Cook, Sean Eberhard, Alireza Golsefidy, Joerg Grande, Ben Green, Dick Gross, Harald Helfgott, Arie Israel, David Joyner, Matthew Kahle, Emmanuel Kowalski, Wolfgang Moens, Vipul Niak, William Orrick, Mikhail Ostrovskii, David Roberts, Misha Rudnev, Alexander Shaposhnikov, Lior Silberman, Vit Tucek, Yilong Yang, and Wei Zhou. These comments can be viewed online at:

terrytao.wordpress.com/category/teaching/254b-expansion-in-groups/

The author was supported by the NSF grant DMS-0649473, the James and Carol Collins Chair, the Mathematical Analysis & Application Research Fund Endowment, and the Simons Foundation.

---

[2] Once we deploy the machinery of nonstandard analysis in Section 6, we will use a closely related, but slightly different, asymptotic notation.

Part 1

Expansion in Cayley Graphs

Chapter 1

Expander graphs: Basic theory

The objective of this text is to present a number of recent constructions of *expander graphs*, which are a type of sparse but "pseudorandom" graph of importance in computer science, the theory of random walks, geometric group theory, and in number theory. The subject of expander graphs and their applications is an immense one, and we will not possibly be able to cover it in full here. For instance, we will say almost nothing about the important applications of expander graphs to computer science, for instance in constructing good pseudorandom number generators, derandomising a probabilistic algorithm, constructing error correcting codes, or in building probabilistically checkable proofs. For such topics, see [**HoLiWi2006**].

Instead of focusing on applications, this text will concern itself much more with the task of *constructing* expander graphs. This is a surprisingly nontrivial problem. On one hand, we shall see that an easy application of the *probabilistic method* shows that a randomly chosen (large, regular, bounded-degree) graph will be an expander graph with very high probability, so expander graphs are extremely abundant. On the other hand, in many applications, one wants an expander graph that is more deterministic in nature (requiring either no or very few random choices to build), and of a more specialised form. For the applications to number theory or geometric group theory, it is of particular interest to determine the expansion properties of a very symmetric type of graph, namely a *Cayley graph*; we will also occasionally work with the more general concept of a *Schreier graph*. It turns out that such questions are related to deep properties of various groups G of Lie type (such as $SL_2(\mathbf{R})$ or $SL_2(\mathbf{Z})$), such as *Kazhdan's property (T)*, the first nontrivial eigenvalue of a Laplacian on a symmetric space G/Γ associated to G, the quasirandomness of G (as measured by the size of irreducible representations), and the product theory of subsets of G. These properties are of intrinsic interest to many other fields of mathematics (e.g., ergodic theory, operator algebras, additive combinatorics, representation theory, finite group theory, number theory, etc.), and it is quite remarkable[1] that a single problem — namely the construction of expander graphs — is so deeply connected with such a rich and diverse array of mathematical topics.

There are also other important constructions of expander graphs that are not related to Cayley or Schreier graphs, such as those graphs constructed by the *zigzag product construction*, but we will not discuss those types of graphs here; again, the reader is referred to [**HoLiWi2006**].

1.1. Expander graphs

We begin by defining formally the concept of an expander graph. As with many fundamentally important concepts in mathematics, there are a number of equivalent definitions of this concept. We will adopt a "spectral" perspective towards expander graphs, defining them in terms of a certain spectral gap, but will relate this formulation of expansion to the more classical notion of edge expansion later in this chapter.

We begin by recalling the notion of a graph. To avoid some very minor technical issues, we will work with undirected, loop-free, multiplicity-free graphs (though later, when we discuss Cayley graphs, we will allow loops and repetition).

Definition 1.1.1. A *graph* is a pair $G = (V, E)$, where V is a set (called the *vertex set* of G), and $E \subset \binom{V}{2}$ is a collection of unordered pairs $\{v, w\}$ of distinct elements v, w of V, known as the *edge set* of E. Elements of V or E are called *vertices* and *edges* of E. A graph is *finite* if the vertex set (and hence the edge set) is finite. If $k \geq 0$ is a natural number, we say that a graph $G = (V, E)$ is *k-regular* if each vertex of V is contained in exactly k edges in E; we refer to k as the *degree* of the regular graph G.

Example 1.1.2. The *complete graph* $(V, \binom{V}{2})$ on a vertex set V has edge set $\binom{V}{2} := \{\{v, w\} : v, w \in V, v \neq w\}$. If V has n elements, the complete graph is $n - 1$-regular.

---

[1] Perhaps this is because so many of these fields are all grappling with aspects of a single general problem in mathematics, namely when to determine whether a given mathematical object or process of interest "behaves pseudorandomly" or not, and how this is connected with the symmetry group of that object or process.

1.1. Expander graphs

In this course, we will mostly be interested in *constant-degree* large finite regular graphs, in which k is fixed (e.g., $k=4$), and the number $n = |V|$ of vertices is going off to infinity.

Given a finite graph $G = (V, E)$, we let $\ell^2(V)$ be the finite-dimensional complex Hilbert space of functions $f: V \to \mathbf{C}$ with norm

$$\|f\|_{\ell^2(V)} := (\sum_{v \in V} |f(v)|^2)^{1/2}$$

and inner product

$$\langle f, g \rangle_{\ell^2(V)} := \sum_{v \in V} f(v)\overline{g(v)}.$$

We then can define the *adjacency operator* $A: \ell^2(V) \to \ell^2(V)$ on functions $f \in \ell^2(V)$ by the formula

$$Af(v) := \sum_{w \in V: \{v,w\} \in E} f(w),$$

thus $Af(v)$ is the sum of f over all of the neighbours of v; this is of course a linear operator. If one enumerates the vertices V as $v_1, \ldots, v_n$ in some fashion, then one can associate A with an $n \times n$ matrix, known as the *adjacency matrix* of G (with this choice of vertex enumeration).

As our graphs are undirected, the adjacency operator A is clearly self-adjoint (and the adjacency matrix is real symmetric). By the spectral theorem, A thus has (counting multiplicity) n real eigenvalues

$$\lambda_1 \geq \ldots \geq \lambda_n.$$

We will write λ_i as $\lambda_i(G)$ whenever we need to emphasise the dependence of the eigenvalues on the graph G.

The largest eigenvalue λ_1 is easily understood for k-regular graphs:

Lemma 1.1.3. *If G is a k-regular graph, then*

$$k = \lambda_1 \geq \lambda_n \geq -k.$$

Proof. Clearly $A\mathbf{1} = k\mathbf{1}$ (where we write $\mathbf{1} \in \ell^2(V)$ for the constant function $v \mapsto 1$), and so k is an eigenvalue of A with eigenvector $\mathbf{1}$. On the other hand, for any $f, g \in \ell^2(V)$ with norm one, one has

$$|\langle Af, g \rangle_{\ell^2(V)}| = \left| \sum_{v,w \in V : \{v,w\} \in E} f(w)\overline{g(v)} \right|$$

$$\leq \frac{1}{2} \sum_{v,w \in V : \{v,w\} \in E} |f(w)|^2 + |g(v)|^2$$

$$\leq \frac{1}{2} k \sum_{w \in V} |f(w)|^2 + \frac{1}{2} k \sum_{v \in V} |g(v)|^2$$

$$= k,$$

and so A has operator norm k (or equivalently, all eigenvalues of A lie between $-k$ and k). The claim follows. □

Now we turn to the next eigenvalue after λ_1.

Definition 1.1.4 (Expander family). Let $\varepsilon > 0$ and $k \geq 1$. A finite k-regular graph is said to be a (one-sided) *ε-expander* if one has

$$\lambda_2 \leq (1 - \varepsilon)k$$

and a *two-sided ε-expander* if one also has

$$\lambda_n \geq -(1 - \varepsilon)k.$$

A sequence $G_i = (V_i, E_i)$ of finite k-regular graphs is said to be a one-sided (resp. two-sided) *expander family* if there is an $\varepsilon > 0$ such that G_i is a one-sided (resp. two-sided) ε-expander for all sufficiently large i.

Remark 1.1.5. The operator $\Delta := 1 - \frac{1}{k} A$ is sometimes known as[2] the *graph Laplacian*. This is a positive semidefinite operator with at least one zero eigenvalue (corresponding to the eigenvector 1). A graph is an ε-expander if and only if there is a *spectral gap* of size ε in Δ, in the sense that the first eigenvalue of Δ exceeds the second by at least ε. The graph Laplacian is analogous to the classical Laplacian in Euclidean space (or the Laplace-Beltrami operator on Riemannian manifolds); see Section 3.3 for a formalisation of this analogy.

The original definition of expander graphs focused on one-sided expanders, but it will be slightly more natural in this text to focus on the two-sided expanders. (But for Cayley graphs the two notions are almost equivalent; see Exercise 5.0.5.)

Strictly speaking, we have not defined the notion of a (one- or two-sided) expander graph in the above definition; we have defined an ε-expander graph

---

[2]It is also common to use the normalisation $k - A$ instead of $1 - \frac{1}{k} A$ in many texts, particularly if one wishes to generalise to graphs that are not perfectly regular.

1.1. Expander graphs

for any given parameter $\varepsilon > 0$, and have defined the notion of an expander family, which is a *sequence* of graphs rather than for an individual graph. One could propose defining an expander graph to be a graph that is a one or two-sided ε-expander for some $\varepsilon > 0$ (or equivalently, a graph G such that the constant sequence $G, G, \ldots$ is an expander family), but this definition collapses to existing concepts in graph theory:

Exercise 1.1.1 (Qualitative expansion). Let $k \geq 1$, and let $G = (V, E)$ be a finite k-regular graph.

(i) Show that $\lambda_2 = k$ if and only if G is not *connected*.

(ii) Show that $\lambda_n = -k$ if and only if G contains a nonempty *bipartite graph* as a connected component.

Thus, a graph is a one-sided expander for some $\varepsilon > 0$ if and only if it is connected, and a two-sided expander for some $\varepsilon > 0$ if and only if it is connected and not bipartite.

To obtain a more interesting theory, it is therefore necessary to either keep a more quantitative track of the ε parameter, or[3] work with expander families (typically involving vertex sets whose cardinality goes to infinity) rather than with individual graphs. Nevertheless, we will often informally drop the ε parameter (or the use of families) and informally refer simply to "expander graphs" in our discussion.

By taking the trace of the adjacency matrix or its square, one obtains some basic identities concerning the eigenvalues of a k-regular graph:

Exercise 1.1.2 (Trace formulae). Let G be a k-regular graph on n vertices for some $n > k \geq 1$.

(i) Show that $\sum_{i=1}^{n} \lambda_i = 0$.

(ii) Show that $\sum_{i=1}^{n} \lambda_i^2 = nk$.

(iii) Show that $\max(|\lambda_2|, |\lambda_n|) \geq \sqrt{k} - o_{n \to \infty; k}(1)$, where $o_{n \to \infty; k}(1)$ denotes a quantity that goes to zero as $n \to \infty$ for fixed k.

Remark 1.1.6. The above exercise places an upper bound on how strong of a two-sided expansion one can obtain for a large k-regular graph. It is not quite sharp; it turns out that one can obtain the improvement

$$\max(|\lambda_2|, |\lambda_n|) \geq 2\sqrt{k-1} - o_{n \to \infty; k}(1),$$

a result of Alon and Boppana; see [**HoLiWi2006**] for a proof. Graphs with $\max(|\lambda_2|, |\lambda_n|) \leq 2\sqrt{k-1}$ are known as *Ramanujan graphs*, and (as the

---
[3]Alternatively, one could adopt a *nonstandard analysis* viewpoint and work with *ultra expander graphs* — i.e., ultraproducts of expander families — but we will postpone using this sort of viewpoint until Section 5.5.

name suggests) have connections to number theory, but we will not discuss this topic here; see for instance [**DaSaVa2003**] for more discussion.

We will give a probabilistic construction of an expander family later, but let us first give an example of a family of regular graphs that is *not* an expander family.

Exercise 1.1.3. For each $n \geq 3$, let G_n be the 2-regular graph whose vertex set is the cyclic group $\mathbf{Z}/n\mathbf{Z}$, and whose edge set is the set of pairs $\{x, x+1\}$ for $x \in \mathbf{Z}/n\mathbf{Z}$. (This is a basic example of a *Cayley graph*; such graphs will be discussed in more depth in Chapter 2.)

  (i) Show that the eigenvalues of the adjacency operator A_n associated to G_n are $2\cos(2\pi j/n)$ for $j = 0, \ldots, n-1$. (*Hint:* You may find the *discrete Fourier transform* to be helpful.)

  (ii) Show that G_n is not a one-sided expander family (and is thus not a two-sided expander family either). This is despite G_n always being connected (and nonbipartite for n odd).

The next exercise shows that the complete graph (Exercise 1.1.2) is an excellent expander; the whole point, though, of expander graph constructions is to come up with much sparser graphs that still have many of the connectivity and expansion properties of the complete graph.

Exercise 1.1.4. Let G be the complete graph on n vertices, which is of course an $n-1$-regular graph. Show that

$$\lambda_2 = \ldots = \lambda_n = -1$$

and so G is a one-sided $1 + \frac{1}{n-1}$-expander and a two-sided $1 - \frac{1}{n-1}$-expander. (This is not a counterexample to Exercise 1.1.2(iii), because the error term $o_{n \to \infty; k}(1)$ is only negligible in the regime when k is either fixed or is a very slowly growing function of n, and this is definitely not the case for the complete graph.)

Exercise 1.1.5. Let $G = (V, E)$ be a k-regular graph on n vertices. Let $G^c = (V, \binom{V}{2} \setminus E)$ be the *complement graph*, consisting of all the edges connecting two vertices in V that are not in E, thus G^c is an $n-k-1$-regular graph. Show that

$$\lambda_i(G^c) = -1 - \lambda_{n+2-i}(G)$$

for all $2 \leq i \leq n$.

Exercise 1.1.6. Let $n \geq 2$ be an even number, and let $G = K_{n/2, n/2}$ be the complete bipartite graph between two sets of $n/2$ vertices each, thus G is an $n/2$-regular graph. Show that $\lambda_n = -n/2$ and $\lambda_2 = \ldots = \lambda_{n-1} = 0$. Thus, G is a one-sided 1-expander, but is not a two-sided expander at all.

Exercise 1.1.7 (Expansion and Poincaré inequality). If $G = (V, E)$ is a k-regular graph and $f \colon V \to \mathbf{C}$ is a function, define the gradient magnitude $|\nabla f| \colon V \to \mathbf{C}$ by the formula

$$|\nabla f(v)| := \left(\sum_{w \in V : \{v,w\} \in E} |f(w) - f(v)|^2 \right)^{1/2}.$$

Show that G is a one-sided ε-expander if and only if one has the *Poincaré inequality*

$$\|\nabla f\|_{\ell^2(V)}^2 \geq 2k\varepsilon \|f\|_{\ell^2(V)}^2$$

whenever f has mean zero.

Exercise 1.1.8 (Connection between one-sided and two-sided expansion). Let $G = (V, E)$ be a k-regular graph, let $\varepsilon > 0$, and let $G' = (V', E')$ be the bipartite version of G in which $V' := V \times \{1, 2\}$ and $E' := \{\{(v,1), (w,2)\} : \{v,w\} \in E\}$. Show that G is a two-sided ε-expander if and only if G' is a one-sided ε-expander. Based on this observation, introduce a notion of two-sided expansion for *directed* graphs, related to the *singular values* of the adjacency matrix, and connect it to one-sided expansion of the (undirected) bipartite version of this graph.

1.2. Connection with edge expansion

The intuition to explain Exercise 1.1.3 should be that while G_n is, strictly speaking, connected, it is not very *strongly* connected; the paths connecting a typical pair of points are quite long (comparable to n, the number of vertices) and it is easy to disconnect the graph into two large pieces simply by removing a handful of edges.

We now make this intuition more precise. Given two subsets F_1, F_2 of the vertex set V in a graph $G = (V, E)$, define $E(F_1, F_2) \subset F_1 \times F_2$ to be the set of all pairs $(v_1, v_2) \in F_1 \times F_2$ such that $\{v_1, v_2\} \in E$. Note that the cardinality of this set can be expressed in terms of the adjacency operator as

$$|E(F_1, F_2)| = \langle A 1_{F_1}, 1_{F_2} \rangle_{\ell^2(V)}.$$

Define the *boundary* ∂F of a subset F of V to be the set $\partial F := E(F, V \setminus F)$, thus ∂F is essentially the set of all edges that connect an element of F to an element outside of F. We define the *edge expansion ratio* $h(G)$ of the graph G to be given by the formula

$$h(G) := \min_{F \subset V : |F| \leq |V|/2} \frac{|\partial F|}{|F|},$$

where F ranges over all subsets of V of cardinality at most[4] $|F| \leq |V|/2$. The quantity $h(G)$ can be interpreted as a type of isoperimetric constant for G, analogous to the *Cheeger constant* [**Ch1970**] of a compact Riemannian manifold, and so $h(G)$ is sometimes known as the *Cheeger constant* of the graph G.

Note that $h(G)$ is nonzero precisely when G is connected. (If G is disconnected, at least one of the components F will have cardinality less than $|V|/2$.) We have an analogous statement for one-sided expansion:

Proposition 1.2.1 (Weak discrete Cheeger inequality). *Let $k \geq 1$, and let G_n be a family of finite k-regular graphs. Then the following are equivalent:*

   (i) *G_n form a one-sided expander family.*

   (ii) *There exists $c > 0$ such that $h(G_n) \geq c$ for all sufficiently large n.*

Proof. Let n be a large number. We abbreviate G_n as $G = (V, E)$.

We first establish the easy direction of this proposition, namely that (i) implies (ii). If n is large enough, then from the hypothesis (i) we have $\lambda_2 \leq (1 - \varepsilon)k$ for some $\varepsilon > 0$ independent of n.

Let F be a subset of V with $|F| \leq |V|/2$. We consider the quantity

$$\langle A 1_F, 1_F \rangle_{\ell^2(V)}. \tag{1.1}$$

We split 1_F into a multiple $\frac{|F|}{|V|} 1$ of the first eigenvector 1, plus the remainder $1_F - \frac{|F|}{|V|} 1$, Using the spectral decomposition of A, we can upper bound (1.1) by

$$k \left\| \frac{|F|}{|V|} 1 \right\|^2_{\ell^2(V)} + (1 - \varepsilon) k \left\| 1_F - \frac{|F|}{|V|} 1 \right\|^2_{\ell^2(V)}$$

which after a brief calculation evaluates to

$$(1 - \varepsilon) k |F| + \varepsilon k \frac{|F|^2}{|V|} \leq (1 - \varepsilon/2) k |F|.$$

On the other hand, (1.1) is also equal to the number of (ordered) pairs of adjacent vertices $v, w \in F$. Since each $v \in F$ is adjacent to exactly k vertices, we conclude that there are at least $\varepsilon k |F|/2$ pairs v, w such that $v \in F$ and $w \notin F$. Thus

$$|\partial F| \geq \varepsilon |F|/2,$$

and so $h(G_n) \geq \varepsilon/2$, and the claim (ii) follows.

Now we establish the harder direction, in which we assume (ii) and prove (i). Thus we may assume that $h(G) \geq c$ for some $c > 0$ independent of n.

---

[4]Some upper bound on F is needed to avoid this quantity from degenerating, since ∂F becomes empty when $F = V$.

1.2. Connection with edge expansion

The difficulty here is basically that the hypothesis (ii) only controls the action $A1_F$ of A on indicator functions 1_F, whereas the conclusion (ii) basically requires us to understand Af for arbitrary functions $f \in \ell^2(V)$. Indeed, from the spectral decomposition one has

$$\lambda_2 = \sup_{f: \|f\|_{\ell^2(V)}=1; \langle f,1\rangle_{\ell^2(V)}=0} \langle Af, f\rangle_{\ell^2(V)},$$

so it suffices to show that

(1.2) $$\langle Af, f\rangle_{\ell^2(V)} \leq (1-\varepsilon)k,$$

whenever $f \in \ell^2(V)$ has norm one and mean zero, and $\varepsilon > 0$ is independent of n. Since A has real matrix coefficients, we may assume without loss of generality that f is real.

The mean zero hypothesis is needed to keep the function f away from 1, but it forces f to change sign. It will be more convenient to first establish a variant of (1.2), namely that

(1.3) $$\langle Af, f\rangle_{\ell^2(V)} \leq (1-c)k\|f\|^2_{\ell^2(V)}$$

whenever f is nonnegative and supported on a set of cardinality at most $|V|/2$.

Let us assume (1.3) for now and see why it implies (1.2). Let $f \in \ell^2(V)$ have norm one and mean zero. We split $f = f_+ - f_-$ into positive and negative parts, where f_+, f_- are nonnegative with disjoint supports. Observe that $\langle Af_+, f_-\rangle_{\ell^2(V)} = \langle Af_-, f_+\rangle_{\ell^2(V)}$ is positive, and so

$$\langle Af, f\rangle_{\ell^2(V)} \leq \langle Af_+, f_+\rangle_{\ell^2(V)} + \langle Af_-, f_-\rangle_{\ell^2(V)}.$$

Also we have

$$1 = \|f_+\|^2_{\ell^2(V)} + \|f_-\|^2_{\ell^2(V)}.$$

At least one of f_+ and f_- is supported on a set of size at most $|V|/2$. By symmetry we may assume that f_- has this small support. Let $\sigma > 0$ be a small quantity (depending on c) to be chosen later. If f_- has $\ell^2(V)$ norm at least σ, then applying (1.3) to f_- (and the trivial bound $\langle Af, f\rangle_{\ell^2(V)} \leq k\|f\|^2_{\ell^2(V)}$ for f_+) we have

$$\langle Af, f\rangle_{\ell^2(V)} \leq (1-c\sigma^2)k,$$

which would suffice. So we may assume that f_- has norm at most σ. By Cauchy-Schwarz, this implies that $\sum_{x \in V} f_-(x) \leq \sigma|V|^{1/2}$, and thus (as f has mean zero) $\sum_{x \in V} f_+(x) \leq \sigma|V|^{1/2}$. Ordering the values $f_+(x)$ and applying Markov's inequality, we see that we can split f_+ as the sum of a function f'_+ supported on a set of size at most $|V|/2$, plus an error of ℓ^2

norm $O(\sigma)$. Applying (1.3) to f'_+ and f_- and using the triangle inequality (and Cauchy-Schwarz) to deal with the error term, we see that
$$\langle Af, f\rangle_{\ell^2(V)} \leq (1 - c + O(\sigma))k,$$
which also suffices (if σ is sufficiently small).

It remains to prove (1.3). We use the "wedding cake" decomposition, writing f as an integral,
$$f = \int_0^\infty 1_{F_t}\, dt,$$
where $F_t := \{x \in V : |f(x)| > t\}$. By construction, all of the F_t have cardinality at most $|V|/2$ and are nonincreasing in t. Also, a computation of the ℓ^2 norm shows that
$$\|f\|_{\ell^2(V)}^2 = \int_0^\infty 2t|F_t|\, dt. \tag{1.4}$$

Expanding $\langle Af, f\rangle_{\ell^2(V)}$ and using symmetry, we obtain
$$2\int_0^\infty \int_0^t \langle A1_{F_s}, 1_{F_t}\rangle\, ds dt.$$

We can bound the integrand in two ways. First, since $A1_{F_s}$ is bounded by k, one has
$$\langle A1_{F_s}, 1_{F_t}\rangle \leq k|F_t|.$$
Second, we may bound $\langle A1_{F_s}, 1_{F_t}\rangle$ by $\langle A1_{F_s}, 1_{F_s}\rangle$.

On the other hand, from the hypothesis $h(G) \geq c$ we see that $|\partial F_s| \geq c|F_s|$, and hence
$$\langle A1_{F_s}, 1_{F_t}\rangle \leq (k - c)|F_s|.$$
We insert the first bound for $s \leq (1-\varepsilon)t$ and the second bound for $(1-\varepsilon)t < s \leq t$, for some $\varepsilon > 0$ to be determined later, and conclude that
$$\langle Af, f\rangle_{\ell^2(V)} \leq 2\int_0^\infty k(1-\varepsilon)t|F_t|\, dt + 2\int_0^\infty \int_{(1-\varepsilon)t}^t (k-c)|F_s|\, ds dt.$$
Interchanging the integrals in the second integral, we conclude that
$$\langle Af, f\rangle_{\ell^2(V)} \leq 2\int_0^\infty k(1-\varepsilon)t|F_t|\, dt + 2\int_0^\infty (k-c)\frac{\varepsilon}{1-\varepsilon}s|F_s|\, ds.$$
For ε small enough, one can check that $k(1-\varepsilon) + (k-c)\frac{\varepsilon}{1-\varepsilon} < k(1-\varepsilon')$ for some $\varepsilon' > 0$ depending only on ε, k, c, and the claim (1.3) then follows from (1.4). $\square$

Example 1.2.2. The graphs in Exercise 1.1.3 contain large sets with small boundary (e.g., $\{1,\ldots,m\} \bmod n$ for $1 \leq m \leq n/2$), which gives a nonspectral way to establish that they do not form an expander family.

1.2. Connection with edge expansion

Exercise 1.2.1. Show that if $k \leq 2$, then the only expander families of k-regular graphs are those families of bounded size (i.e., the vertex sets V_n have cardinality bounded in n).

Remark 1.2.3. There is a more precise relationship between the edge expansion ratio $h(G)$ and the best constant ε that makes G a one-sided ε-expander, namely the *discrete Cheeger inequality*

$$\frac{\varepsilon}{2} k \leq h(G) \leq \sqrt{2\varepsilon} k, \tag{1.5}$$

first proven in [**Do1984**] and [**AlMi1985**] (see also [**Al1996**]), based on the continuous isoperimetric inequalities in [**Ch1970**], [**Bu1982**]. The first inequality in (1.5) is already implicit in the proof of the above lemma, but the second inequality is more difficult to establish. However, we will not use this more precise inequality here.

There is an analogous criterion for two-sided expansion, but it is more complicated to state. Here is one formulation that is quite useful:

Exercise 1.2.2 (Expander mixing lemma). Let $G(V, E)$ be a k-regular graph on n vertices which is a two-sided ε-expander. Show that for any subsets F_1, F_2 of V, one has

$$\left| |E(F_1, F_2)| - \frac{k}{n} |F_1||F_2| \right| \leq (1-\varepsilon) k \sqrt{|F_1||F_2|}.$$

(Actually, the factors $|F_1|, |F_2|$ on the right-hand side can be refined slightly to $|F_1| - \frac{|F_1|^2}{n}$ and $|F_2| - \frac{|F_2|^2}{n}$ respectively.)

Thus, two-sided expanders behave analogously to the pseudorandom graphs that appear in the *Szemerédi regularity lemma* [**Sz1978**] (but with the caveat that expanders are usually sparse graphs, whereas pseudorandom graphs are usually dense).

Here is a variant of the above lemma that more closely resembles Proposition 1.2.1.

Exercise 1.2.3. Let $k \geq 1$, and let G_n be a family of finite k-regular graphs. Show that the following are equivalent:
  (i) G_n form a two-sided expander family.
  (ii) There exists $c > 0$ such that whenever n is sufficiently large and F_1, F_2 are subsets of V_n of cardinality at most $|V_n|/2$, then
  $$|E(F_1, F_2)| \leq (k-c)\sqrt{|F_1||F_2|}.$$

The exercises below connect expansion to some other graph-theoretic properties. On a connected graph G, one can define the *graph metric* $d \colon G \times$

$G \to \mathbf{R}^+$ by defining $d(v, w)$ to be the length of the shortest path from v to w using only edges of G. This is easily seen to be a metric on G.

Exercise 1.2.4 (Expanders have low diameter). Let G be a k-regular graph on n vertices that is a one-sided ε-expander for some $n > k \geq 1$ and $\varepsilon > 0$. Show that there is a constant $c > 0$ depending only on k and ε such that for every vertex $v \in V$ and any radius $r \geq 0$, the ball $B(v, r) := \{w \in V : d(v, w) \leq r\}$ has cardinality

$$|B(v, r)| \geq \min((1 + c)^r, n).$$

(*Hint:* First establish the weaker bound $|B(v, r)| \geq \min((1 + c)^r, n/2)$.) In particular, G has diameter $O(\log n)$, where the implied constant can depend on k and ε.

Exercise 1.2.5 (Expanders have high connectivity). Let G be a k-regular graph on n vertices that is a one-sided ε-expander for some $n > k \geq 1$ and $\varepsilon > 0$. Show that if one removes m edges from G for some $m \geq 0$, then the resulting graph has a connected component of size at least $n - Cm$, where C depends only on k and ε.

Exercise 1.2.6 (Expanders have high chromatic number). Let G be a k-regular graph on n vertices that is a two-sided ε-expander for some $n > k \geq 1$ and $\varepsilon > 0$.

(i) Show that any *independent set*[5] in G has cardinality at most $(1 - \varepsilon)n$.

(ii) Show that the *chromatic number*[6] of G is at least $\frac{1}{1-\varepsilon}$. (Of course, this bound only becomes nontrivial for ε close to 1; however, it is still useful for constructing bounded-degree graphs of high chromatic number and large *girth*[7].)

Exercise 1.2.7 (Expansion and concentration of measure). Let G be a k-regular graph on n vertices that is a one-sided ε-expander for some $n > k \geq 1$ and $\varepsilon > 0$. Let $f \colon G \to \mathbf{R}$ be a function which is Lipschitz with some Lipschitz constant K, thus $|f(v) - f(w)| \leq K d(v, w)$ for all $v, w \in V$. Let M be a median value of f (thus $f(v) \geq M$ for at least half of the vertices

---

[5] A set of vertices in a graph G is *independent* if there are no edges in G that connect two elements in this set.

[6] The *chromatic number* of a graph G is the fewest number of colours needed to colour the vertices of the graph in such a way that no two vertices of the same colour are connected by an edge.

[7] The *girth* of a graph is the length of the shortest cycle in the graph, or infinity if the graph does not contain any cycles.

v, and $f(v) \leq M$ for at least half the vertices v; note that the median may be nonunique in some cases). Show that
$$|\{v \in V : |f(v) - M| \geq \lambda K\}| \leq Cn\exp(-c\lambda)$$
for all $\lambda > 0$ and some constants $C, c > 0$ depending only on k, ε.

1.3. Random walks on expanders

We now discuss a connection between expanders and random walks, which will be of particular importance in this text as a tool for demonstrating expansion. Given a k-regular graph G for some $k \geq 1$, and an initial vertex $v_0 \in G$, we define the random walk on G starting at v_0 to be a random sequence $v_0, v_1, v_2, \ldots$ of vertices in G defined recursively by setting, once $v_0, \ldots, v_i$ have been chosen, v_{i+1} to be one of the k neighbours of v_i, chosen at random[8]. For each i, let $\mu^{(i)} \colon G \to \mathbf{R}^+$ be the probability distribution of v_i, thus
$$\mu^{(i)}(v) = \mathbf{P}(v_i = v).$$
Thus $\mu^{(0)}$ is the Dirac mass δ_{v_0} at v_0, and we have the recursion
$$\mu^{(i+1)} = \frac{1}{k}A\mu^{(i)}$$
and thus
$$\mu^{(i)} = k^{-i}A^i\delta_{v_0}.$$

Among other things, this shows that the quantity $\|\mu^{(i)} - \frac{1}{|V_n|}\|_{\ell^2(V_n)}$, which measures the extent to which v_i is uniformly distributed, is nonincreasing in i. The rate of this decrease is tied to the expansion properties of the graph:

Exercise 1.3.1. Let $k \geq 1$, and let $G_n = (V_n, E_n)$ be a family of finite k-regular graphs. Let $\alpha > 1/2$. Show that the following are equivalent:

(i) The G_n are a two-sided expander family.
(ii) There is a $C > 0$ independent of n, such that for all sufficiently large n one has $\|\mu^{(i)} - \frac{1}{|V_n|}\|_{\ell^2(V_n)} \leq |V_n|^{-\alpha}$ for all $i \geq C\log|V_n|$, and all choices of initial vertex v_0.

Informally, the above exercise asserts that a two-sided expander on n vertices is one for which random walks (from an arbitrary starting point) become very close to uniform in just $O(\log n)$ steps. (Compare with, say,

---

[8]The existence of such a random process can be easily justified by using the *Kolmogorov extension theorem*; see, e.g., [**Ta2011**, Theorem 2.4.3]. Alternatively, one can select a random real number from $[0, 1]$ and express it in base k, obtaining an infinite string of digits in $\{0, \ldots, k-1\}$ that can be used (after arbitrarily ordering the edges emanating from each vertex in G) to generate the random sequence $v_0, v_1, \ldots$.

the graphs in Exercise 1.1.3, in which the random walks do not come close to mixing until time well beyond n^2, as indicated by the central limit theorem.) This rapid mixing is useful for many applications; for instance, it can be used in computer science to generate almost perfectly uniformly distributed random elements of various interesting sets (e.g., elements of a finite group); see [**HoLiWi2006**] for more discussion. It is also useful in number theory to facilitate certain sieving estimates, as will be discussed in Chapter 7.

Remark 1.3.1. In the above exercise, we assumed that the initial vertex v_0 was deterministic rather than random. However, it is easy to see (from Minkowski's inequality) that the exercise also holds if we permit v_0 to be drawn from an arbitrary probability distribution on V_n, rather than being a single deterministic vertex. Thus one can view the uniform distribution in a two-sided expander to be a very strong attractor for all the other probability distributions on the vertex set.

Remark 1.3.2. The ℓ^2 norm is the most convenient norm to use when using the spectral theorem, but one can certainly replace this norm if desired by other norms (e.g., the ℓ^1 norm, which in this finite setting is the same thing as the *total variation norm*), after adjusting the lower bound of A slightly, since all norms on a finite-dimensional space are equivalent (though one typically has to concede some powers of $|V_n|$ to attain this equivalency). One can also use some other norm-like quantities here to measure distance to uniformity, such as *Shannon entropy*, although we will not do so here.

Note that for graphs that are only one-sided expanders instead of two-sided expanders, the random walk is only partially mixing, in that the probability distribution tends to flatten out rapidly, but not converge as rapidly (or at all) to the uniform distribution. For instance, in the case of the complete bipartite graph $K_{n/2,n/2}$, it is clear that the random walk simply alternates between the two vertex sets of size $n/2$ in the bipartite graph (although it is uniformly distributed in each set). But this lack of rapid mixing can be dealt with by replacing the random walk with the *lazy random walk* $w_0, w_1, w_2, \ldots$, which is defined similarly to the random walk $v_0, v_1, v_2, \ldots$ except that w_{i+1} is only set equal to a randomly chosen neighbour of w_i with probability $1/2$, and remains equal to w_i with probability $1/2$. (Here, one can also select probabilities other than $1/2$, as this will not significantly affect the exercise below.) Indeed, we have:

Exercise 1.3.2. Let $k \geq 1$, and let $G_n = (V_n, E_n)$ be a family of finite k-regular graphs. Let $\alpha > 1/2$. Show that the following are equivalent:

(i) The G_n are a one-sided expander family.

(ii) There is a $C > 0$ independent of n, such that for all sufficiently large n one has $\|\nu^{(i)} - \frac{1}{|V_n|}\|_{\ell^2(V_n)} \leq |V_n|^{-\alpha}$ for all $i \geq C \log n$, and all choices of initial vertex w_0, where $\nu^{(i)}$ is the law of the random walk w_i.

1.4. Random graphs as expanders

We now turn to the task of constructing expander families of k-regular graphs. The first result in this direction was by Pinsker [**Pi1973**] (with a closely related result also established by Barzdin and Kolmogorov [**KoBa1967**]), who showed that if one chose a k-regular graph on n vertices randomly, and then sent n to infinity along a sufficiently sparse sequence, the resulting sequence would be almost surely[9] be an expander family. We will not quite prove this result here (because it requires some understanding of the probability distribution of k-regular graphs, which has some subtleties as the parity obstruction already indicates), but establish a closely related result, in which k is restricted to be even (to avoid parity problems) and sufficiently large (for convenience) and the k-regular graph is drawn from a slightly nonuniform distribution.

Before we do this, though, let us perform a heuristic computation as to why, when k is fixed but large, and n goes to infinity, one expects a "random" k-regular graph $G = (V, E)$ on n vertices to be an expander. For simplicity, we work with the one-sided expansion condition. By Proposition 1.2.1, we would then like to say that with high probability, one has

$$|\partial F| \geq c|F|$$

for all $F \subset V$ with cardinality at most $n/2$, and some $c > 0$ independent of n. An equivalent formulation would be to say that the neighbourhood $N(F)$ of F has to have cardinality at least $(1 + c')|F|$ for all $F \subset V$ with cardinality at most $n/2$, and some $c' > 0$ independent of n. Thus, we wish to exclude the possibility that there are sets $F \subset F' \subset V$ with $|F| \leq n/2$ and $|F'| \leq (1 + c')|F|$, for which all the edges starting from F end up in F'.

To bound this failure probability, we use the union bound: we try each pair F, F' in turn, bound the probability that the claim follows for that particular value of F, F', and then sum in F, F'. The goal is to obtain a total probability bound of $o_{n \to \infty; k}(1)$, that goes to zero as $n \to \infty$ for fixed k.

Accordingly, pick $1 \leq r \leq n/2$, and then pick $F \subset V$ of cardinality r, and then $F \subset F' \subset V$ of cardinality $r + r'$, where $r' := \lfloor c'r \rfloor + 1$ for some small constant $c' > 0$ to be chosen later. For each fixed r, there are $\frac{n!}{r! r'! (n-r-r')!}$

---

[9]Here, of course, one needs to avoid the parity obstruction that one cannot have a k-regular graph on n vertices if k and n are both odd.

choices of F and F'. For each F, F', there are kr edges emanating from F. Intuitively, if we choose the graph randomly, each edge has a probability about $\frac{r+r'}{n}$ of landing back in F', so the probability that they all do is about $(\frac{r+r'}{n})^{kr}$. So the failure rate should be about

$$\sum_{1 \leq r \leq n/2} \frac{n!}{r!r'!(n-r-r')!} \left(\frac{r+r'}{n}\right)^{kr}.$$

We can bound $\frac{n!}{r!r'!(n-r-r')!}$ somewhat crudely as $\frac{n^{r+r'}}{(r+r')!}O(1)^r$. Applying Stirling's formula, we obtain a bound of

$$\sum_{1 \leq r \leq n/2} O(1)^r \left(\frac{r+r'}{n}\right)^{kr-r-r'}.$$

For c small enough, $(r+r')/n$ is less than 0.6 (say), and then (for k large enough) we see that this series is bounded by $o_{n \to \infty; k}(1)$ as required.

Now we turn to the rigorous construction of random expander graphs. We will assume that k is a large (but fixed) even integer, $k = 2l$. To build a $2l$-regular graph on n vertices $\{1, \ldots, n\}$, what we will do is pick l permutations $\pi_1, \ldots, \pi_l : \{1, \ldots, n\} \to \{1, \ldots, n\}$, and let G be the graph formed by connecting v to $\pi_i(v)$ for all $v \in \{1, \ldots, n\}$ and $i = 1, \ldots, l$. This is not always a $2l$-regular graph, but we will be able to show the following two claims (if k is large enough):

Proposition 1.4.1 (*G can be k-regular*)**.** *The graph G is $2l$-regular with probability at least $c - o_{n \to \infty; k}(1)$, where $c > 0$ depends only on k.*

Proposition 1.4.2 (*G usually expands*)**.** *There is an $\varepsilon > 0$ depending only on $k = 2l$, such that the probability that G is $2l$-regular but not a one-sided ε-expander is $o_{n \to \infty; k}(1)$.*

Putting these two propositions together, we conclude:

Corollary 1.4.3. *With probability at least $c - o_{n \to \infty; k}(1)$, G is a $2l$-regular one-sided ε-expander.*

In particular, this allows us to construct one-sided expander families of k-regular graphs for any fixed large even k.

Remark 1.4.4. If one allowed graphs to have multiple edges and loops, then it would be possible to dispense with the need for Proposition 1.4.1, and show that G (now viewed as a $2l$-regular graph with multiple edges and loops) is a one-sided ε-expander with probability $1 - o_{n \to \infty; k}(1)$. (This requires extending results such as the weak discrete Cheeger inequality to the case when there are multiple edges and loops, but this turns out to

1.4. Random graphs as expanders

be straightforward.) However, we will not do so here as it requires one to introduce a slight amount of additional notation.

Let us prove Proposition 1.4.2 first, which will follow the informal sketch at the beginning of this section. By Proposition 1.2.1, it suffices to show that there is a $c > 0$ depending only on k such that the probability that G is $2l$-regular with $h(G) \leq c$ is $o_{n \to \infty; k}(1)$. As in the sketch, we first bound for each $1 \leq r \leq n/2$ and $F \subset F' \subset \{1, \ldots, n\}$ with $|F| = r$ and $|F'| = r + r'$, where $r' := \lfloor cr \rfloor + 1$, the probability that all the edges from F end up in F'. A necessary condition for this to occur is that $\pi_i(F) \subset F'$ for each $i = 1, \ldots, k$. For each i, and for fixed r, F, F', the probability that the random permutation π_i does this, is

$$\frac{\binom{r+r'}{r}}{\binom{n}{r}} \leq \left(\frac{r+r'}{n}\right)^r,$$

so the total failure probability can be bounded by

$$\sum_{1 \leq r \leq n/2} \frac{n!}{r! r'! (n-r-r')!} \left(\frac{r+r'}{n}\right)^{kr/2},$$

which is acceptable as discussed previously.

Now we turn to Proposition 1.4.1. We observe that G will be $2l$-regular unless there are distinct i, j and a vertex $v \in \{1, \ldots, n\}$ such that either $\pi_i(v) = \pi_j(v)$, $\pi_i(v) = \pi_j^{-1}(v)$, $\pi_i(v) = v$, or $\pi_i(v) = \pi_i^{-1}(v)$ (as such cases lead to repeated edges or loops). Unfortunately, each of these events can occur with a fairly sizeable probability (e.g., for each i, j, the probability that $\pi_i(v) = \pi_j(v)$ for some v is about $1 - 1/e$, by the classical theory of *derangements*), so the union bound is not enough here. Instead, we will proceed by an interpolant between the union bound and the *inclusion-exclusion formula*, known as the *Bonferroni inequalities*:

Exercise 1.4.1 (Bonferroni inequalities).

(i) Show that if $n, k \geq 0$ are natural numbers, that

$$\sum_{j=0}^{k} (-1)^j \binom{n}{j} \geq 1_{n=0}$$

when k is even, and

$$\sum_{j=0}^{k} (-1)^j \binom{n}{j} \leq 1_{n=0}$$

when k is odd, where we adopt the convention that $\binom{n}{j} = 0$ when $j > n$.

(ii) Show that if $N, k \geq 0$ and $E_1, \ldots, E_N$ are events, then

$$\sum_{j=0}^{k} (-1)^j \sum_{1 \leq i_1 < \ldots < i_j \leq N} \mathbf{P}(E_{i_1} \cap \ldots \cap E_{i_j}) \geq \mathbf{P}\left(\bigcup_{i=1}^{N} E_i\right)$$

when k is even, and

$$\sum_{j=0}^{k} (-1)^j \sum_{1 \leq i_1 < \ldots < i_j \leq N} \mathbf{P}(E_{i_1} \cap \ldots \cap E_{i_j}) \leq \mathbf{P}\left(\bigcup_{i=1}^{N} E_i\right)$$

when k is odd. (Note that the $k = 1$ case of this inequality is essentially the union bound, and the $k = N$ case is the inclusion-exclusion formula.) Here we adopt the convention that the empty intersection occurs with probability 1.

We return to the proof of Proposition 1.4.1. By a conditioning argument, it suffices to show the following:

Proposition 1.4.5. *Let $1 \leq i \leq k$, and suppose that the permutations $\pi_1, \ldots, \pi_{i-1}$ have already been chosen (and are now viewed as fixed deterministic objects). Let $\pi_i \colon \{1, \ldots, n\} \to \{1, \ldots, n\}$ be a permutation chosen uniformly at random. Then with probability at least $c - o_{n \to \infty; k}(1)$ for some $c > 0$ depending only on k, one has $\pi_i(v) \neq \pi_j(v), v, \pi_i^{-1}(v)$ for all $v \in \{1, \ldots, n\}$ and $j = 1, \ldots, i-1$.*

It remains to establish Proposition 1.4.5. We modify an argument of Bollobas [**Bo2001**], based on some technical counting asymptotics which we defer to the exercises. Consider the set $\Omega \subset \{1, \ldots, n\} \times \{1, \ldots, n\}$ of pairs

$$\Omega := \{(v, \pi_j(v)) : v = 1, \ldots, n; j = 1, \ldots, i-1\}$$
$$\cup \{(v, \pi_j(v)^{-1}) : v = 1, \ldots, n; j = 1, \ldots, i-1\}$$
$$\cup \{(v, v) : v = 1, \ldots, n\}.$$

Each pair (v, w) in Ω gives rise to a bad event $E_{(v,w)} := (\pi_i(v) = w)$. Each pair $(v, w) \in \{1, \ldots, n\}^2 \backslash \Omega$ gives rise to another bad event

$$E_{(v,w)} := (\pi_i(v) = w) \wedge (\pi(w) = v).$$

Note that $E_{(v,w)} = E_{(w,v)}$ for $(v,w) \in \{1, \ldots, n\}^2 \backslash \Omega$. To avoid this collision problem, we work in the space

$$\Omega' := \Omega \cup \{(v, w) \in \{1, \ldots, n\}^2 \backslash \Omega : v < w\}.$$

Our task is now to show that

$$\mathbf{P}\left(\overline{\bigcup_{s \in \Omega'} E_s}\right) \geq c - o_{n \to \infty; k}(1).$$

1.4. Random graphs as expanders

It is easy to check (by direct counting arguments) that

$$(1.6) \qquad \mathbf{P}(E_s) = \frac{1}{n} + O\left(\frac{1}{n^2}\right)$$

for $s \in \Omega$, and

$$(1.7) \qquad \mathbf{P}(E_s) = \frac{1}{n^2} + O\left(\frac{1}{n^3}\right)$$

for $s \notin \Omega'$. In particular, if we set

$$\alpha := \sum_{s \in \Omega'} \mathbf{P}(E_s),$$

then we have

$$\alpha = 2i - \frac{1}{2} + O\left(\frac{1}{n}\right).$$

More generally, for any fixed j, the same sort of counting arguments (which we leave as an exercise) gives the approximate independence property

$$(1.8) \qquad \mathbf{P}(E_{s_1} \cap \ldots \cap E_{s_j}) = (1 + O_{j,k}\left(\frac{1}{n}\right))\mathbf{P}(E_{s_1})\ldots\mathbf{P}(E_{s_j})$$

whenever $s_1 = (v_1, w_1), \ldots, s_j = (v_j, w_j)$ are distinct elements of Ω' with the disjointness property $\{v_i, w_i\} \cap \{v_{i'}, w_{i'}\} = \emptyset$ for all $1 \leq i < i' \leq j$. If the disjointness property fails, we can still obtain the weaker bound

$$(1.9) \qquad \mathbf{P}(E_{s_1} \cap \ldots \cap E_{s_j}) = O_{j,k}(\mathbf{P}(E_{s_1})\ldots\mathbf{P}(E_{s_j})).$$

(The subscripts in the $O()$-notation indicate that the implied constant in that notation can depend on the subscripted parameters.) Summing this, we conclude that

$$(1.10) \qquad \sum_{s_1,\ldots,s_j \in \Omega', \text{ distinct}} \mathbf{P}(E_{s_1} \cap \ldots \cap E_{s_j}) = \alpha^j + O_{j,k}\left(\frac{1}{n}\right)$$

and thus by the Bonferroni inequalities

$$\mathbf{P}(\bigcup_{s=1}^{\overline{2in}} E_s) \geq \sum_{j=1}^{m}(-1)^{j-1}\frac{\alpha^j}{j!} + O_{m,k}\left(\frac{1}{n}\right)$$

for any odd m. But from Taylor series expansion, the sum on the right-hand side converges to the positive quantity $e^{-\alpha}$, and the claim follows by taking m to be a sufficiently large odd number depending on k.

Exercise 1.4.2. Verify the estimates (1.6), (1.8), (1.9), (1.10).

Remark 1.4.6. Another way to establish Proposition 1.4.1, via the "swapping method", was pointed out to me by Brendan McKay. The key observation is that if l permutations $\pi_1, \ldots, \pi_l$ have m problematic edges (i.e., edges that are either repeated or loops), and then one applies m random transpositions to the $\pi_1, \ldots, \pi_l$ (as selected at random), then with probability $\gg_m n^{-m}$ (if n is sufficiently large depending on m), all problematic edges are erased and one obtains a random regular graph. Conversely, if one starts with a random regular graph and applies m random transpositions, the probability of obtaining m problematic edges as a result is $O_m(n^{-m})$. Combining the two facts, we see that p_m is the probability of having m problematic edges, then $p_0 \gg p_m$ for each fixed m (and n sufficiently large depending on m). Since the expected number of problematic edges is bounded, the desired bound $p_0 \gg 1$ then follows from Markov's inequality and the pigeonhole principle.

Exercise 1.4.3. Show that "one-sided" can be replaced with "two-sided" in Proposition 1.4.2 and hence in Corollary 1.4.3.

Remark 1.4.7. It turns out that the random k-regular graphs formed by taking l permutations as indicated above, and conditioning on the event that there are no "collisions" (so that one genuinely gets a k-regular graph) does not quite give a uniform distribution on the k-regular graphs. However, it is close enough to one that any property which is true with probability $1 - o_{n \to \infty; k}(1)$ for this model of random k-regular graph, is also true with probability $1 - o_{n \to \infty; k}(1)$ for uniform k-regular graphs, and conversely. This fact (known as *contiguity* of the two random models, and analogous to the concept of mutually absolutely continuous measures in measure theory) is established, for instance, in [**Wo1999**]. As a consequence of this fact (and a more refined version of the above analysis), one can show that $1 - o_{n \to \infty; k}(1)$ of all k-regular graphs on n vertices are ε-expanders for some $\varepsilon = \varepsilon_k > 0$ if $k \geq 3$ (assuming of course the parity requirement that nk be even, otherwise there are no k-regular graphs at all).

Chapter 2

Expansion in Cayley graphs, and Kazhdan's property (T)

In Chapter 1 we introduced the notion of expansion in arbitrary k-regular graphs. For the rest of the text, we will now focus attention primarily to a special type of k-regular graph, namely a *Cayley graph*.

Definition 2.0.1 (Cayley graph)**.** Let $G = (G, \cdot)$ be a group, and let S be a finite subset of G. We assume that S is symmetric (thus $s^{-1} \in S$ whenever $s \in S$) and does not contain the identity 1 (this is to avoid loops). Then the (right-invariant) *Cayley graph* $\operatorname{Cay}(G, S)$ is defined to be the graph with vertex set G and edge set $\{\{sx, x\} : x \in G, s \in S\}$, thus each vertex $x \in G$ is connected to the $|S|$ elements sx for $s \in S$, and so $\operatorname{Cay}(G, S)$ is a $|S|$-regular graph.

Example 2.0.2. The graph in Exercise 1.1.3 is the Cayley graph on $\mathbf{Z}/N\mathbf{Z}$ with generators $S = \{-1, +1\}$.

Remark 2.0.3. We call the above Cayley graphs right-invariant because every right translation $x \mapsto xg$ on G is a graph automorphism of $\operatorname{Cay}(G, S)$. This group of automorphisms acts transitively on the vertex set of the Cayley graph. One can thus view a Cayley graph as a *homogeneous space* of G, as it "looks the same" from every vertex. One could of course also consider *left-invariant Cayley graphs*, in which x is connected to xs rather than sx. However, the two such graphs are isomorphic using the inverse map $x \mapsto x^{-1}$, so we may without loss of generality restrict our attention throughout to left Cayley graphs.

Remark 2.0.4. For minor technical reasons, it will be convenient later on to allow S to contain the identity and to come with multiplicity (i.e., it will be a multiset rather than a set). If one does so, of course, the resulting Cayley graph will now contain some loops and multiple edges.

For the purposes of building expander families, we would of course want the underlying group G to be finite. However, it will be convenient at various times to "lift" a finite Cayley graph up to an infinite one, and so we permit G to be infinite in our definition of a Cayley graph.

We will also sometimes consider a generalisation of a Cayley graph, known as a *Schreier graph*:

Definition 2.0.5 (Schreier graph). Let G be a finite group that *acts* (on the left) on a space X, thus there is a map $(g, x) \mapsto gx$ from $G \times X$ to X such that $1x = x$ and $(gh)x = g(hx)$ for all $g, h \in G$ and $x \in X$. Let S be a symmetric subset of G which acts freely on X in the sense that $sx \neq x$ for all $s \in S$ and $x \in X$, and $sx \neq s'x$ for all distinct $s, s' \in S$ and $x \in X$. Then the *Schreier graph* $\mathrm{Sch}(X, S)$ is defined to be the graph with vertex set X and edge set $\{\{sx, x\} : x \in X, s \in S\}$.

Example 2.0.6. Every Cayley graph $\mathrm{Cay}(G, S)$ is also a Schreier graph $\mathrm{Sch}(G, S)$, using the obvious left-action of G on itself. The k-regular graph formed from l permutations $\pi_1, \ldots, \pi_l \in S_n$ that were studied in Chapter 1 is also a Schreier graph provided that $\pi_i(v) \neq v, \pi_i^{-1}(v), \pi_j(v)$ for all distinct $1 \leq i, j \leq l$, with the underlying group being the permutation group S_n (which acts on the vertex set $X = \{1, \ldots, n\}$ in the obvious manner), and $S := \{\pi_1, \ldots, \pi_l, \pi_1^{-1}, \ldots, \pi_l^{-1}\}$.

Exercise 2.0.1. If k is an even integer, show that every k-regular graph is a Schreier graph involving a set S of generators of cardinality $k/2$. (*Hint:* First show that every k-regular graph can be decomposed into $k/2$ unions of cycles, each of which partition the vertex set, then use the previous example.)

We return now to Cayley graphs. It is easy to characterise qualitative expansion properties of Cayley graphs:

Exercise 2.0.2 (Qualitative expansion). Let $\mathrm{Cay}(G, S)$ be a finite Cayley graph.

(i) Show that $\mathrm{Cay}(G, S)$ is a one-sided ε-expander for G for some $\varepsilon > 0$ if and only if S generates G.

(ii) Show that $\operatorname{Cay}(G,S)$ is a two-sided ε-expander for G for some $\varepsilon > 0$ if and only if S generates G, and furthermore S intersects each index 2 subgroup of G.

We will, however, be interested in more quantitative expansion properties, in which the expansion constant ε is independent of the size of the Cayley graph, so that one can construct nontrivial expander families $\operatorname{Cay}(G_n, S_n)$ of Cayley graphs.

One can analyse the expansion of Cayley graphs in a number of ways. For instance, by taking the edge expansion viewpoint, one can study Cayley graphs combinatorially, using the product set operation

$$A \cdot B := \{ab : a \in A, b \in B\}$$

of subsets of G.

Exercise 2.0.3 (Combinatorial description of expansion). Let $\operatorname{Cay}(G_n, S_n)$ be a family of finite k-regular Cayley graphs. Show that $\operatorname{Cay}(G_n, S_n)$ is a one-sided expander family if and only if there is a constant $c > 0$ independent of n such that $|E_n \cup E_n S_n| \geq (1+c)|E_n|$ for all sufficiently large n and all subsets E_n of G_n with $|E_n| \leq |G_n|/2$.

Remark 2.0.7. Note that $|E_n \cap E_n s| = |E_n \cap E_n s^{-1}|$ for all $s \in G$. Thus one can define combinatorial expansion for asymmetric generating sets S_n as well, but it is equivalent to combinatorial expansion for $S_n \cup S_n^{-1}$ and so the theory is essentially the same. (See also Exercise 1.1.8, which is a related observation connecting directed expansion with undirected expansion.)

One can also give a combinatorial description of two-sided expansion for Cayley graphs; see Exercise 5.0.5.

Exercise 2.0.4 (Abelian groups do not expand). Let $\operatorname{Cay}(G_n, S_n)$ be a family of finite k-regular Cayley graphs, with the G_n all abelian, and the S_n generating G_n. Show that $\operatorname{Cay}(G_n, S_n)$ are a one-sided expander family if and only if the Cayley graphs have bounded cardinality (i.e., $\sup_n |G_n| < \infty$). (*Hint:* Assume for contradiction that $\operatorname{Cay}(G_n, S_n)$ is a one-sided expander family with $|G_n| \to \infty$, and show by two different arguments that $\sup_n |S_n^m|$ grows at least exponentially in m and also at most polynomially in m, giving the desired contradiction.)

The left-invariant nature of Cayley graphs also suggests that such graphs can be profitably analysed using some sort of Fourier analysis; as the underlying symmetry group is not necessarily abelian, one should use the Fourier analysis of nonabelian groups, which is better known as (unitary) *representation theory*. The Fourier-analytic nature of Cayley graphs can be highlighted by recalling the operation of *convolution* of two functions $f, g \in \ell^2(G)$, defined by the formula

$$f * g(x) := \sum_{y \in G} f(y) g(y^{-1} x) = \sum_{y \in G} f(xy^{-1}) g(y).$$

This convolution operation is bilinear and associative[1] (at least when one imposes a suitable decay condition on the functions, such as compact support), but is not commutative unless G is abelian. The adjacency operator A on a Cayley graph $\mathrm{Cay}(G, S)$ can then be viewed as a convolution

$$Af = |S| \mu * f,$$

where μ is the probability density

(2.1) $$\mu := \frac{1}{|S|} \sum_{s \in S} \delta_s$$

where δ_s is the Kronecker delta function on s. Using the spectral definition of expansion, we thus see that $\mathrm{Cay}(G, S)$ is a one-sided expander if and only if

(2.2) $$\langle f, \mu * f \rangle \leq (1 - \varepsilon) \|f\|_{\ell^2(G)}^2$$

whenever $f \in \ell^2(G)$ is orthogonal to the constant function 1, and is a two-sided expander if

(2.3) $$\|\mu * f\|_{\ell^2(G)} \leq (1 - \varepsilon) \|f\|_{\ell^2(G)}$$

whenever $f \in \ell^2(G)$ is orthogonal to the constant function 1.

We remark that the above spectral definition of expansion can be easily extended to symmetric sets S which contain the identity or have multiplicity (i.e., are multisets). (We retain symmetry, though, in order to keep the operation of convolution by μ self-adjoint.) In particular, one can say (with some slight abuse of notation) that a set of elements $s_1, \ldots, s_l$ of G (possibly with repetition, and possibly with some elements equaling the identity) generates a one-sided or two-sided ε-expander if the associated symmetric probability density

$$\mu := \frac{1}{2l} \sum_{i=1}^{l} \delta_{s_i} + \delta_{s_i^{-1}}$$

obeys either (2.2) or (2.3).

We saw in Chapter 1.3 that expansion can be characterised in terms of random walks. One can of course specialise this characterisation to the Cayley graph case:

---

[1] If one is more algebraically minded, one can also identify $\ell^2(G)$ (when G is finite, at least) with the *group algebra* $\mathbf{C}G$, in which case convolution is simply the multiplication operation in this algebra.

Exercise 2.0.5 (Random walk description of expansion). Let $\mathrm{Cay}(G_n, S_n)$ be a family of finite k-regular Cayley graphs, and let μ_n be the associated probability density functions. Let $A > 1/2$ be a constant.

(i) Show that the $\mathrm{Cay}(G_n, S_n)$ are a two-sided expander family if and only if there exists a $C > 0$ such that for all sufficiently large n, one has $\|\mu_n^{*m} - \frac{1}{|G_n|}\|_{\ell^2(G_n)} \leq \frac{1}{|G_n|^A}$ for some $m \leq C \log |G_n|$, where
$$\mu_n^{*m} := \mu_n * \ldots * \mu_n$$
denotes the convolution of m copies of μ_n.

(ii) Show that the $\mathrm{Cay}(G_n, S_n)$ are a one-sided expander family if and only if there exists a $C > 0$ such that for all sufficiently large n, one has
$$\|(\tfrac{1}{2}\delta_1 + \tfrac{1}{2}\mu_n)^{*m} - \tfrac{1}{|G_n|}\|_{\ell^2(G_n)} \leq \frac{1}{|G_n|^A}$$
for some $m \leq C \log |G_n|$.

In this chapter, we will connect expansion of Cayley graphs to an important property of certain infinite groups, known as *Kazhdan's property (T)* (or property (T) for short). In [**Ma1973**], Margulis exploited this property to create the first known *explicit* and deterministic examples of expanding Cayley graphs. As it turns out, property (T) is somewhat overpowered for this purpose; in particular, we now know that there are many families of Cayley graphs for which the associated infinite group does not obey property (T) (or weaker variants of this property, such as property τ). In later chapters we will therefore turn to other methods of creating Cayley graphs that do not rely on property (T). Nevertheless, property (T) is of substantial intrinsic interest, and also has many connections to other parts of mathematics than the theory of expander graphs, so it is worth spending some time to discuss it here.

The material here is based in part on [**BedeVa2008**].

2.1. Kazhdan's property (T)

Kazhdan's property (T) is a representation-theoretic property of groups. Although we will primarily be interested in finite groups (such as $\mathrm{SL}_d(\mathbf{Z}/p\mathbf{Z})$), or at least discrete, finitely generated groups (such as $\mathrm{SL}_d(\mathbf{Z})$), it will be convenient to work in the more general category of *locally compact groups* which includes discrete finitely generated groups, but also Lie groups (such as $\mathrm{SL}_d(\mathbf{R})$) as examples. For minor technical reasons we will restrict attention to locally compact groups G that are Hausdorff, second countable, and compactly generated (so that there is a compact set S that generates G as a group, as is for instance, the case for discrete finitely generated groups). We shall therefore abuse notation and abbreviate "locally compact second countable Hausdorff compactly generated group" as "locally compact group". One can extend the study of property (T) to more general classes of groups than these, but this class will be sufficient for our applications, while avoiding some technical subtleties that are not relevant for our purposes.

We will focus on a particular type of representation of locally compact groups, namely a unitary representation.

Definition 2.1.1 (Unitary representation). Let G be a locally compact group. A *unitary representation* (or *representation*, for short) of G is a (complex) *separable* Hilbert space H (possibly infinite-dimensional), together with a homomorphism $\rho\colon G \to U(H)$ from G to the group $U(H)$ of unitary transformations on H. Furthermore, we require ρ to be continuous, where we give $U(H)$ the *strong operator topology*; thus the map $g \mapsto \rho(g)v$ is continuous for each $v \in H$. We often abuse notation and refer to ρ (rather than the pair (ρ, H)) as the representation of G.

Two unitary representations $\rho\colon G \to U(H)$, $\rho'\colon G \to U(H')$ are *isomorphic* if there is a Hilbert space isomorphism $\phi\colon H \to H'$ such that $\rho'(g) \circ \phi = \phi \circ \rho(g)$ for all $g \in G$. When one has such an isomorphism, we write $\rho \equiv \rho'$.

Note that if G is a discrete group, then the continuity hypothesis is automatic and can be omitted. One could easily turn the space of unitary representations of a fixed group G into a *category* by defining the notion of a morphism between two unitary representations that generalises the notion of an isomorphism given above (basically by replacing "Hilbert space isomorphism" with "Hilbert space isometry"), but we will not need to do so here. One could also consider inseparable Hilbert space representations, but for minor technical reasons it will be convenient to restrict attention to the separable case[2].

Example 2.1.2 (Trivial representation). Any locally compact group G acts trivially on any Hilbert space H by defining $\rho(g)$ to be the identity on H for all $g \in G$.

Example 2.1.3 (Regular representation). If G is a discrete group, then we have the (left) *regular representation* $\tau\colon G \to U(\ell^2(G))$ of G, in which the Hilbert space is $\ell^2(G)$, and the action is given by the formula

$$\tau(g)f(x) := f(g^{-1}x) = \delta_g * f(x)$$

---

[2]Note that G, being second countable, is separable, and so any orbit $\{gv : g \in G\}$ of a vector v in a Hilbert space is automatically contained in a separable subspace. As such, one can usually restrict without loss of generality to separable Hilbert spaces in applications.

2.1. Kazhdan's property (T)

for $g \in G$ and $x \in G$. (Note that the g^{-1} here is needed to make τ a homomorphism.) This is easily verified to be a unitary representation. One can also consider the right-regular representation that takes $f(x)$ to $f(xg)$, but this is easily seen to be isomorphic to the left-regular representation and will not be needed here.

More generally, if G is a locally compact group equipped with a left-invariant *Haar measure* μ (that is to say, a Radon measure which is invariant with respect to left translations), one can define the (left) regular representation τ on[3] $L^2(G, d\mu)$ by setting $\tau(g)f(x) := f(g^{-1}x)$ for all $f \in L^2(G, d\mu)$.

Example 2.1.4 (Quasiregular representation). If (X, μ) is a measure space that G acts on in a transitive measure-preserving fashion, then we have the (left) *quasiregular representation* $\tau_X \colon G \to U(L^2(X, \mu))$, in which the Hilbert space is $L^2(X, \mu)$, and the action is given by the formula

$$\tau_X(g)f(x) := f(g^{-1}x)$$

for $g \in G$ and $x \in X$. Of course, the regular representation can be viewed as a special case of a quasiregular representation, as can the one-dimensional trivial representation.

Example 2.1.5 (Direct sum). If $\rho_1 \colon G \to U(H_1)$ and $\rho_2 \colon G \to U(H_2)$ are unitary representations of a locally compact group G, then their *direct sum* $\rho_1 \oplus \rho_2 \colon G \to U(H_1 \oplus H_2)$ is also a unitary representation, where $H_1 \oplus H_2$ is the Hilbert space of all formal sums $v_1 \oplus v_2$ with $v_1 \in H_1$ and $v_2 \in H_2$ with the inner product

$$\langle v_1 \oplus v_2, w_1 \oplus w_2 \rangle_{H_1 \oplus H_2} := \langle v_1, w_1 \rangle_{H_1} + \langle v_2, w_2 \rangle_{H_2}$$

and the representation $\rho_1 \oplus \rho_2$ is given by the formula

$$(\rho_1 \oplus \rho_2)(g)(v_1 \oplus v_2) := (\rho_1(g)v_1) \oplus (\rho_2(g)v_2).$$

One can also form the direct sum of finitely many or even countably[4] many unitary representations in a similar manner; we leave the details to the reader.

Example 2.1.6 (Subrepresentation). If $\rho \colon G \to U(H)$ is a unitary representation of a locally compact group G, and H' is a closed subspace of H which is G-invariant (thus $\rho(g)H' \subset H'$ for all $g \in G$), then we can form the restriction $\rho \restriction_{H'} \colon G \to U(H')$ of ρ to H', defined by setting $\rho \restriction_{H'} (g)v := \rho(g)v$ for all $g \in G$ and $v \in H'$. This is easily seen to also be a unitary representation, and is known as a *subrepresentation* of ρ. By unitarity, we see that

---

[3]Note that $L^2(G, d\mu)$ will be separable because we are assuming G to be second countable. For a review of the theory of Haar measure (and in particular, a proof that any locally compact group has a Haar measure, unique up to scalar multiplication); see [**Ta2013**, §1.4].

[4]There is also a construction to combine uncountably many such representations, known as the *direct integral*, but we will not need it here.

the orthogonal complement $(H')^\perp$ of H' in H is an invariant space, leading to the complementary subrepresentation $\rho\restriction_{(H')^\perp}$. One easily verifies the isomorphism

$$\rho \equiv \rho\restriction_{H'} \oplus \rho\restriction_{(H')^\perp}.$$

Example 2.1.7 (Invariant vectors)**.** If $\rho\colon G \to U(H)$ is a unitary representation of a locally compact group G, then the space $H^G := \{v \in H : \rho(g)v = v \text{ for all } g \in G\}$ of G-invariant vectors in H is a closed invariant subspace of H. By the preceding example, we thus have the decomposition

$$\rho \equiv \rho\restriction_{H^G} \oplus \rho\restriction_{(H^G)^\perp}$$

into a trivial representation $\rho\restriction_{H^G}$, and a representation $\rho\restriction_{(H^G)^\perp}$ with no nontrivial invariant vectors. (Indeed, this is the only such decomposition up to isomorphism; we leave this as an exercise to the reader.) For instance, if G is a finite group and we consider the regular representation τ (so $H = \ell^2(G)$), then H^G is the one-dimensional space of constants $\mathbf{C}$, while $(H^G)^\perp$ is the space $\ell^2(G)_0$ of functions of mean zero, so we have the decomposition

$$\tau \equiv \mathbf{C} \oplus \tau\restriction_{\ell^2(G)_0}.$$

Note that if G is an *infinite* discrete group, then there are already no nontrivial invariant vectors in $\ell^2(G)$ (why?), so the decomposition in this case is trivial:

$$\tau \equiv 0 \oplus \tau\restriction_{\ell^2(G)}.$$

We now make a key definition of a *Kazhdan constant*, which is analogous to the expansion constant of a Cayley graph.

Definition 2.1.8 (Kazhdan constant)**.** Let $\rho\colon G \to U(H)$ be a unitary representation of a locally compact group G, and let S be a compact subset of G. The *Kazhdan constant* $\mathrm{Kaz}(G, S, \rho)$ of S and ρ is then the supremum of all the constants $\varepsilon \geq 0$ for which one has the bound

$$\sup_{s \in S} \|\rho(s)v - v\|_H \geq \varepsilon \|v\|_H$$

for all $v \in H$. Thus, for instance, $\mathrm{Kaz}(G, S, \rho)$ vanishes whenever the representation ρ contains nontrivial invariant vectors. The Kazhdan constant $\mathrm{Kaz}(G, S)$ of S is defined as

$$\mathrm{Kaz}(G, S) := \inf_{\rho} \mathrm{Kaz}(G, S, \rho),$$

where ρ ranges over all unitary representations of G with no nontrivial invariant vectors. A group G is said to have *Kazhdan property (T)* if one has $\mathrm{Kaz}(G, S) > 0$ for at least one compact set S.

2.1. Kazhdan's property (T)

Thus, if G has Kazhdan property (T), then one can find at least one compact set S and some $\varepsilon > 0$ with the property that for every representation unitary $\rho\colon G \to U(H)$ with no nontrivial invariant vectors, and every $v \in H$, one can find $s \in S$ such that $\|\rho(s)v - v\|_H \geq \varepsilon \|v\|_H$. Thus, we have a dichotomy: either a representation contains invariant vectors, or every vector in the representation is moved by a nontrivial amount by some "small" element of G.

Example 2.1.9. Later on in this text, we will show that $\mathrm{SL}_d(\mathbf{Z})$ and $\mathrm{SL}_d(\mathbf{R})$ have property (T) if and only if $d \geq 3$. We will also see that a free nonabelian group F_k on k generators will never have property (T), and also no finitely generated infinite abelian group will have property (T).

Exercise 2.1.1. Show that every finite group G has property (T). (*Hint:* First show that every vector v in a unitary representation is contained in a subrepresentation of dimension at most $|G|$, and that all the unitary maps $\rho(g)$ have order at most $|G|$.) Later on, we shall establish the stronger statement that every *compact* group has property (T).

We record some basic properties of the Kazhdan constants:

Exercise 2.1.2. Let G be a locally compact group, let $\rho\colon G \to U(H)$ be a representation, and let S, S' be compact subsets of G.

(i) If $S \subset S'$, then $\mathrm{Kaz}(G, S, \rho) \leq \mathrm{Kaz}(G, S', \rho)$ and $\mathrm{Kaz}(G, S) \leq \mathrm{Kaz}(G, S')$.

(ii) One has $\mathrm{Kaz}(G, S, \rho) = \mathrm{Kaz}(G, S^{-1}, \rho) = \mathrm{Kaz}(G, S \cup S^{-1}, \rho)$ and $\mathrm{Kaz}(G, S) = \mathrm{Kaz}(G, S^{-1}) = \mathrm{Kaz}(G, S \cup S^{-1})$.

(iii) One has $\mathrm{Kaz}(G, S, \rho) = \mathrm{Kaz}(G, S \cup \{1\}, \rho)$ and $\mathrm{Kaz}(G, S) = \mathrm{Kaz}(G, S \cup \{1\})$.

(iv) One has $\mathrm{Kaz}(G, S^m, \rho) \leq m\,\mathrm{Kaz}(G, S, \rho)$ and $\mathrm{Kaz}(G, S^m) \leq m\,\mathrm{Kaz}(G, S)$ for all $m \geq 1$.

(v) If S generates G as a group (thus every element of G is a finite word in S), show that G has Kazhdan property (T) if and only if $\mathrm{Kaz}(G, S) > 0$.

From the above exercise, we see that the precise choice of compact set S needed to establish the Kazhdan property is not important, so long as it generates the group (and by hypothesis, every locally compact group G has at least one compact generating set S).

Remark 2.1.10. In our conventions, we are only considering locally compact groups that are compactly generated. However, in some applications it is important to note that the compact generation hypothesis is in fact automatic if one has Kazhdan's property (T). Indeed, if $\mathrm{Kaz}(G, S) > 0$ for some compact S (which we may assume without loss of generality to be a neighbourhood of the identity), and G' is the (necessarily open) subgroup of G generated by S, then the Dirac delta δ_1 in the Hilbert space $\ell^2(G/G')$ is G'-invariant and thus (by the hypothesis $\mathrm{Kaz}(G, S) > 0$) $\ell^2(G/G')$ has a G-invariant vector, which forces G/G' to be finite, and so G is compactly generated.

Exercise 2.1.3. Let G be a locally compact group, let S be a compact subset of G. Show that the following assertions are equivalent:

  (i) $\mathrm{Kaz}(G, S) = 0$.
  (ii) There exists a unitary representation $\rho \colon G \to U(H)$ with no non-trivial invariant vectors such that $\mathrm{Kaz}(G, S, \rho) = 0$.

(*Hint:* If $\mathrm{Kaz}(G, S) = 0$, take a sequence of unitary representations of G with no nontrivial invariant vectors but a sequence of increasingly S-approximately-invariant vectors, and form their (infinite) direct sum.)

Exercise 2.1.4. Let G be a locally compact group, let S be a compact subset of G, and let $\varepsilon > 0$. Show that the following assertions are equivalent:

  (i) $\mathrm{Kaz}(G, S) \geq \varepsilon$.
  (ii) For any unitary representation $\rho \colon G \to U(H)$ (possibly containing invariant vectors), and any $v \in H$, one has

$$\mathrm{dist}(v, H^G) \leq \frac{1}{\varepsilon} \sup_{s \in S} \|\rho(s)v - v\|_H.$$

Exercise 2.1.5. Let G be a locally compact group. Show that exactly one of the following is true:

  (i) G has property (T).
  (ii) There exists a sequence $\rho_n \colon G \to U(H_n)$ of representations and a sequence of unit vectors $v_n \in H_n$ such that $\|sv_n - v_n\|_{H_n} \to 0$ for all $s \in G$, but such that each of the H_n contains no nontrivial invariant vectors.

Remark 2.1.11. Informally, the statement in (ii) of the preceding exercise shows that one can reach the trivial representation as a "limit" of a sequence of representations with no nontrivial invariant vectors. As such, property (T) can be viewed as the assertion that the trivial representation T is isolated from the other representations in some sense, which is the origin for the term "property (T)". This intuition can be formalised by introducing *Fell's topology* on unitary representations; see [**BedeVa2008**] for details.

Now we show why Kazhdan constants are related to expansion in Cayley graphs.

Exercise 2.1.6. Let $\mathrm{Cay}(G,S)$ be a finite k-regular Cayley graph, and let $\varepsilon > 0$. Let $\rho = \tau \restriction_{\ell^2(G)_0}$ be the restriction of the regular representation of G to the functions of mean zero.

(i) If $\mathrm{Kaz}(G,S,\rho) \geq \varepsilon$, show that $\mathrm{Cay}(G,S)$ is a one-sided c-expander for some $c = c(\varepsilon, k) > 0$.

(ii) Conversely, if $\mathrm{Cay}(G,S)$ is a one-sided ε-expander, show that $\mathrm{Kaz}(G,S,\rho) \geq c$ for some $c = c(\varepsilon, k) > 0$.

(iii) Show that $\mathrm{Kaz}(G,S) \leq \mathrm{Kaz}(G,S,\rho) \leq C_k \mathrm{Kaz}(G,S)$ for some $C_k > 0$ depending only on k.

(*Hint:* Use the triangle inequality and the *cosine rule*: if v, w are unit vectors with $\|v-w\|_H^2 = 2 - 2\langle v,w\rangle_H$. Part (iii) can be strengthened to the exact identity $\mathrm{Kaz}(G,S) = \mathrm{Kaz}(G,S,\rho)$, but this requires more effort to prove; see [**Ne2006**].)

Thus, to show that a sequence $\mathrm{Cay}(G_n, S_n)$ of k-regular Cayley graphs forms a one-sided expander family, it suffices to obtain a lower bound on the Kazhdan constants $\mathrm{Kaz}(G_n, S_n, \rho_n)$. There is a similar criterion for two-sided expansion:

Exercise 2.1.7. Let $\mathrm{Cay}(G,S)$ be a finite k-regular Cayley graph, and let $\varepsilon > 0$. Let $\rho = \tau \restriction_{\ell^2(G)_0}$ be the restriction of the regular representation of G to the functions of mean zero.

(i) If $\mathrm{Kaz}(G,S^2,\rho) \geq \varepsilon$, show that $\mathrm{Cay}(G,S)$ is a two-sided c-expander for some $c = c(\varepsilon, k) > 0$.

(ii) Conversely, if $\mathrm{Cay}(G,S)$ is a two-sided ε-expander, show that $\mathrm{Kaz}(G,S^2,\rho) \geq c$ for some $c = c(\varepsilon, k) > 0$.

One advantage of working with Kazhdan constants instead of expansion constants is that they behave well with respect to homomorphisms:

Exercise 2.1.8. Let G, G' be locally compact groups, and suppose that there is a continuous surjective homomorphism $\pi \colon G \to G'$ from G to G'. Let S be a compact subset of G. Show that for any unitary representation $\rho' \colon G' \to U(H)$ of G', one has $\mathrm{Kaz}(G', \pi(S), \rho') = \mathrm{Kaz}(G, S, \rho' \circ \pi)$. Conclude that $\mathrm{Kaz}(G', \pi(S)) \geq \mathrm{Kaz}(G, S)$. In particular, if G has property (T), then G' does also.

As a corollary of the above results, we can use Kazhdan's property (T) to generate expander families:

Exercise 2.1.9. Let G be a finitely generated discrete group, and let S be a symmetric subset of G that generates G. Let N_n be a sequence of finite index normal subgroups of G, and let $\pi_n \colon G \to G/N_n$ be the quotient maps.

Suppose that for all sufficiently large n, π_n is injective on $S \cup \{1\}$. Show that if G has property (T), then $\text{Cay}(G/N_n, \pi_n(S))$ for sufficiently large n is an expander family.

Thus, it is of interest to find ways to demonstrate property (T) or, in other words, to create invariant vectors from almost invariant vectors. The next few exercises will develop some tools for this purpose.

Exercise 2.1.10. Let $\rho \colon G \to U(H)$ be a unitary representation of a locally compact group G. Suppose that there is a closed convex set K in H that contains an orbit $\{gv_0 : g \in G\}$ of some element v_0 in K. Show that K contains an invariant vector. (*Hint:* Show that the set $K' := \{v \in H : gv \in K \text{ for all } g \in G\}$ is closed, convex, and G-invariant; now study an element of K' of minimal norm.)

Exercise 2.1.11. Show that every compact group has property (T). (*Hint:* Use Exercise 2.1.10.)

Exercise 2.1.12 (Direct products and property (T)). Let G, G' be locally compact groups. Show that the product group $G \times G'$ (with the product topology, of course) has property (T) if and only if G and G' both separately have property (T). (*Hint:* One direction follows from Exercise 2.1.8. To obtain the other direction, start with an approximately invariant vector v for $G \times G'$ and use Exercise 2.1.4 (and Exercise 2.1.2(v)) to show that the $G \times G'$-orbit of v stays close to v, then use Exercise 2.1.10.)

Exercise 2.1.13 (Short exact sequences and property (T)). Let G be a locally compact group, and let N be a closed normal subgroup of G; then N and G/N are also locally compact. Show that if N and G/N have property (T), then G also has property (T).

Exercise 2.1.14. Let G be an infinite discrete finitely generated group, generated by a finite set S. Show that the following assertions are equivalent:

(i) There exists a sequence F_n of finite nonempty subsets in G with the property that $|sF_n \Delta F_n|/|F_n| \to 0$ as $n \to \infty$ for each $s \in S$. (Such a sequence of sets is known as a *Følner sequence* for G.)

(ii) One has $\text{Kaz}(G, S, \tau) = 0$, where τ is the regular representation of G.

(*Hint:* You may wish to mimic the proof of the weak discrete Cheeger inequality.)

Infinite, finitely generated groups G with property (i) or (ii) of the above exercise are known as *amenable groups*; amenability is an important property in ergodic theory, operator algebras, and many other areas of mathematics, but will not be discussed extensively in this course. The notion of amenability can also be extended to other locally compact groups, but we again will not discuss these matters here. From the above exercise, we see that an infinite amenable finitely generated group cannot have property (T). The next example shows, though, that it is also possible for nonamenable groups (such as the *free group* on two or more generators) to not have property (T).

Exercise 2.1.15.

(i) Show that the integers $\mathbf{Z}$ do not have property (T).

(ii) Show that any infinite discrete abelian finitely generated group does not have property (T).

(iii) Show that any finitely generated group G with infinite abelianisation $G/[G,G]$ does not have property (T).

(iv) Show that for any $k \geq 1$, the free group on k generators does not have property (T).

Exercise 2.1.16 (Property (T) and group cohomology)**.** The purpose of this exercise is to link property (T) to some objects of interest in *group cohomology*, and in particular, to demonstrate some "rigidity" properties of groups with property (T). (This exercise will not be needed in the rest of the text.) The results here originate from the work of Delorme [**De1977**] and Guichardet [**Gu1980**]; see [**BedeVa2008**] or [**Sh2000**] for a further discussion of the rigidity (and superrigidity) properties of groups with property (T).

Let $\rho\colon G \to U(H)$ be a unitary representation of a locally compact group G. Define a ρ-*cocycle* to be a continuous function $c\colon G \to H$ obeying[5] the *cocycle equation*
$$c(gh) = c(g) + \rho(g)c(h)$$
for all $g, h \in G$. Define a ρ-*coboundary* to be a function $c\colon G \to H$ of the form
$$c(g) = v_0 - \rho(g)v_0$$
for some fixed vector $v_0 \in H$ (or equivalently, an affine isometric action of G on H with a common fixed point v_0); observe that every ρ-coboundary is automatically a ρ-cocycle.

(i) Show that a ρ-cocycle $c\colon G \to H$ is a ρ-coboundary if and only if it is bounded. (*Hint:* If c is a bounded ρ-cocycle, then for any vector v the orbit $\{\rho(g)v + c(g) : g \in G\}$ of v lies in a ball. Exploit convexity to construct a shrinking sequence of such balls and use the completeness of H to pass the limit.)

---
[5]Equivalently, a ρ-cocycle determines an affine isometric action $v \mapsto \rho(g)v + c(g)$ of G on H with ρ as the unitary component of the action.

(ii) Show that if G has property (T), then every ρ-cocycle is a ρ-coboundary. (*Hint:* The main difficulty is to lift the affine isometric action to a unitary action. One way to do this is to work in the Hilbert space $\tilde{H}$ that is the completion of the finitely supported measures on H, using an inner product such as $\langle \delta_v, \delta_w \rangle_{\tilde{H}} := e^{-\varepsilon \|v-w\|^2}$ for some sufficiently small $\varepsilon > 0$ (one needs to verify that this does indeed determine a positive definite inner product, which can be seen for instance by working in finite dimensions and using either Fourier transforms or Gaussian identities). Note that the separability of H will imply the separability of $\tilde{H}$.)

(iii) Conversely, show that if for every unitary representation ρ, every ρ-cocycle is a ρ-coboundary, then G has property (T). (*Hint:* First show (by taking a direct sum of counterexamples, as in Exercise 2.1.3) that if S is a compact neighbourhood of the identity that generates G, $\rho\colon G \to U(H)$ is any unitary representation, and c is a ρ-cocycle, then $\sup_{g \in G} \|c(g)\|_H \leq C \sup_{s \in S} \|c(s)\|_H$, where C depends on S but is independent of ρ or c. Apply this fact to a coboundary generated by an approximate vector.)

Exercise 2.1.17 (Groups with expanding Cayley graphs have few low-dimensional representations)**.** Let $\mathrm{Cay}(G, S)$ be a finite k-regular Cayley graph which is a two-sided ε-expander for some $\varepsilon > 0$. Let H be a finite-dimensional Hilbert space.

(i) [**Wa1991**] If $\rho\colon G \to U(H)$ and $\tilde{\rho}\colon G \to U(H)$ are two nonisomorphic irreducible representations of G, show that

$$\|\frac{1}{k}\sum_{s \in S} \rho(s) A \tilde{\rho}(s)^*\|_{\mathrm{HS}(H)} \leq (1-\delta)\|A\|_{\mathrm{HS}(H)}$$

for some $\delta > 0$ depending only on ε, and all linear transformations $A\colon H \to H$ where $\|A\|_{\mathrm{HS}(H)} := \mathrm{tr}(AA^*)^{1/2}$ is the Hilbert-Schmidt norm of A. (*Hint:* Use an appropriate unitary action of G on $\mathrm{HS}(H)$, use Schur's lemma to exclude invariant vectors, and the Peter-Weyl theorem to relate this representation to the regular representation.)

(ii) [**deRoVa1993**] If $\rho_i\colon G \to U(H)$ are pairwise nonisomorphic irreducible representations of G for $i = 1, \ldots, m$, show that $m \leq \exp(Ck \dim(H)^2)$, where C depends only on ε. (*Hint:* Use (i) to show that the vectors $(\rho_i(s))_{s \in S}$ for $i = 1, \ldots, m$ are separated from each other in the Hilbert space $\mathrm{HS}(H)^k$, then use a volume packing argument.)

For further results of this type (limiting the representations of groups with expanding Cayley graphs), see [**MeWi2004**].

2.2. Induced representations and property (T)

Let $\tilde{G}$ be a locally compact group, and let G be a subgroup of $\tilde{G}$ which is closed (and thus also locally compact). Clearly, every unitary representation $\tilde{\rho}\colon \tilde{G} \to U(H)$ of $\tilde{G}$ can be restricted to form a unitary representation $\operatorname{Res}^G_{\tilde{G}} \rho\colon G \to U(H)$ of G. It is then natural to ask whether the converse is also true, that is to say whether any unitary representation $\rho\colon G \to U(H)$ of H can be extended to a representation $\tilde{\rho}\colon \tilde{G} \to U(H)$ of $\tilde{G}$ on the same Hilbert space H.

In general, the answer is no. For instance, if $\tilde{G}$ is the Heisenberg group $\tilde{G} = \begin{pmatrix} 1 & \mathbf{Z} & \mathbf{Z} \\ 0 & 1 & \mathbf{Z} \\ 0 & 0 & 1 \end{pmatrix}$, and $G = [\tilde{G}, \tilde{G}]$ is the commutator group $G = \begin{pmatrix} 1 & 0 & \mathbf{Z} \\ 0 & 1 & 0 \\ 0 & 0 & 1 \end{pmatrix}$, then any one-dimensional representation $\tilde{\rho}\colon \tilde{G} \to U_1(\mathbf{C})$ must annihilate the commutator G, but there are clearly nontrivial one-dimensional representations $\rho\colon G \to U_1(\mathbf{C})$ of G which thus cannot be extended to a representation of G.

However, there is a fundamental construction that (under some mild hypotheses) takes a representation $\rho\colon G \to U(H)$ of G and converts it to an *induced representation* $\tilde{\rho} := \operatorname{Ind}^{\tilde{G}}_G \rho\colon \tilde{G} \to U(\tilde{H})$ of $\tilde{G}$, that acts on a somewhat larger Hilbert space $\tilde{H}$ than H (in particular, the induced representation construction is *not* an inverse of the restricted representation construction). This construction will provide an important link between the representation theories of G and $\tilde{G}$, and in particular, will connect property (T) for G to property (T) for $\tilde{G}$.

To motivate the induced representation construction, we work for simplicity in the case when G and $\tilde{G}$ are discrete, consider the regular representations $\tau\colon G \to U(H)$ and $\tilde{\tau}\colon \tilde{G} \to U(\tilde{H})$ of G and $\tilde{G}$ respectively, where $H := \ell^2(G)$ and $\tilde{H} := \ell^2(\tilde{G})$. We wish to view the $\tilde{G}$-representation $\tilde{\tau}$ as somehow being induced from the G-representation τ:

$$\tilde{\tau} = \operatorname{Ind}^{\tilde{G}}_G \tau.$$

To do this, we must somehow connect H with $\tilde{H}$, and τ with $\tilde{\tau}$.

One natural way to proceed is to express $\tilde{G}$ as the union of cosets kG of G for k in some set K of coset representatives. We can then split $\ell^2(\tilde{G})$ as a direct sum $\ell^2(\tilde{G}) = \bigoplus_{k \in K} \ell^2(kG)$ (at least in the model case when K is finite), and each factor space $\ell^2(kG)$ can be viewed as a shift $\ell^2(kG) = \tilde{\rho}(k)\ell^2(G)$ of $\ell^2(G)$. This does indeed give enough of a relationship between τ and $\tilde{\tau}$ to generalise to other representations, but it is a somewhat inelegant "coordinate-dependent" formalism because it initially depends on making a somewhat arbitrary choice of coset representatives K (though one can show, at the end of the construction, that the choice of K was ultimately irrelevant). Also, this method develops some technical complications when the quotient space $\tilde{G}/G$ is not discrete.

Because of this, we shall work instead with a more "coordinate-free" construction that does not require an explicit construction of coset representatives. Instead, we rely on the orthogonal projection $\pi \colon \tilde{H} \to H$ of the larger Hilbert space $\tilde{H} = \ell^2(\tilde{G})$ to the smaller Hilbert space $\ell^2(G)$, which in the case of the regular representation is just the restriction map, $\pi(f) := f \restriction_G$.

Observe that given any vector $f \in \tilde{H}$ and group element $g \in \tilde{G}$, one can form the projection $F(g) := \pi(\tilde{\rho}(g^{-1})f)$ in H, which can be written explicitly as
$$F(g)(x) = f(gx)$$
for $g \in G$ and $x \in \tilde{G}$. Thus F is a function from $\tilde{G}$ to H which obeys the symmetry

(2.4) $$F(gh) = \rho(h^{-1})F(g)$$

for all $g \in \tilde{G}$ and $h \in G$. Conversely, any function $F \colon \tilde{G} \to H$ obeying the symmetry (2.4) arises from an element f of $\tilde{H}$ in this manner. Thus, we may identify $\tilde{H}$ (as a vector space, at least), with the space of functions $F \colon \tilde{G} \to H$ obeying (2.4). Furthermore, the Hilbert space norm $\|f\|_{\tilde{H}} = \|f\|_{\ell^2(\tilde{G})}$ of $\tilde{G}$ can be expressed in terms of F via the identity
$$\|f\|_{\tilde{H}} = \Big(\sum_{g \in \tilde{G}/G} \|F(g)\|_H^2 \Big)^{1/2},$$
where we use the fact (from (2.4)) that $\|F(gh)\|_H = \|F(g)\|_H$ for all $h \in G$ to view (by abuse of notation) $\|F(\cdot)\|_H$ as a function of $\tilde{G}/G$ rather than of $\tilde{G}$. Similarly, given two vectors $f, f' \in \tilde{H}$ with associated functions $F, F' \colon \tilde{G} \to H$, the inner product $\langle f, f' \rangle_{\tilde{H}}$ can be recovered from F, F' by the formula
$$\langle f, f' \rangle_{\tilde{H}} = \sum_{g \in \tilde{G}/G} \langle F(g), F'(g) \rangle_H,$$
where we adopt the same abuse of notation as before.

Motivated by this example, we now have the following construction (essentially due to Frobenius).

Definition 2.2.1 (Induced representation). Let $\tilde{G}$ be a locally compact group, and let G be a subgroup of $\tilde{G}$ which is closed (and thus also locally compact). Suppose that there is a nonzero Radon measure $\mu_{\tilde{G}/G}$ on the quotient space $\tilde{G}/G$ which is invariant under the left-action of $\tilde{G}$ (i.e., it is a *Haar measure* of $\tilde{G}/G$). Let $\rho \colon G \to U(H)$ be a unitary representation of G. Then we define the *induced representation* $\tilde{\rho} = \operatorname{Ind}_G^{\tilde{G}} \rho \colon \tilde{G} \to U(\tilde{H})$ as

2.2. Induced representations and property (T)

follows. We define $\tilde{H}$ to be the (pre-)Hilbert space of all functions $F\colon \tilde{G} \to H$ obeying (2.4) for all $g \in \tilde{G}$ and $h \in G$, that are weakly measurable in the sense that $\lambda(F)$ is Borel measurable for all bounded linear functionals $\lambda\colon H \to \mathbf{C}$, and such that the norm

$$\|F\|_{\tilde{H}} := (\int_{g \in \tilde{G}/G} \|F(g)\|_H^2 \, d\mu_{\tilde{G}/G}(g))^{1/2}$$

is finite, where we abuse notation as before and view $\|F(\cdot)\|_H$ as a function of $\tilde{G}/G$. (Note from the separability of H that the function $\|F(\cdot)\|_H$ is measurable.) We also define the inner product on $\tilde{H}$ by

$$\langle F, F' \rangle_{\tilde{H}} := \int_{g \in \tilde{G}/G} \langle F(g), F'(g) \rangle_H \, d\mu_{\tilde{G}/G}(g),$$

and identify together elements of $\tilde{H}$ whose difference has norm zero, to obtain a genuine Hilbert space rather than a pre-Hilbert space. (We leave it to the reader to verify that this space is in fact complete; this is a "G-space" version of the standard argument establishing that $L^2(X, \mu)$ is complete for any measure space (X, μ).) We then define the representation $\tilde{\rho}$ on $\tilde{H}$ by the formula

$$\tilde{\rho}(g)F(x) := F(g^{-1}x)$$

for all $g, x \in \tilde{G}$; one can verify that this is a well-defined unitary representation.

Remark 2.2.2. Given a Haar measure $\mu_{\tilde{G}/G}$ on $\tilde{G}/G$ and a Haar measure μ_G on G, one can build a Haar measure $\mu_{\tilde{G}}$ on $\tilde{G}$ via the *Riesz representation theorem* and the integration formula

$$\int_{\tilde{G}} f(x) \, d\mu_{\tilde{G}}(x) := \int_{\tilde{G}/G} \left(\int_G f(yz) \, d\mu_G(z) \right) d\mu_{\tilde{G}/G}(y)$$

for $f \in C_c(\tilde{G})$, where we abuse notation by noting that the integrand is a G-right-invariant function of $y \in \tilde{G}$, and can thus be viewed as a function on the quotient space $\tilde{G}/G$. As Haar measures on groups are determined up to constants (as shown for instance in [**Ta2013**, §1.3]), we conclude that Haar measures on quotient spaces $\tilde{G}/G$, if they exist, are also determined up to constants. As such, the induced representation construction given above does not depend (up to isomorphism) on the choice of Haar measure on $\tilde{G}/G$. However, it is possible to have quotient spaces for which no Haar measure is available; for instance, the group of affine transformations $x \mapsto ax + b$ acts on $\mathbf{R}$ (which is then a quotient of the affine group by the stabiliser of a point), but without any invariant measure. It is possible to extend the induced representation construction to this setting also, but this is more technical; see [**Fo1995**] for details.

Example 2.2.3. If G is an open subgroup of $\tilde{G}$, and $\mu_{\tilde{G}/G}$ is counting measure, then the induced representation of the trivial one-dimensional representation of G is the quasiregular representation of $\tilde{G}$ on $\tilde{G}/G$, and the induced representation of the regular representation of G is the regular representation of $\tilde{G}$.

Exercise 2.2.1 (Transitivity of induction). Let $G_3 \leq G_2 \leq G_1$ be locally compact groups, such that G_3 has finite index in G_2, and G_2 has finite index in G_1. Show that for any unitary representation $\rho\colon G_3 \to U(H)$ of G_3, one has $\operatorname{Ind}_{G_2}^{G_1} \operatorname{Ind}_{G_3}^{G_2} \rho \equiv \operatorname{Ind}_{G_1}^{G_3} \rho$. (A similar statement is also true in the infinite index case, but is more technical to establish.)

As a first application of the induced representation construction (and the much simpler restricted representation construction), we show that one can pass from a group to a finite index subgroup as far as property (T) is concerned.

Proposition 2.2.4. Let $\tilde{G}$ be a locally compact group, and let G be a finite index closed subgroup of $\tilde{G}$. Then $\tilde{G}$ has property (T) if and only if G has property (T).

Proof. Suppose first that G has property (T). Let S be a compact generating subset of G. As G has finite index in $\tilde{G}$, we may find a finite set K in $\tilde{G}$ such that $KG = \tilde{G}$. Let $\rho\colon \tilde{G} \to U(H)$ be a unitary representation, and suppose that H has a unit vector v such that $\|sv - v\|_H \leq \varepsilon$ for all $s \in S \cup K$, where $\varepsilon > 0$ is a small quantity (independent of ρ) to be determined later. We will show that v lies within distance $O(\varepsilon)$ from a $\tilde{G}$-invariant vector (where the implied constant can depend on $S, G, \tilde{G}$ but not on ε), which will give the claim for ε small enough.

By Exercise 2.1.2(v) applied to G, we have $\operatorname{Kaz}(G, S) > 0$. By Exercise 2.1.4, we thus see that v lies within $O(\varepsilon)$ from a G-invariant vector, so by the triangle inequality we may assume without loss of generality (and adjusting ε slightly) that v is G-invariant. If $g \in \tilde{G}$ is arbitrary, we may write $g = k\gamma$ for some $k \in K$ and $\gamma \in G$. Then $\|gv - v\|_H = \|kv - v\|_H = O(\varepsilon)$. Thus the $\tilde{G}$-orbit of v lies in a ball of radius $O(\varepsilon)$ centred at v, and so by Exercise 2.1.10 this ball contains an invariant vector as required.

Conversely, suppose that $\tilde{G}$ has property (T). Let S be a compact generating subset of G, which we may assume without loss of generality to be a neighbourhood of the identity. Let $\rho\colon G \to U(H)$ be a unitary representation, and let v_0 be a unit vector such that

(2.5) $$\|sv_0 - v_0\|_H \leq \varepsilon$$

for all $s \in S$, where $\varepsilon > 0$ is a small quantity independent of ρ to be chosen later. Our task is to show that v_0 lies within $O(\varepsilon)$ of a G-invariant

2.2. Induced representations and property (T) 41

vector. Now we form the induced representation $\tilde{\rho} := \mathrm{Ind}_G^{\tilde{G}} \rho \colon \tilde{G} \to U(\tilde{H})$ (using counting measure for $\mu_{\tilde{G}/G}$). By construction, $\tilde{H}$ is the space of all functions $\tilde{v} \colon \tilde{G} \to H$ such that $\tilde{v}(x\gamma) := \rho(\gamma)^{-1}\tilde{v}(x)$ for all $\gamma \in G$ and $x \in \tilde{G}$. If we let K be a finite set consisting of one representative of each left-coset of G, it is easy to see that each element $\tilde{v} \colon \tilde{G} \to H$ of $\tilde{H}$ is determined by its restriction to K, and conversely every function from $\tilde{v} \colon K \to H$ can be extended uniquely to an element of $\tilde{H}$; thus $\tilde{H}$ can be identified with the direct sum $\bigoplus_{k \in K} H$ of $|K|$ copies of H. Also, $\tilde{H}$ contains H as a G-invariant subspace, by identifying each vector $v \in H$ with the function $\tilde{v}$ defined by setting $\tilde{v}(\gamma) = \rho(\gamma)^{-1}v$ for $\gamma \in G$ and $\tilde{v}(x) = 0$ for $x \notin G$. The actions of $\tilde{\rho}$ and ρ are then compatible in the sense that $\tilde{\rho}(\gamma)v = \rho(\gamma)v$ for all $\gamma \in G$ and $v \in H$.

Now consider the vector
$$\tilde{v} := \sum_{k \in K} \tilde{\rho}(k)v_0.$$

We now study its invariance properties of this vector with respect to $\tilde{S} := S \cup K$ (which generates $\tilde{G}$). For any $s \in \tilde{S}$ and $k \in K$, sk lies in a compact subset of $\tilde{G}$, and thus $sk = k's'$ for some $k' = k'(k,s)$ in K and $s' = s'(k,s)$ for some s' in a compact subset of G. Also, for fixed $s \in \tilde{S}$, the map $k \mapsto k'(k,s)$ is a permutation of K. Since S is a compact neighbourhood of the identity generating G, we see from compactness that there is a finite m such that $s'(k,s) \in S^m$ for all k, s. In particular, from (2.5) and the triangle inequality we have
$$\|\rho(s')v_0 - v_0\|_H \le m\varepsilon.$$
Since
$$\tilde{\rho}(s)\tilde{v} - \tilde{v} = \sum_{k \in K} \tilde{\rho}(k'(k,s))\left(\rho(s'(k,s))v_0 - v_0\right)$$
we conclude from the triangle inequality that
$$\|\tilde{\rho}(s)\tilde{v} - \tilde{v}\|_H = O(\varepsilon),$$
where the implied constant can depend on $m, S, K, G, \tilde{G}$ but not on ρ or ε. As $\tilde{G}$ has property (T), we conclude (for ε small enough) using Exercise 2.1.4 that $\tilde{v}$ lies within $O(\varepsilon)$ of an $\tilde{G}$-invariant vector w. In particular, as $\tilde{G}$-invariant vectors are also G-invariant, $w(1)$ is a G-invariant vector in H which is within $O(\varepsilon)$ of $\tilde{v}(1) = v_0$, as desired. □

Actually, with a bit more effort, one can generalise the above proposition from the finite index case to the finite covolume case, as was first observed by Kazhdan [**Ka1967**]. There is, however, a technical issue; once $\tilde{G}/G$ is not discrete, the original Hilbert space H does not embed naturally into the induced Hilbert space $\tilde{H}$ (basically because G now has measure zero in $\tilde{G}$, so there is no obvious way to embed $L^2(G)$ into $L^2(\tilde{G})$). This issue, however, can be dealt with by "convolving" with a suitable approximation to the identity $f \in C_c(\tilde{G})$, where $C_c(\tilde{G})$ is the space of continuous functions $f \colon \tilde{G} \to \mathbf{C}$ with compact support. More precisely, we have the following definition:

Definition 2.2.5. Let $\tilde{G}$ be a locally compact group, and let G be a subgroup of $\tilde{G}$ that is also a locally compact group. Let $\mu_{\tilde{G}/G}$ be a Haar measure on $\tilde{G}/G$, and let μ_G be a Haar measure on G. Let $\rho \colon G \to U(H)$ be a unitary representation, and let $\tilde{\rho} \colon \tilde{G} \to U(\tilde{H})$ be the induced representation $\tilde{\rho} = \operatorname{Ind}_G^{\tilde{G}} \rho$. Let $v \in H$, and let $f \in C_c(\tilde{G})$. Then we define the convolution $C_f(v) \in \tilde{H}$ of v by f to be the function $C_f(v) \colon \tilde{G} \to H$ given by the formula
$$C_f(v)(g) := \int_G f(gh)\rho(h)v\, d\mu_G(h).$$
One easily verifies that $C_f(v)$ is well defined and lies in $\tilde{H}$.

Related to the convolution operation will be the projection $\Pi_{\tilde{G}/G}(f) \in C_c(\tilde{G}/G)$ of a function $f \in C_c(\tilde{G})$, defined by the formula
$$\Pi_{\tilde{G}/G}(f)(g) = \int_G f(gh)\, d\mu_G(h),$$
where the right-hand side is right G-invariant in g, and can thus be viewed as a function of $\tilde{G}/G$ rather than $\tilde{G}$. We have the following key technical fact:

Lemma 2.2.6 (Surjectivity)**.** *The projection operator $\Pi_{\tilde{G}/G} \colon C_c(\tilde{G}) \to C_c(\tilde{G}/G)$ is surjective. Furthermore, given a nonnegative function F in $C_c(\tilde{G}/G)$, we may find a nonnegative function f' in $C_c(\tilde{G})$ with $\Pi_{\tilde{G}/G}(f') = F$.*

Proof. It suffices to prove the second claim. Let $F \colon \tilde{G}/G \to \mathbf{R}^+$ be a nonnegative continuous function supported on some compact subset K of $\tilde{G}/G$. By compactness, one can find a compact subset $\tilde{K}$ of $\tilde{G}$ which covers K in the sense that for every coset gG in the preimage of K, the set $\tilde{K} \cap gG$ is nonempty and open in gG (or equivalently, $g^{-1}\tilde{K} \cap G$ is open in G). By Urysohn's lemma, we may find a nonnegative function $f \in C_c(\tilde{G})$ which equals 1 on $\tilde{K}$, and then $\Pi_{\tilde{G}/G}(f)$ is nonzero on K. Thus we may write $F = \Pi_{\tilde{G}/G}(f) F'$ for some $F' \in C_c(\tilde{G}/G)$. If we let $\pi \colon \tilde{G} \to \tilde{G}/G$ be the projection map, then we easily verify that the function $f' := f(F' \circ \pi)$ lies in $C_c(\tilde{G})$ and is nonnegative with $F = \Pi_{\tilde{G}/G}(f')$, and the claim follows. $\square$

Now we generalise Proposition 2.2.4 to the finite covolume case:

2.2. Induced representations and property (T) 43

Proposition 2.2.7. *Let $\tilde{G}$ be a locally compact group, and let G be a locally compact subgroup of $\tilde{G}$. Suppose that G has finite covolume in $\tilde{G}$, which means that there exists a finite Haar measure $\mu_{\tilde{G}/G}$ on $\tilde{G}/G$. Then $\tilde{G}$ has property (T) if and only if G has property (T).*

Proof. We may normalise the Radon measure μ to be a probability measure. Suppose first that G has property (T). Let S be a compact generating subset of G, and let ε be chosen later. As $\mu(\tilde{G}/G) = 1$ and $\tilde{G}$ is compactly generated, we may use inner regularity and find a compact subset K of $\tilde{G}$ such that $\mu(\pi(K)) \geq 1 - \varepsilon$, where $\pi \colon \tilde{G} \to \tilde{G}/G$ is the projection map.

Now let $\rho \colon \tilde{G} \to U(H)$ be a unitary representation, and suppose that H has a unit vector v such that $\|sv - v\|_H \leq \varepsilon$ for all $s \in S \cup K$. As before, the goal is to show that v lies within distance $O(\varepsilon)$ from a $\tilde{G}$-invariant vector, which will give the claim for ε small enough.

By Exercise 2.1.2(v) and Exercise 2.1.4 as in the previous argument, we may assume that v is G-invariant. The function $g \mapsto \rho(g)v$ then descends from a bounded H-valued function on $\tilde{G}$ to a function on $F \colon \tilde{G}/G \to H$. We may thus form the average

$$\bar{v} := \int_{\tilde{G}/G} F(x) \, d\mu_{\tilde{G}/G}(x),$$

where we can define the H-valued integral in the weak sense, using bounded linear functionals $\lambda \colon H \to \mathbf{C}$ on H and the Riesz representation theorem for Hilbert spaces, noting that F is weakly integrable in the sense that $\lambda(F)$ is absolutely integrable for all bounded linear functionals λ. By the left-invariance of $\mu_{\tilde{G}/G}$ we see that $\bar{v}$ is $\tilde{G}$-invariant. Also, by construction, we have $\|F(x)\|_H = O(1)$ for all $x \in \tilde{G}/G$, and $\|F(x) - x\|_H = O(\varepsilon)$ for all $x \in \pi(K)$, which has measure $1 - \varepsilon$. As such we see that $\|\bar{v} - v\|_H = O(\varepsilon)$, and the claim follows.

Now suppose instead that $\tilde{G}$ has property (T). Let S be a compact generating subset of G, which we may assume without loss of generality to be a neighbourhood of the identity. Let $\rho \colon G \to U(H)$ be a unitary representation, and let v_0 be a unit vector obeying (2.5) for all $s \in S$, and some small $\varepsilon > 0$ (depending only on $S, G, \tilde{G}$ and to be chosen later). We will suppose for contradiction that H contains no nontrivial G-invariant vectors.

As before, the first step is to build the induced representation $\tilde{\rho} := \operatorname{Ind}_G^{\tilde{G}} \rho \colon \tilde{G} \to U(\tilde{H})$ of ρ, using the given Haar measure $\mu_{\tilde{G}/G}$. Let $\delta > 0$ be a sufficiently small quantity (depending on $S, G, \tilde{G}$, but not depending on ε) to be chosen later. By inner regularity, we can find a compact subset K of $\tilde{G}/G$ of measure $\mu_{\tilde{G}/G}(K) \geq 1 - \delta$. By Urysohn's lemma followed by Lemma 2.2.6, we may find a function $f \in C_c(\tilde{G})$ such that $\Pi_{\tilde{G}/G}(f)$ is bounded by

1 and equals 1 on K; in particular, it differs by 1 only on a set of measure $O(\delta)$. Now we consider the vector $v \in \tilde{H}$ defined by the formula

$$v := C_f(v_0),$$

thus

$$v(g) = \int_G f(gh)\rho(h)v_0 d\mu_G(h).$$

By the triangle inequality one has

(2.6) $$\|v(g)\|_H \leq \int_G f(gh)d\mu_G(h) = \Pi_{\tilde{G}/G}(f)(f) \leq 1$$

for all $g \in \tilde{G}$.

Let K' be a compact subset of $\tilde{G}$ with $\pi(K')$ containing K. For $g \in K'$ and $h \in G$ with gh in the support of f, we see from the approximate invariance of v_0 that

$$\|\rho(h)v_0 - v_0\|_H \ll_\delta \varepsilon$$

and thus

$$\|v(g) - \Pi_{\tilde{G}/G}f(g)v_0\|_H \ll_\delta \varepsilon.$$

In particular, we see that $\|v(g)\|_H$ is equal to $1 - O_\delta(\varepsilon)$ for all $g \in K$ (again abusing notation and descending from $\tilde{G}$ to $\tilde{G}/G$). From this and (2.6) we see that

$$1 \ll \|v\|_H \leq 1$$

if δ is small enough (and ε sufficiently small depending on δ).

Now we investigate the approximate invariance properties of v. Let $\tilde{S}$ be a compact generating subset of $\tilde{G}$ (not depending on δ or ε). For $g \in K'$ and $s \in \tilde{S}$, the preceding argument gives

$$\|v(g) - \Pi_{\tilde{G}/G}f(g)v_0\|_H \ll_\delta \varepsilon$$

and

$$\|v(s^{-1}g) - \Pi_{\tilde{G}/G}f(s^{-1}g)v_0\|_H \ll_\delta \varepsilon$$

and thus (by choice of $\Pi_{\tilde{G}/G}f$)

$$\|v(g) - v(s^{-1}g)\|_H \ll_\delta \varepsilon$$

whenever $g \in K \cap sK$ (again abusing notation). This is a measure $1 - O(\delta)$ subset of $\tilde{G}/G$. From this and (2.6) we see that

$$\|v - \tilde{\rho}(s)v\|_{\tilde{H}} \leq O_\delta(\varepsilon) + O(\delta^{1/2}).$$

Since $\tilde{G}$ has property (T), we conclude (if δ is small enough, and ε sufficiently small depending on δ) that there exists a nonzero $\tilde{G}$-invariant vector $w \in \tilde{H}$. Thus, for all $g \in \tilde{G}$, one has $w(gx) = w(x)$ for almost all $x \in \tilde{G}$ (using Haar measure on $\tilde{G}$, of course). By the Fubini-Tonelli theorem, this implies that for almost all x, one has $w(gx) = w(x)$ for almost all $g \in \tilde{G}$. Fixing such a

2.2. Induced representations and property (T)

x, we conclude in particular that w is almost everywhere equal to a constant $w_0 \in H$. By another application of Fubini-Tonelli, this implies the existence of a coset xG on which w is almost everywhere equal to w_0 (this time using the Haar measure coming from G). By (2.4), this makes w_0 G-invariant, and thus zero by choice of H. But this makes w zero almost everywhere, contradicting the nonzero nature of w. □

A discrete subgroup Γ of a locally compact group G with finite covolume is known as a *lattice*. Thus, for instance, $\mathbf{Z}^d$ is a lattice in $\mathbf{R}^d$. Here is another important example of a lattice, involving the *special linear groups* SL_d:

Proposition 2.2.8. *For any $d \geq 1$, $\mathrm{SL}_d(\mathbf{Z})$ (the group of $d \times d$ integer matrices of determinant one) is a lattice in $\mathrm{SL}_d(\mathbf{R})$ (the group of $d \times d$ real matrices of determinant one).*

Proof. Clearly, $\mathrm{SL}_d(\mathbf{Z})$ is discrete, so it suffices to show that there is a finite Haar measure on $\mathrm{SL}_d(\mathbf{R})/\mathrm{SL}_d(\mathbf{Z})$. It will suffice to show that there is a subset E of $\mathrm{SL}_d(\mathbf{R})$ of finite measure (with respect to a Haar measure on $\mathrm{SL}_d(\mathbf{R})$, of course) whose projection onto $\mathrm{SL}_d(\mathbf{R})/\mathrm{SL}_d(\mathbf{Z})$ is surjective. Indeed, if this is the case, then one can construct a fundamental domain K in E by selecting, for each left coset $g\,\mathrm{SL}_d(\mathbf{Z})$ of $\mathrm{SL}_d(\mathbf{Z})$, a single element of $g\,\mathrm{SL}_d(\mathbf{Z}) \cap E$ in some measurable fashion (noting that this set is discrete and so has finite intersection with every compact set, allowing one to locate minimal elements with respect to some measurable ordering on $\mathrm{SL}_d(\mathbf{R})$). As the translates of K by the countable group $\mathrm{SL}_d(\mathbf{Z})$ cover $\mathrm{SL}_d(\mathbf{R})$, K must have positive measure; and one can then construct a Haar measure on $\mathrm{SL}_d(\mathbf{R})/\mathrm{SL}_d(\mathbf{Z})$ by *pushing forward* the Haar measure on K.

One can interpret $\mathrm{SL}_d(\mathbf{R})$ as the space of all lattices in $\mathbf{R}^d$ generated by d marked generators $v_1, \ldots, v_d \in \mathbf{R}^d$ which are *unimodular* in the sense that $\det(v_1, \ldots, v_d) = 1$. The quotient space $\mathrm{SL}_d(\mathbf{R})/\mathrm{SL}_d(Z)$ can then be viewed as the space of all unimodular lattices *without* marked generators (since the action of $\mathrm{SL}_d(\mathbf{Z})$ simply serves to move one set of generators to another). Our task is thus to find a finite measure set of unimodular lattices with marked generators, which cover all unimodular lattices.

It will be slightly more convenient to work with $\mathrm{GL}_d(\mathbf{R})$ instead of $\mathrm{SL}_d(\mathbf{R})$; i.e., the space of all lattices (not necessarily unimodular) with marked generators $v_1, \ldots, v_d$. The reason for this is that there is a very simple Haar measure on $\mathrm{GL}_d(\mathbf{R})$, namely the Lebesgue measure $dv_1 \ldots dv_d$ on the generators (or equivalently, the measure induced from the open embedding $\mathrm{GL}_d(\mathbf{R}) \subset \mathbf{R}^{d^2}$); this is easily verified to be a Haar measure. One can then use dilation to convert a Haar measure on $\mathrm{GL}_d(\mathbf{R})$ to one on $\mathrm{SL}_d(\mathbf{R})$, for instance, by declaring the $\mathrm{SL}_d(\mathbf{R})$ Haar measure of a set $E \subset \mathrm{SL}_d(\mathbf{R})$ to be the $\mathrm{GL}_d(\mathbf{R})$ Haar measure of the set $\{tA : t \in [1,2]; A \in E\}$. Our task is now to find a finite measure set of lattices with marked generators, which cover all lattices of covolume in the interval $[1, 2^d]$. Clearly one can replace $[1, 2^d]$ here by any other compact interval in the positive real line.

This claim is trivial for $d = 1$, so suppose inductively that $d > 1$, and that the claim has already been proven for $d-1$. From *Minkowski's theorem* (see, e.g., [**TaVu2006**, Theorem 3.28]), every lattice Γ of covolume in $[1, 2^d]$ contains a nonzero vector v_d of norm $O(1)$, where we allow implied constants to depend on d. We may assume this vector to be irreducible, so that Γ is generated by v_d and $d-1$ other generators $v_1, \ldots, v_{d-1}$ that are independent of v_d. By subtracting or adding an integer multiple of v_d to these other generators, we may assume that they take the form $v_i = w_i + t_i n_d$ for some w_i orthogonal to v_d and some $t_i \in [0, |v_d|]$ for each $i = 1, \ldots, d-1$, where $n_d := v_d/|v_d|$ is the direction vector of v_d. Furthermore, $w_1, \ldots, w_{d-1}$ span a lattice in the $d-1$-dimensional space $n_d^\perp$ of covolume comparable to $1/|v_d|$.

For each fixed v_d, the parameters $t_1, \ldots, t_{d-1}$ range over a cube of $d-1$-dimensional Lebesgue measure $|v_d|^{d-1}$. By induction hypothesis and a rescaling argument, the $w_1, \ldots, w_{d-1}$ can be made (after identifying $n_d^\perp$ arbitrarily with $\mathbf{R}^{d-1}$) to range over a set of $\mathrm{GL}_{d-1}(\mathbf{R})$ of Haar measure $O(1/|v_d|)$. By the Fubini-Tonelli theorem (and the rotation-invariance of Lebesgue measure), we may thus cover all the lattices of covolume in $[1, 2^d]$ in $\mathbf{R}^d$ by a subset of $\mathrm{GL}_d(\mathbf{R})$ of measure at most

$$\int_{v_d \in \mathbf{R}^d : |v_d| = O(1)} |v_d|^{d-1} O(1/|v_d|) \, dv_d$$

which evaluates to $O(1)$, and the claim follows. □

Combining this fact with Proposition 2.2.7, we obtain

Corollary 2.2.9. *For any $d \geq 1$, $\mathrm{SL}_d(\mathbf{R})$ has property (T) if and only if $\mathrm{SL}_d(\mathbf{Z})$ has property (T).*

The usefulness of this corollary lies in the fact that there is a certain asymptotic conjugation argument of Mautner and Moore which is available for connected Lie groups such as $\mathrm{SL}_d(\mathbf{R})$, but not for discrete groups such as $\mathrm{SL}_d(\mathbf{Z})$, and allows one to boost the invariance properties of a vector; see Proposition 2.3.9 below.

We will now study the property (T) nature of the special linear group.

Remark 2.2.10. Another consequence of Proposition 2.2.7 (and Remark 2.1.10) is that if a locally compact group $\tilde{G}$ has property (T), then all lattices G in $\tilde{G}$ are finitely generated; this was one of Kazhdan's original applications[6] of property (T) in [**Ka1967**]. It is surprisingly difficult to replicate this result for, say, $SL_3(\mathbf{R})$, without using property (T) (or something very close to it).

2.3. The special linear group and property (T)

The purpose of this section is to prove the following theorem of Kazhdan [**Ka1967**]:

Theorem 2.3.1. $SL_d(\mathbf{R})$ *has property (T) if and only if* $d \neq 2$.

Combining this theorem with Corollary 2.2.9 and Exercise 2.1.9, we obtain some explicit families of expanders:

Corollary 2.3.2 (Margulis' expander construction). *If $d \geq 3$ and S is a symmetric set of generators of $SL_d(\mathbf{Z})$ that does not contain the identity, then the Cayley graphs $\mathrm{Cay}(SL_d(\mathbf{Z}/n\mathbf{Z}), \pi_n(S))$ form an expander family, where $\pi_n \colon SL_d(\mathbf{Z}) \to SL_d(\mathbf{Z}/n\mathbf{Z})$ is the obvious projection homomorphism.*

We now prove this theorem. We first deal with the $d = 1, 2$ cases. The group $SL_1(\mathbf{R})$ is trivial and thus has property (T). As for $SL_2(\mathbf{R})$, we may rule out property (T) by using the following basic fact:

Lemma 2.3.3. $SL_2(\mathbf{R})$ *contains a lattice isomorphic to the free group* F_2 *on two generators.*

Indeed, from this lemma, Proposition 2.2.7, and Exercise 2.1.15, we conclude that $SL_2(\mathbf{R})$ does not have property (T).

A proof of the above lemma is given in the exercise below.

Exercise 2.3.1. Let Γ be the subgroup of $SL_2(\mathbf{Z})$ (and hence of $SL_2(\mathbf{R})$) generated by the elements $a := \begin{pmatrix} 1 & 2 \\ 0 & 1 \end{pmatrix}$ and $b := \begin{pmatrix} 1 & 0 \\ 2 & 1 \end{pmatrix}$.

  (i) If $A := \{(x,y) \in \mathbf{R}^2 : |x| < |y|\}$ and $B := \{(x,y) \in \mathbf{R}^2 : |x| > |y|\}$, show that $a^n A \subset B$ and $b^n B \subset A$ for any nonzero integer n, where $SL_2(\mathbf{R})$ acts on $\mathbf{R}^2$ in the obvious manner.
  (ii) Show that Γ is a free group on two generators. (*Hint:* Use (i) to show that any *reduced word*[7] of a, b that both begins and ends with $a^{\pm 1}$, or begins and ends with $b^{\pm 1}$, is not equal to the identity. This argument is a variant of the *ping-pong lemma* argument used in [**Ti1972**].)

---

[6]Strictly speaking, one has to modify the proof of Proposition 2.2.7 to obtain this application, because one is not allowed to assume that G is compactly generated any more; however, if one inspects the proof, one sees that the set S in that proof does not need to generate all of G, but merely needs to generate those h for which gh lies in the support of f for some $g \in K'$. As this is already a compact set, we can remove the hypothesis that G is compactly generated.

[7]A word in a, b, a^{-1}, b^{-1} is *reduced* if a and a^{-1} do not appear adjacent to each other, and similarly for b, b^{-1}.

(iii) Show that Γ has finite index in $\mathrm{SL}_2(\mathbf{Z})$. (*Hint:* Show that given an element of $\mathrm{SL}_2(\mathbf{Z})$ with columns $v, w \in \mathbf{Z}^2$, one can multiply this element on the left by some word in a, b to minimise the magnitude $|v \cdot w|$ of the dot product, until one reaches a point where $|v \cdot w| \leq \|v\|^2, \|w\|^2$. Now use the Lagrange identity $|v \cdot w|^2 + 1 = \|v\|^2 \|w\|^2$ to conclude that v, w have bounded size.)

(iv) Establish Lemma 2.3.3.

Now we turn to the higher dimensional cases $d \geq 3$. The idea is to first use Fourier analysis[8] to understand the action of various simpler subgroups of $\mathrm{SL}_d(\mathbf{R})$ acting on a space H with approximately invariant vectors, and obtain nontrivial vectors that are invariant with respect to those simpler subgroups. Then, we will use an asymptotic conjugation trick of Mautner to boost this invariance up to increasingly larger groups, until we obtain a nontrivial vector invariant under the whole group $\mathrm{SL}_d(\mathbf{R})$.

We begin with a preliminary result, reminiscent of property (T) but in the category of probability measures.

Lemma 2.3.4. *Let S be a compact neighbourhood of the identity in $\mathrm{SL}_2(\mathbf{R})$, and let $\varepsilon > 0$. Suppose that μ is a probability measure on $\mathbf{R}^2$ with the property that*
$$\|s_* \mu - \mu\|_{TV} \leq \varepsilon$$
for all $s \in S$, where s acts on $\mathbf{R}^2$ in the obvious manner, $s_* \mu$ is the pushforward of μ by s (thus $s_* \mu(E) := \mu(s^{-1}(E))$ for all measurable E), and $\|\|_{TV}$ denotes the total variation norm of a measure. Then $\mu(\{0\}) = 1 - O(\varepsilon)$, where the implied constant can depend on S.

Proof. We modify[9] the argument used to establish Exercise 2.3.1. Let A, B, a, b be as in that exercise. Then
$$\|a_* \mu - \mu\|_{TV} = O(\varepsilon)$$
and thus
$$\mu(B) \geq \mu(aA) = \mu(A) + O(\varepsilon)$$
and similarly
$$\mu(A) \geq \mu(bB) = \mu(B) + O(\varepsilon).$$
Putting these estimates together, we conclude that
$$\mu(B) = \mu(aA) + O(\varepsilon).$$

---

[8] As such, this section will presuppose some familiarity with Fourier analysis, as reviewed, for instance, in [**Ta2010**, §1.12].

[9] There are many other proofs available, but this one has the advantage of extending without difficulty to the integer setting of $\mathrm{SL}_2(\mathbf{Z})$.

2.3. The special linear group and property (T)

Similarly one has
$$\mu(B) = \mu(a^2 A) + O(\varepsilon).$$
Since
$$B \backslash a^2 A \supset \{(x,y) \in \mathbf{R}^2 : |y| < |x| \leq 3|y|\},$$
we conclude that the set $\{(x,y) \in \mathbf{R}^2 : |y| < |x| \leq 3|y|\}$ has measure $O(\varepsilon)$. Translating this set around by a finite number of explicit elements of $SL_2(\mathbf{R})$, we conclude that $\mu(\mathbf{R}^2 \backslash \{0\}) = O(\varepsilon)$, and the claim follows. □

Now we use Fourier analysis to pass from probability measures back to Hilbert spaces. We will need (a special case of) a fundamental result from abstract harmonic analysis, namely *Bochner's theorem*:

Proposition 2.3.5 (Bochner's theorem for $\mathbf{R}^d$). *Let $f : \mathbf{R}^d \to \mathbf{C}$ be a bounded continuous function which is positive semidefinite, in the sense that $f(x) = \overline{f(-x)}$ for all $x \in \mathbf{R}^d$, and*

$$(2.7) \qquad \int_{\mathbf{R}^d} \int_{\mathbf{R}^d} f(x-y) \, d\nu(x) d\overline{\nu}(y) \geq 0$$

for all finite complex measures ν. Then there exists a nonnegative finite measure μ on $\mathbf{R}^d$ such that f is the inverse Fourier transform of μ, in the sense that

$$f(x) = \int_{\mathbf{R}^d} e^{2\pi i x \cdot \xi} \, d\mu(\xi)$$

for all $x \in \mathbf{R}^d$.

Proof. Suppose first that f was square-integrable. Then by Plancherel's theorem, there is a square-integrable Fourier transform $\hat{f}$ for which one has

$$f(x) = \int_{\mathbf{R}^d} e^{2\pi i x \cdot \xi} \hat{f}(\xi) \, d\xi$$

in the sense of tempered distributions (or in an L^2 approximation sense). From (2.7) (applied to a smooth measure $d\nu(x) = g(x) \, dx$) and standard Fourier identities, one then has

$$\int_{\mathbf{R}^d} \hat{f}(\xi) |\hat{g}(\xi)|^2 \, d\xi \geq 0$$

for any Schwartz function $\hat{g}$. From this and the Lebesgue differentiation theorem we see that $\hat{f}$ is nonnegative almost everywhere. By testing f against an approximation $R^d \phi(Rx)$ to the identity, we also see from the continuity of f that

$$f(0) = \lim_{R \to \infty} \int_{\mathbf{R}^d} \hat{f}(\xi) \hat{\phi}(\xi/R) \, d\xi$$

which (after choosing ϕ to have nonnegative Fourier transform) we see that $\hat{f}$ is absolutely integrable. Setting $d\mu(\xi) := \hat{f}(\xi) \, d\xi$, one obtains the claim.

Now we consider the general case. We consider a truncation $f_R(x) := f(x)\psi(x/R)$ of f for some large $R > 0$, where $\psi = \eta * \eta$ and η is a real even Schwartz function with unit L^2 norm. From the identity

$$\int_{\mathbf{R}^d} \int_{\mathbf{R}^d} f_R(x-y) \, d\nu(x) d\overline{\nu}(y)$$
$$= \int_{\mathbf{R}^d} |\hat{\eta}(\xi)|^2 \int_{\mathbf{R}^d} f(x-y) \, e^{2\pi i \xi \cdot x} d\nu(x) e^{-2\pi i \xi \cdot y} d\overline{\nu}(y)$$

we see that f_R is also positive semidefinite, and is thus the Fourier transform of a finite nonnegative measure μ_R, with $\mu_R(\mathbf{R}^d) = f_R(0) = f(0)$. Since the f_R converge in the sense of tempered distributions to f, μ_R must converge in distribution to the distributional Fourier transform of f. In particular, by the Riesz representation theorem, $\hat{f}$ must be another finite nonnegative measure, and the claim follows. $\square$

Remark 2.3.6. Bochner's theorem can be extended to arbitrary locally compact abelian groups, and this fact can be used to build[10] the foundation of Fourier analysis on such groups; see for instance [**Ru1990**] for details. There are several substitutes for Fourier analysis that can serve this purpose, such as spectral theory or the Gelfand theory of C^* algebras, but we will not discuss these topics here.

We can use Bochner's theorem to analyse[11] unitary representations $\rho \colon \mathbf{R}^d \to U(H)$ of Euclidean groups. Given a vector v in the Hilbert space H, we consider the associated *autocorrelation function* $f_{v,v} \colon \mathbf{R}^d \to \mathbf{C}$ defined by the formula

$$f_{v,v}(x) := \langle \rho(x)v, v \rangle_H.$$

This is a continuous bounded function of $\mathbf{R}^d$, and from the identity

$$\int_{\mathbf{R}^d} \int_{\mathbf{R}^d} f_{v,v}(x-y) \, d\nu(x) d\overline{\nu}(y) = \left\| \int_{\mathbf{R}} \rho(x)v \, d\nu(x) \right\|_H^2$$

we see that it is positive semidefinite. Thus, by Bochner's theorem, there exists a nonnegative finite measure $\mu_{v,v}$ whose inverse Fourier transform is $f_{v,v}$, thus

(2.8) $$\langle \rho(x)v, v \rangle_H = \int_{\mathbf{R}^d} e^{2\pi i x \cdot \xi} \, d\mu_{v,v}(\xi).$$

In particular, $\mu_{v,v}$ has total mass $\|v\|_H^2$. From a quantum mechanical viewpoint, one can view $\mu_{v,v}$ as the probability distribution of the momentum of v (now viewed as a quantum state, and normalising v to be a unit vector).

---

[10] Note though that in order to do this without circularity, one needs a different proof than the one above, which presupposes Plancherel's theorem, which in the case of general locally compact abelian groups is usually proven using Bochner's theorem.

[11] Actually, much the same analysis will apply to unitary representations of arbitrary locally compact abelian groups, but we will only need to work with $\mathbf{R}^d$ (and more specifically, $\mathbf{R}^2$) here.

2.3. The special linear group and property (T)

By depolarisation, we can then assign a complex finite measure $\mu_{v,w}$ to any pair of vectors $v, w \in H$ such that

$$\langle \rho(x)v, w \rangle_H = \int_{\mathbf{R}^d} e^{2\pi i x \cdot \xi} \, d\mu_{v,w}(\xi).$$

Indeed, one can explicitly define $\mu_{v,w}$ using the *polarisation identity* as

$$\mu_{v,w} := \frac{1}{4}(\mu_{v+w,v+w} - \mu_{v-w,v-w} + i\mu_{v+iw,v+iw} - i\mu_{v-iw,v-iw}).$$

By the uniqueness of Fourier inversion, we see that $\mu_{v,w}$ is uniquely determined by v, w, and is sesquilinear with respect to these inputs. By depolarising with the right normalisations, we see that this measure has total mass $O(\|v\|_H \|w\|_H)$.

Exercise 2.3.2 (Functional calculus). Show that for any bounded Borel-measurable function $m\colon \mathbf{R}^d \to \mathbf{C}$, there is a bounded operator $m(\rho)\colon H \to H$ such that

$$\langle m(\rho)v, w \rangle = \int_{\mathbf{R}^d} m(\xi) d\mu_{v,w}(\xi)$$

for all unit vectors v, w, with the operator norm of $m(\rho)$ bounded by the supremum norm of m. Furthermore, show the map $m \mapsto m(\rho)$ is a *-homomorphism of *-algebras, thus it is a (complex) algebra homomorphism that also preserves the conjugation operation. In particular, for any Borel set E, the operator $\mu(E) := 1_E(\rho)$ is an orthogonal projection on H. Show that μ is a countably additive measure taking values as orthogonal projections on H, with $\mu(\mathbf{R}^d)$ equal to the identity operator on H. Define a notion of integration with respect to such measures in such a way that one has the identities

$$\rho(x) = \int_{\mathbf{R}^d} e^{2\pi i x \cdot \xi} \, d\mu(\xi)$$

and

$$m(\rho) = \int_{\mathbf{R}^d} m(\xi) \, d\mu(\xi)$$

for all $x \in \mathbf{R}^d$ and bounded Borel-measurable m. (This exercise is not explicitly used in the sequel, though the functional calculus perspective is definitely lurking beneath the surface in the arguments below. One can use the results of this exercise to establish *Stone's theorem on one-parameter groups* [**St1932**]; see Theorem 10.3.3.)

Now we can obtain a relative version of property (T), relating the Euclidean group $\mathbf{R}^2$ with the semidirect product $\mathrm{SL}_2(\mathbf{R}) \ltimes \mathbf{R}^2$, defined in the obvious manner.

Proposition 2.3.7. *Let S be a compact neighbourhood of the identity in $\mathrm{SL}_2(\mathbf{R}) \ltimes \mathbf{R}^2$, and let $\rho\colon \mathrm{SL}_2(\mathbf{R}) \ltimes \mathbf{R}^2 \to U(H)$ be a unitary representation.*

If $\mathrm{Kaz}(\mathrm{SL}_2(\mathbf{R}) \ltimes \mathbf{R}^2, S, H)$ *is sufficiently small depending on* S, *then* H *contains a nontrivial* $\mathbf{R}^2$*-invariant vector.*

Remark 2.3.8. Another way of stating the conclusion of this proposition is that the pair $(\mathrm{SL}_2(\mathbf{R}) \ltimes \mathbf{R}^2, \mathbf{R}^2)$ of locally compact groups has relative property (T). See [BedeVa2008] for a more thorough discussion of this property.

Proof. Suppose that $\mathrm{Kaz}(\mathrm{SL}_2(\mathbf{R}) \ltimes \mathbf{R}^2, S, H) < \varepsilon$ for some sufficiently small $\varepsilon > 0$, then there is a unit vector v in H such that $\|\rho(s)v - v\|_H \leq \varepsilon$ for all $s \in S$. Now let g be an element of $\mathrm{SL}_2(\mathbf{R})$ (which we can view as a subgroup of $\mathrm{SL}_2(\mathbf{R}) \ltimes \mathbf{R}^2$, and similarly for $\mathbf{R}^2$). Observe that

$$\langle \rho(x)\rho(g)v, \rho(g)v \rangle_H = \langle \rho(g(x))v, v \rangle_H$$

for all $x \in \mathbf{R}^2$. Comparing this with the Fourier inversion formula (2.8) we see that

$$\mu_{\rho(g)v, \rho(g)v} = (g^*)_* \mu_{v,v},$$

where $(g^*)_*$ is the pushforward by the adjoint g^* of g. By the sesquilinearity and boundedness of $\mu_{v,w}$, we thus see that

$$\|(g^*)_* \mu_{v,v} - \mu_{v,v}\|_{TV} \ll \|\rho(g)v - v\|_H.$$

By Lemma 2.3.4, we conclude that

$$\mu_{v,v}(\{0\}) \geq 1 - O(\varepsilon)$$

which, by (2.8), implies that

$$\langle \rho(x)v, v \rangle_H = 1 - O(\varepsilon)$$

for all $x \in \mathbf{R}^2$. Using Exercise 2.1.10, we conclude that H has a nontrivial $\mathbf{R}^2$-invariant vector, as claimed. $\square$

The above proposition gives a vector which is invariant with respect to the action of an abelian group, namely $\mathbf{R}^2$. The next lemma, using an argument of Moore exploiting an asymptotic conjugation idea of Mautner, shows how to boost invariance from a small abelian group to a larger nonabelian group. We will only need this argument in the context of $\mathrm{SL}_2(\mathbf{R})$, and using the three subgroups

$$U^+ := \{u_+(t) : t \in \mathbf{R}\},$$
$$D := \{d(t) : t \in \mathbf{R}\},$$
$$U^- := \{u_-(t) : t \in \mathbf{R}\}$$

2.3. The special linear group and property (T)

of $SL_2(\mathbf{R})$, where

$$u_+(t) := \begin{pmatrix} 1 & t \\ 0 & 1 \end{pmatrix},$$

$$d(t) := \begin{pmatrix} e^t & 0 \\ 0 & e^{-t} \end{pmatrix},$$

$$u_-(t) := \begin{pmatrix} 1 & 0 \\ t & 1 \end{pmatrix}.$$

Proposition 2.3.9 (Mautner phenomenon). *Let $\rho\colon SL_2(\mathbf{R}) \to U(H)$ be a unitary representation. Then any vector v which is U^+-invariant, is also $SL_2(\mathbf{R})$ invariant.*

Proof. The main idea (due to Moore [**Mo1966**]) is to show that D can be approximated by double cosets $U^+ u_-(\varepsilon) U^+$ for ε arbitrarily small. More precisely, we will use the identity

$$\begin{pmatrix} e^t & 0 \\ \varepsilon & e^{-t} \end{pmatrix} = u_+(\frac{e^t - 1}{\varepsilon}) u_-(\varepsilon) u_+(\frac{e^{-t} - 1}{\varepsilon}) \in U^+ u_-(\varepsilon) U^+$$

for any $t \in \mathbf{R}$ and $\varepsilon > 0$. In particular, from the U^+-invariance of v, we have

$$\left\langle \rho\left(\begin{pmatrix} e^t & 0 \\ \varepsilon & e^{-t} \end{pmatrix}\right) v, v \right\rangle_H = \langle u_-(\varepsilon) v, v \rangle_H.$$

Sending $\varepsilon \to 0$ we conclude that

$$\langle d(t) v, v \rangle_H = \langle v, v \rangle_H;$$

since $d(t)v$ has the same length as v, we conclude that $d(t)v = v$, thus v is D-invariant.

Now we use a similar argument of Mautner [**Ma1957**] to finish up. Starting with the identity

$$d(t) u_-(s) d(-t) = u_-(e^{-t} s)$$

for $s, t \in \mathbf{R}$, we see from the D-invariance of v that

$$\langle u_-(s) v, v \rangle = \langle u_-(e^{-t} s) v, v \rangle.$$

Sending $t \to \infty$ and arguing as before we conclude that v is also U^--invariant. Since U^+, D, U^- generate $SL_2(\mathbf{R})$, the claim follows. $\square$

Remark 2.3.10. As a corollary of the above proposition, we see that if a probability space (X, μ) with a measure-preserving action of $SL_2(\mathbf{R})$ is ergodic with respect to the $SL_2(\mathbf{R})$ action (in the sense that all $SL_2(\mathbf{R})$-invariant sets have either full measure or zero measure, or equivalently that $L^2(X, \mu)_0$ has no nontrivial $SL_2(\mathbf{R})$-invariant vectors), then it is necessarily ergodic with respect to the U^+ action as well. Thus, for instance, the action of U^+ on $\mathrm{SL}_2(\mathbf{R})/\mathrm{SL}_2(\mathbf{Z})$ (which is known as the *horocycle flow*) is ergodic. This is a special case of an ergodic theorem of Moore [**Mo1966**].

We can now establish that $\mathrm{SL}_3(\mathbf{R})$ has property (T) by navigating between various subgroups of that group. Indeed, let S be a compact neighbourhood of the identity in $\mathrm{SL}_3(\mathbf{R})$, and suppose that $\rho\colon \mathrm{SL}_3(\mathbf{R}) \to U(H)$ is a unitary representation with $\mathrm{Kaz}(\mathrm{SL}_3(\mathbf{R}), S, \rho)$ sufficiently small. We need to show that H contains an $\mathrm{SL}_3(\mathbf{R})$-invariant vector. To do this, we first note that $\mathrm{SL}_3(\mathbf{R})$ contains a copy of the semidirect product $\mathrm{SL}_2(\mathbf{R}) \ltimes \mathbf{R}^2$, namely the space of all matrices in $\mathrm{SL}_3(\mathbf{R})$ of the form

$$\begin{pmatrix} * & * & * \\ * & * & * \\ 0 & 0 & 1 \end{pmatrix}$$

where the entries marked $*$ are unconstrained (beyond the SL_3 requirement that the entire matrix have determinant 1). Applying Proposition 2.3.7, we conclude that H contains a nontrivial vector v which is invariant with respect to the matrices of the form

(2.9) $$\begin{pmatrix} 1 & 0 & * \\ 0 & 1 & * \\ 0 & 0 & 1 \end{pmatrix}.$$

Now we work with a copy of $\mathrm{SL}_2(\mathbf{R})$ in $\mathrm{SL}_3(\mathbf{R})$, namely the matrices in $\mathrm{SL}_3(\mathbf{R})$ of the form

(2.10) $$\begin{pmatrix} * & 0 & * \\ 0 & 1 & 0 \\ * & 0 & * \end{pmatrix}.$$

The associated copy of U^+ here is a subgroup of the matrices of the form (2.9). Applying Proposition 2.3.9, we see that v is invariant under the matrices in $\mathrm{SL}_3(\mathbf{R})$ of the form (2.9). A similar argument shows that v is also invariant with respect to matrices in $\mathrm{SL}_3(\mathbf{R})$ of the form

(2.11) $$\begin{pmatrix} 1 & 0 & 0 \\ 0 & * & * \\ 0 & * & * \end{pmatrix}.$$

But it is easy to see (e.g., by working with the Lie algebras) that (2.10), (2.11) generate all of $\mathrm{SL}_3(\mathbf{R})$, and the claim follows.

Exercise 2.3.3. Adapt the above argument to larger values of d to finish the proof of Theorem 2.3.1. (*Hint:* One either has to extend Lemma 2.3.4 to higher dimensions, or else use a version of the Mautner argument to boost invariance with respect to, say, a copy of $\mathrm{SL}_3(\mathbf{R})$, to invariance with respect to larger subgroups of $\mathrm{SL}_d(\mathbf{R})$.)

2.4. A more elementary approach

In Corollary 2.3.2 we constructed an explicit family of expander graphs, but the verification of the expander graph property was quite complicated, involving for instance the theory of induced representations, Bochner's theorem, the Riesz representation theorem, and many other tools besides. It turns out that this is overkill; if all one wants to do is construct expanders (as opposed to establishing property (T) for various groups), one can skip much of the above theory and establish expansion by more elementary methods (one still needs some Fourier analysis, but now just for finite abelian groups). In this section we outline this approach, following the work of Gabber-Galil [**GaGa1981**] and Jimbo-Maruocka [**JiMa1987**], as presented in [**HoLiWi2006**].

To avoid the need to exploit Mautner's phenomenon, the example is based on the semidirect product $SL_2(\mathbf{R}) \ltimes \mathbf{R}^2$ (or more accurately, the lattice $SL_2(\mathbf{Z}) \ltimes \mathbf{Z}^2$) rather than $SL_d(\mathbf{R})$ or $SL_d(\mathbf{Z})$. More precisely, we show

Theorem 2.4.1. *Let S be a symmetric finite set generating $SL_2(\mathbf{Z}) \ltimes \mathbf{Z}^2$. Then the Schreier graphs*[12] $Sch((\mathbf{Z}/n\mathbf{Z})^2, \pi_n(S))$ *form a one-sided expander family, where* $\pi_n \colon SL_2(\mathbf{Z}) \ltimes \mathbf{Z}^2 \to SL_2(\mathbf{Z}/n\mathbf{Z}) \ltimes (\mathbf{Z}/n\mathbf{Z})^2$ *is the obvious projection homomorphism.*

One can deduce this theorem from Proposition 2.3.7 and (a relative version of) Proposition 2.2.7. We will not do so here, but instead establish the theorem directly. We first need a discrete variant of Lemma 2.3.4:

Exercise 2.4.1. Let S be a finite generating subset of $SL_2(\mathbf{Z})$, and let $\varepsilon > 0$. Suppose that μ is a probability measure on $\mathbf{Z}^2$ with the property that

$$\|s_*\mu - \mu\|_{TV} \leq \varepsilon$$

for all $s \in S$, where s acts on $\mathbf{Z}^2$ in the obvious manner. Show that $\mu(\{0\}) = 1 - O(\varepsilon)$, where the implied constant can depend on S.

Now we prove the theorem. Suppose for contradiction that the Schreier graphs do not form a one-sided expander family. Then we obtain a family of nearly invariant vectors:

Exercise 2.4.2. With the above assumption, show that after passing to a subsequence of n's, one can find a sequence $f_n \in \ell^2((\mathbf{Z}/n\mathbf{Z})^2)$ of mean zero functions of ℓ^2 norm 1 such that

$$\sup_{s \in S} \|\rho_n(s)f_n - f_n\|_{\ell^2((\mathbf{Z}/n\mathbf{Z})^2)} = o(1),$$

---

[12]Here, we allow the Schreier graphs to contain loops or repeated edges; one has to check that the theory of expander graphs used here extends to this setting, but this is routine and will be glossed over here.

where ρ_n is the quasiregular representation of $SL_2(\mathbf{Z}) \ltimes \mathbf{Z}^2$ on $(\mathbf{Z}/n\mathbf{Z})^2$, and $o(1)$ denotes a quantity that goes to zero as $n \to \infty$.

Let f_n be as in the above exercise. If we let e_1, e_2 be the generators of the translation group $\mathbf{Z}^2$, we see, in particular, that
$$\|\rho_n(e_j)f_n - f_n\|_{\ell^2((\mathbf{Z}/n\mathbf{Z})^2)} = o(1)$$
for $j = 1, 2$. If we then introduce the finite Fourier transform
$$\hat{f}_n(\xi_1, \xi_2) := \frac{1}{n} \sum_{x_1, x_2 \in \mathbf{Z}/n\mathbf{Z}} f_n(x_1, x_2) e^{-2\pi i (x_1\xi_1 + x_2\xi_2)/n},$$
normalised to be an isometry on $\ell^2((\mathbf{Z}/n\mathbf{Z})^2)$, we conclude from Plancherel's theorem that
$$\|(e^{-2\pi i \xi_j/n} - 1)\hat{f}_n(\xi_1, \xi_2)\|_{\ell^2_{\xi_1, \xi_2}((\mathbf{Z}/n\mathbf{Z})^2)} = o(1).$$
In particular, we can find a ball B_n of radius $o(n)$ centred at the origin in $(\mathbf{Z}/n\mathbf{Z})^2$ on which $\hat{f}_n$ concentrates almost all of its ℓ^2 mass:
$$\|\hat{f}_n\|_{\ell^2((\mathbf{Z}/n\mathbf{Z})^2 \setminus B_n)} = o(1).$$
Let $g_n \in \ell^2(\mathbf{Z}^2)$ be the restriction of $\hat{f}_n$ to B_n, which one then identifies with a subset of $\mathbf{Z}^2$. Then we have
$$\|g_n\|_{\ell^2(\mathbf{Z}^2)} = 1 - o(1).$$
If s is any fixed element of $SL_2(\mathbf{Z})$, then we have
$$\|\rho_n(s)f_n - f_n\|_{\ell^2((\mathbf{Z}/n\mathbf{Z})^2)} = o(1)$$
and thus by Fourier duality
$$\|\hat{f}_n \circ s^* - \hat{f}_n\|_{\ell^2((\mathbf{Z}/n\mathbf{Z})^2)} = o(1).$$
Restricting to B_n and then embedding into $\mathbf{Z}^2$, we conclude that
$$\|g_n \circ s^* - g_n\|_{\ell^2(\mathbf{Z}^2)} = o(1).$$
Applying Exercise 2.4.1, we conclude that
$$g_n(0) = 1 - o(1).$$
But as the f_n have mean zero, we have $g_n(0) = 0$, giving the desired contradiction.

Remark 2.4.2. One advantage of this more elementary approach is that it is easier to obtain explicit bounds on the expansion constant of these graphs; see [**JiMa1987**] for details.

Chapter 3

Quasirandom groups

In Chapter 2 we saw how a representation-theoretic property of groups, namely *Kazhdan's property (T)*, could be used to demonstrate expansion in Cayley graphs. In this chapter we discuss a different representation-theoretic property of groups, namely *quasirandomness*, which is also useful for demonstrating expansion in Cayley graphs, though in a somewhat different way to property (T). For instance, whereas property (T), being qualitative in nature, is only interesting for infinite groups such as $SL_d(\mathbf{Z})$ or $SL_d(\mathbf{R})$, and only creates Cayley graphs after passing to a finite quotient, quasirandomness is a quantitative property which is directly applicable to finite groups, and is able to deduce expansion in a Cayley graph, provided that random walks in that graph are known to become sufficiently "flat" in a certain sense.

The definition of quasirandomness is easy enough to state:

Definition 3.0.1 (Quasirandom groups). Let G be a finite group, and let $D \geq 1$. We say that G is *D-quasirandom* if all nontrivial unitary representations $\rho\colon G \to U(H)$ of G have dimension at least D. (Recall a representation is *trivial* if $\rho(g)$ is the identity for all $g \in G$.)

Exercise 3.0.1. Let G be a finite group, and let $D \geq 1$. A unitary representation $\rho\colon G \to U(H)$ is said to be *irreducible* if H has no G-invariant subspaces other than $\{0\}$ and H. Show that G is D-quasirandom if and only if every nontrivial irreducible representation of G has dimension at least D.

Remark 3.0.2. The terminology "quasirandom group" was introduced explicitly (though with slightly different notational conventions) in [**Go2008**]; the name arises because dense Cayley graphs in quasirandom groups are quasirandom graphs in the sense of Chung, Graham, and Wilson in [**ChGrWi1989**], as we shall see below. This property had already been used implicitly to construct expander graphs in [**SaXu1991**], [**Ga2002**], [**BoGa2008**]. One can of course define quasirandomness for more general locally compact groups[1] than the finite ones, but we will only need this concept in the finite case.

Quasirandomness behaves fairly well with respect to quotients and short exact sequences:

Exercise 3.0.2. Let $0 \to H \to G \to K \to 0$ be a short exact sequence of finite groups H, G, K.

  (i) If G is D-quasirandom, show that K is D-quasirandom also. (Equivalently, any quotient of a D-quasirandom finite group is again a D-quasirandom finite group.)

  (ii) Conversely, if H and K are both D-quasirandom, show that G is D-quasirandom also. (In particular, the direct or semidirect product of two D-quasirandom finite groups is again a D-quasirandom finite group.)

Informally, we will call G *quasirandom* if it is D-quasirandom for some "large" D, though the precise meaning of "large" will depend on context. For applications to expansion in Cayley graphs, "large" will mean "$D \geq |G|^c$ for some constant $c > 0$ independent of the size of G", but other regimes of D are certainly of interest.

In the way we have set things up, the trivial group $G = \{1\}$ is infinitely quasirandom (i.e., it is D-quasirandom for every D). This is, however, a degenerate case and will not be discussed further here. In the nontrivial case, a finite group can only be quasirandom if it is large and has no large subgroups:

Exercise 3.0.3. Let $D \geq 1$, and let G be a finite D-quasirandom group.

  (i) Show that if G is nontrivial, then $|G| \geq D + 1$. (*Hint:* Use the mean zero component $\tau \lfloor_{\ell^2(G)_0}$ of the *regular representation* $\tau : G \to U(\ell^2(G))$, see Example 2.1.3.) In particular, nontrivial finite groups cannot be infinitely quasirandom.

  (ii) Show that any proper subgroup H of G has index $[G : H] \geq D+1$. (*Hint:* Use the mean zero component of the *quasiregular representation*, see Example 2.1.4.)

The following exercise shows that quasirandom groups have to be quite nonabelian, and in particular, *perfect*:

---

[1] For instance, one can view the paper [**KuSt1960**] as exploiting the quasirandomness properties of the locally compact group $SL_2(\mathbf{R})$ to obtain mixing estimates in that group.

Exercise 3.0.4 (Quasirandomness, abelianness, and perfection). Let G be a finite group.

(i) If G is abelian and nontrivial, show that G is not 2-quasirandom. (*Hint:* Use Fourier analysis or the *classification of finite abelian groups*.)

(ii) Show that G is 2-quasirandom if and only if it is *perfect*, i.e., the *commutator group* $[G, G]$ is equal to G. (Equivalently, G is 2-quasirandom if and only if it has no nontrivial abelian quotients.)

(iii) If G is a perfect group, $Z(G)$ is the *centre*[2] of G, and $G/Z(G)$ is D^2-1-quasirandom for some $D \geq 1$, show that G is D-quasirandom. (*Hint:* Starting from a unitary action of G on a finite-dimensional Hilbert space H, consider the conjugation action of $G/Z(G)$ on $\mathfrak{sl}(H)$.)

Later on we shall see that there is a converse to the two exercises above; any nontrivial perfect finite group with no large subgroups will be quasirandom.

Exercise 3.0.5. Let G be a finite D-quasirandom group. Show that for any subgroup G' of G, G' is $D/[G : G']$-quasirandom, where $[G : G'] := |G|/|G'|$ is the index of G' in G. (*Hint:* Use induced representations.)

Exercise 3.0.6. Let $D \geq 1$, and let G be a finite simple group. Show that if *any* nontrivial subgroup of G is D-quasirandom, then G is D-quasirandom also. This suggests that simple groups are quite likely to be rather quasirandom; this intuition will be confirmed in the specific examples of simple (or almost simple) groups to be discussed shortly.

Now we give an example of a more quasirandom group.

Lemma 3.0.3 (Frobenius lemma). *If $\mathbf{F}_p$ is a field of some prime order p, then $\mathrm{SL}_2(\mathbf{F}_p)$ is $\frac{p-1}{2}$-quasirandom.*

This should be compared with the cardinality $|\mathrm{SL}_2(\mathbf{F}_p)|$ of the special linear group, which is easily computed to be $(p^2 - 1) \times p = p^3 - p$.

Proof. We may of course take p to be odd. Suppose for contradiction that we have a nontrivial representation $\rho: \mathrm{SL}_2(\mathbf{F}_p) \to U_d(\mathbf{C})$ on a unitary group of some dimension d with $d < \frac{p-1}{2}$. Set a to be the group element

$$a := \begin{pmatrix} 1 & 1 \\ 0 & 1 \end{pmatrix},$$

---
[2] The centre $Z(G) := \{g \in G : gh = hg \text{ for all } h \in G\}$ of a group G consists of all the elements of G which commute with (or *centralise*) all the other elements of G.

and suppose first that $\rho(a)$ is nontrivial. Since $a^p = 1$, we have $\rho(a)^p = 1$; thus all the eigenvalues of $\rho(a)$ are p^{th} roots of unity. On the other hand, by conjugating a by diagonal matrices in $\text{SL}_2(\mathbf{F}_p)$, we see that a is conjugate to a^m (and hence $\rho(a)$ conjugate to $\rho(a)^m$) whenever m is a quadratic residue mod p. As such, the eigenvalues of $\rho(a)$ must be permuted by the operation $x \mapsto x^m$ for any quadratic residue mod p. Since $\rho(a)$ has at least one nontrivial eigenvalue, and there are $\frac{p-1}{2}$ distinct quadratic residues, we conclude that $\rho(a)$ has at least $\frac{p-1}{2}$ distinct eigenvalues. But $\rho(a)$ is a $d \times d$ matrix with $d < \frac{p-1}{2}$, a contradiction. Thus a lies in the kernel of ρ. By conjugation, we then see that this kernel contains all unipotent matrices. But these matrices generate $\text{SL}_2(\mathbf{F}_p)$ (see exercise below), and so ρ is trivial, a contradiction. $\square$

Exercise 3.0.7. Show that for any prime p, the unipotent matrices

$$\begin{pmatrix} 1 & t \\ 0 & 1 \end{pmatrix}, \begin{pmatrix} 1 & 0 \\ t & 1 \end{pmatrix}$$

for t ranging over $\mathbf{F}_p$ generate $\text{SL}_2(\mathbf{F}_p)$ as a group.

Exercise 3.0.8. Let G be a finite group, and let $D \geq 1$. If G is generated by a collection $G_1, \ldots, G_k$ of D-quasirandom subgroups, show that G is itself D-quasirandom.

Exercise 3.0.9. Show that $\text{SL}_d(\mathbf{F}_p)$ is $\frac{p-1}{2}$-quasirandom for any $d \geq 2$ and any prime p. (This is not sharp; the optimal bound here is $\gg_d p^{d-1}$, which follows from the results in [**LaSe1974**].)

As a corollary of the above results and Exercise 3.0.2, we see that the *projective special linear group* $P\text{SL}_d(\mathbf{F}_p)$ is also $\frac{p-1}{2}$-quasirandom.

Remark 3.0.4. One can ask whether the bound $\frac{p-1}{2}$ in Lemma 3.0.3 is sharp, assuming of course that p is odd. Noting that $\text{SL}_2(\mathbf{F}_p)$ acts linearly on the plane $\mathbf{F}_p^2$, we see that it also acts projectively on the projective line $P\mathbf{F}_p^1 := (\mathbf{F}_p^2 \backslash \{0\})/\mathbf{F}_p^\times$, which has $p+1$ elements. Thus $\text{SL}_2(\mathbf{F}_p)$ acts via the quasiregular representation on the $p+1$-dimensional space $\ell^2(P\mathbf{F}_p^1)$, and also on the p-dimensional subspace $\ell^2(P\mathbf{F}_p^1)_0$; this latter representation (known as the *Steinberg representation*) is irreducible. This shows that the $\frac{p-1}{2}$ bound cannot be improved beyond p. More generally, given any character $\chi\colon \mathbf{F}_p^\times \to S^1$, $\text{SL}_2(\mathbf{F}_p)$ acts on the $p+1$-dimensional space V_χ of functions $f \in \ell^2(\mathbf{F}_p^2 \backslash \{0\})$ that obey the twisted dilation invariance $f(tx) = \chi(t)f(x)$ for all $t \in \mathbf{F}_p^\times$ and $x \in \mathbf{F}_p^2 \backslash \{0\}$; these are known as the *principal series representations*. When χ is the trivial character, this is the quasiregular representation discussed earlier. For most other characters, this is an irreducible representation, but it turns out that when χ is the quadratic representation (thus taking values in $\{-1,+1\}$ while being nontrivial), the principal series representation splits into the direct sum of two $\frac{p+1}{2}$-dimensional representations, which comes very close to matching the bound in Lemma 3.0.3. There is a parallel series of representations to the principal series (known as the *discrete series*) which is more complicated to describe (roughly speaking, one has to embed $\mathbf{F}_p$ in a quadratic extension $\mathbf{F}_{p^2}$ and then use a rotated version of the above construction, to change a split torus into a nonsplit torus), but can generate irreducible representations of dimension $\frac{p-1}{2}$, showing that the bound in Lemma 3.0.3 is in fact exactly sharp. These constructions can be generalised to arbitrary finite simple groups of Lie type (as defined in Section 12.3) using *Deligne-Luzstig theory*, but this is beyond the scope of this text.

Exercise 3.0.10. Let q be a power of an odd prime. Show that $\mathrm{SL}_2(F_q)$ is $\frac{q-1}{2}$-quasirandom. (*Hint:* Diagonalise an irreducible representation of $\mathrm{SL}_2(F_q)$ relative to the action of the unipotent (and abelian) group

$$\left\{ \begin{pmatrix} 1 & t \\ 0 & 1 \end{pmatrix} : t \in \mathbf{F}_q \right\}$$

and then study how this action is conjugated by the action of diagonal matrices in $\mathrm{SL}_2(F_q)$.)

Exercise 3.0.11. Let p be an odd prime. Show that for any $n \geq p+2$, the *alternating group* A_n (i.e., the group of even permutations on n elements) is $p-1$-quasirandom. (*Hint:* Show that all cycles of order p in A_n are conjugate to each other in A_n (and not just in the symmetric group S_n); in particular, a cycle is conjugate to its jth power for all $j = 1, \ldots, p-1$. Also, as $n \geq 5$, A_n is simple, and so the cycles of order p generate the entire group.)

Remark 3.0.5. By using more precise information on the representations of the alternating group (using the theory of *Specht modules* and *Young tableaux*), one can show the slightly sharper statement that A_n is $n-1$-quasirandom for $n \geq 6$ (but is only 3-quasirandom for $n = 5$ due to *icosahedral symmetry*, and 1-quasirandom for $n \leq 4$ due to lack of perfectness). Using Exercise 3.0.3 with the index n subgroup A_{n-1}, we see that the bound $n-1$ cannot be improved. Thus, A_n (for large n) is not as quasirandom as the special linear groups $\mathrm{SL}_d(\mathbf{F}_p)$ (for p large and d bounded), because in the latter case the quasirandomness is as strong as a power of the size of the group, whereas in the former case it is only logarithmic in size.

If one replaces the alternating group A_n with the slightly larger *symmetric group* S_n, then quasirandomness is destroyed (since S_n, having the abelian quotient S_n/A_n, is not perfect); indeed, S_n is 1-quasirandom and no better.

Remark 3.0.6. Thanks to the monumental achievement of the *classification of finite simple groups*, we know that apart from a finite number (26, to be precise) of *sporadic exceptions*, all finite simple groups (up to isomorphism) are either a cyclic group $\mathbf{Z}/p\mathbf{Z}$, an alternating group A_n, or is a *finite simple group of Lie type* such as $P\operatorname{SL}_d(\mathbf{F}_p)$; see Section 12.3 for more discussion of the latter family of groups. In the case of finite simple groups G of Lie type with bounded rank $r = O(1)$, it is known (see [**LaSe1974**]) that such groups are $\gg |G|^c$-quasirandom for some $c > 0$ depending only on the rank. On the other hand, by the previous remark, the large alternating groups do not have this property, and one can show that the finite simple groups of Lie type with large rank also do not have this property. Thus, we see using the classification that if a finite simple group G is $|G|^c$-quasirandom for some $c > 0$ and $|G|$ is sufficiently large depending on c, then G is a finite simple group of Lie type with rank $O_c(1)$. It would be of interest to see if there was an alternate way to establish this fact that did not rely on the classification, as it may lead to an alternate approach to proving the classification (or perhaps a weakened version thereof).

A key reason why quasirandomness is desirable for the purposes of demonstrating expansion is that quasirandom groups happen to be rapidly mixing at large scales, as we shall see below the fold. As such, quasirandomness is an important tool for demonstrating expansion in Cayley graphs, though because expansion is a phenomenon that must hold at all scales, one needs to supplement quasirandomness with some additional input that creates mixing at small or medium scales also before one can deduce expansion. As an example of this technique of combining quasirandomness with mixing at small and medium scales, we present a proof (due to [**SaXu1991**], and simplified in [**Ga2002**]) of a weak version of the famous "3/16 theorem" of Selberg [**Se1965**] on the least nontrivial eigenvalue of the Laplacian on a modular curve, which among other things can be used to construct a family of expander Cayley graphs in $\operatorname{SL}_2(\mathbf{Z}/N\mathbf{Z})$; compare this with the property (T)-based methods in Chapter 2, which could construct expander Cayley graphs in $\operatorname{SL}_d(\mathbf{Z}/N\mathbf{Z})$ for any fixed $d \geq 3$.

3.1. Mixing in quasirandom groups

Let G be a finite group. Given two functions $f, g \in \ell^2(G)$, we can define the *convolution* $f * g \in \ell^2(G)$ by the formula

$$f * g(x) := \sum_{y \in G} f(y)g(y^{-1}x) = \sum_{y \in G} f(xy^{-1})g(y).$$

This operation is bilinear and associative, but is not commutative unless G is abelian. From the Cauchy-Schwarz inequality one has

$$\|f * g\|_{\ell^\infty(G)} \leq \|f\|_{\ell^2(G)} \|g\|_{\ell^2(G)}$$

3.1. Mixing in quasirandom groups

and hence
$$\|f * g\|_{\ell^2(G)} \leq |G|^{1/2} \|f\|_{\ell^2(G)} \|g\|_{\ell^2(G)}.$$
This inequality is sharp in the sense that if we set f and g to both be constant-valued, then the left-hand side and right-hand side match. For abelian groups, one can also see this example is sharp when f and g are multiples of the same character.

It turns out, though, that if one restricts one of f or g (or both) to be of mean zero, and G is quasirandom, then one can improve this inequality, which first appeared explicitly in [**BaNiPy2008**]:

Proposition 3.1.1 (Mixing inequality). *Let G be a finite D-quasirandom group, and let $f, g \in \ell^2(G)$. If at least one of f, g has mean zero, then*
$$\|f * g\|_{\ell^2(G)} \leq D^{-1/2} |G|^{1/2} \|f\|_{\ell^2(G)} \|g\|_{\ell^2(G)}.$$

Proof. By subtracting a constant from f or g, we may assume that f or g both have mean zero.

Observe that $f * g$ (being a superposition of right-translates of f) also has mean zero. Thus, we see that we may define an operator $T_g \colon \ell^2(G)_0 \to \ell^2(G)_0$ by setting $T_g f := f * g$. It thus suffices to show that the operator norm of T_g is at most $D^{-1/2} |G|^{1/2} \|g\|_{\ell^2(G)}$.

Fix g. We can view T_g as a $|G| - 1 \times |G| - 1$ matrix. We apply the *singular value decomposition* to this matrix to obtain singular values
$$\sigma_1 \geq \ldots \geq \sigma_{|G|-1} \geq 0$$
of T_g, together with associated singular vectors. The operator norm of T_g is the largest singular value σ_1. The operator $T_g^* T_g$ is then a self-adjoint operator (or matrix) with eigenvalues $\sigma_1^2, \ldots, \sigma_{|G|-1}^2$. In particular, we have
$$\operatorname{tr} T_g^* T_g = \sigma_1^2 + \ldots + \sigma_{|G|-1}^2.$$
Now, a short computation shows that $T_g^* T_g f = f * g * \tilde{g}$, where $\tilde{g}(x) := \overline{g(x^{-1})}$, and (by embedding $\ell^2(G)_0$ in $\ell^2(G)$, and noting that $T_g^* T_g$ annihilates constants) the trace can computed as
$$\operatorname{tr} T_g^* T_g = |G| g * \tilde{g}(0) = |G| \|g\|_{\ell^2(G)}^2.$$
Thus, if V is the eigenspace of $T_g^* T_g$ corresponding to the eigenvalue σ_1^2 (so that the dimension of V is the multiplicity of σ_1), we have
$$\dim(V) \sigma_1^2 \leq |G| \|g\|_{\ell^2(G)}^2.$$
Now observe that if $\tau \colon G \to U(\ell^2(G)_0)$ is the left-regular representation (restricted to $\ell^2(G)_0$), then
$$T_g^* T_g \tau(h) f = \tau(h) T_g^* T_g f$$

for any $f \in \ell^2(G)_0$ and $h \in G$ (this is a special case of the associativity of convolution). In particular, we see that V is invariant under τ. Since τ has no nontrivial invariant vectors in $\ell^2(G)_0$, we conclude from quasirandomness that V has dimension at least D, and the claim follows. □

Remark 3.1.2. One can also establish the above inequality using the non-abelian Fourier transform (which is based on the *Peter-Weyl theorem* combined with *Schur's lemma*, and is developed for instance in [**Ta2013**, Section 2.8]); we leave this as an exercise for the interested reader.

Exercise 3.1.1. Let A, B, C be subsets of a finite D-quasirandom group G. Show that
$$\left\| 1_A * 1_B - \frac{|A||B|}{|G|} \right\|_{\ell^2(G)} \leq D^{-1/2} |G|^{1/2} |A|^{1/2} |B|^{1/2}$$
and
$$\left\| 1_A * 1_B * 1_C - \frac{|A||B||C|}{|G|} \right\|_{\ell^\infty(G)} \leq D^{-1/2} |G|^{1/2} |A|^{1/2} |B|^{1/2} |C|^{1/2}.$$
Conclude, in particular, that
$$|AB| \geq |G| - \frac{|G|^3}{D|A||B|}$$
(with the convention that $\frac{|G|^3}{D|A||B|} = +\infty$ if A or B is empty), and that $ABC = G$ whenever $|A||B||C| > |G|^3/D$. (The bounds here are not quite sharp, but are simpler than the optimal bounds, and suffice for most applications.)

Thus, for instance, if A is a subset of a finite D-quasirandom group G of density $|A|/|G|$ more than $D^{-1/3}$, then A^2 will be most of G (with fewer than $|A|$ elements omitted), and A^3 will be all of G; thus large subsets of a quasirandom group rapidly expand to fill out the whole group. In the converse direction, we have

Exercise 3.1.2. Let $D \geq 1$, and let G be a finite group which is *not D-quasirandom*. Show that there exists a subset A of G with $|A|/|G| \geq C^{-D^2}$ for some absolute constant $C > 1$, such that $A^3 \subsetneq G$. (*Hint:* By hypothesis, one has a nontrivial unitary representation $\rho: G \to U(H)$ of dimension at most D. Show that $\rho(G)$ contains an element at distance $\gg 1$ from the identity in operator norm, and take A to be the preimage of a suitable ball around the identity in the operator norm, and use a pigeonhole (or Dirichlet box principle) argument to obtain the lower bound on A.)

One can improve this result by using a quantitative form of *Jordan's theorem*; see Section 3.2 below.

Exercise 3.1.3 (Mixing inequality for actions). Let G be a finite D-quasirandom group acting (on the left) on a discrete set X. Given functions $f \in \ell^2(G)$ and $g \in \ell^2(X)$, one can define the convolution $f * g \in \ell^2(X)$ in much the same way as before:
$$f * g(x) := \sum_{h \in G} f(h) g(h^{-1} x).$$
Show that
$$\|f * g\|_{\ell^2(X)} \leq D^{-1/2} |G|^{1/2} \|f\|_{\ell^2(G)} \|g\|_{\ell^2(X)}$$
whenever f has mean zero, or whenever g has mean zero on every orbit of G.

One can use quasirandomness to show that Cayley graphs of very large degree k in a quasirandom group are expanders:

Exercise 3.1.4. Let $\text{Cay}(G, S)$ be a k-regular Cayley graph in a finite D-quasirandom group G on n vertices.

(i) If A, B are subsets of G, show that
$$\left| E(A, B) - \frac{k}{n} |A||B| \right| \leq \sqrt{\frac{kn|A||B|}{D}}$$
(compare with the expander mixing lemma, Exercise 1.2.2).

(ii) Show that $\text{Cay}(G, S)$ is a two-sided ε-expander whenever
$$\varepsilon \leq 1 - \sqrt{\frac{n}{Dk}}.$$

Unfortunately, the above result is only nontrivial in the regime $k \gg n/D$, whereas for our applications we are interested instead in the regime when $k = O(1)$. We record a tool for this purpose.

Proposition 3.1.3 (Using quasirandomness to demonstrate expansion). Let $\text{Cay}(G, S)$ be a k-regular Cayley graph in a finite group G. Assume the following:

(i) *(Quasirandomness)* G is $c|G|^\alpha$-quasirandom for some $c, \alpha > 0$.

(ii) *(Flattening of random walk)* One has
(3.1) $$\|\mu^{*n}\|_{\ell^2(G)} \leq C|G|^{-1/2+\beta}$$
for some $C, \beta, n > 0$ with $\beta < \alpha/2$ and $n \leq C \log |G|$, where $\mu := \frac{1}{|S|} \sum_{s \in S} \delta_s$ and μ^{*n} is the n-fold convolution of μ.

Then G is a two-sided ε-expander for some $\varepsilon > 0$ depending only on c, α, C, β, k, if $|G|$ is sufficiently large depending on these quantities. If we replace μ by $\nu := \frac{1}{2}\delta_1 + \frac{1}{2}\mu$ in the flattening hypothesis, then G is a one-sided ε-expander instead.

Proof. We allow implied constants to depend on c, α, C, β. We will just prove the first claim, as the second claim is similar. By Exercise 2.0.5, it will suffice to show that

$$\left\| \mu^{*m} - \frac{1}{|G|} \right\|_{\ell^2(G)} \ll |G|^{-1}$$

(say) for some $m = O(\log|G|)$. But from Proposition 3.1.1 (with $f := \mu^{*m} - \frac{1}{|G|}$ and $g := \mu^{*n}$) and the hypotheses we have

$$\left\| \mu^{*(m+n)} - \frac{1}{|G|} \right\|_{\ell^2(G)} \ll |G|^{-\alpha/2+\beta} \left\| \mu^{*m} - \frac{1}{|G|} \right\|_{\ell^2(G)}$$

for any $m \geq 0$. Iterating this (starting from, say, $m = n$, and advancing in steps of n $O(1)$ times) we obtain the claim. $\square$

Exercise 3.1.5. Obtain an alternate proof of the above result that proceeds directly from the spectral decomposition of the adjacency operator $Af := f * \mu$ into eigenvalues and eigenvectors and quasirandomness, rather than through Exercise 2.0.5 and Proposition 3.1.1. (This alternate approach is closer in spirit to the arguments of [**SaXu1991**] and [**BoGa2008**], though the two approaches are largely equivalent in the final analysis.)

Informally, the flattening hypothesis in Proposition 3.1.3 asserts that by time $O(\log|G|)$, the random walk has expanded to the point where it is covering a large portion of the group G (roughly speaking, it is spread out over a set of size at least $|G|^{1-2\beta}$). The point is that the scale of this set is large enough for the quasirandomness properties of the group G to then mix the random walk rapidly towards the uniform distribution. However, this proposition provides no tools with which to *prove* this flattening property; this task will be a focus of subsequent chapters.

The following exercise extends some of the above theory from quasirandom groups to "virtually quasirandom" groups, which have a bounded index subgroup that is quasirandom, but need not themselves be quasirandom.

Exercise 3.1.6 (Virtually quasirandom groups). Let G be a finite group that contains a normal D-quasirandom subgroup G' of index at most K.

(i) If $f, g \in \ell^2(G)$, and at least one of f, g has mean zero on every coset of G', show that

$$\|f * g\|_{\ell^2(G)} \leq D^{-1/2} |G|^{1/2} \|f\|_{\ell^2(G)} \|g\|_{\ell^2(G)}.$$

(ii) If $|D| \geq c|G|^\alpha$ for some $c, \alpha > 0$, and $\mathrm{Cay}(G, S)$ is a connected k-regular Cayley graph obeying (3.1) for some $C, \beta, n > 0$ with $\beta < \alpha/2$ and $n \leq C \log|G|$, show that G is a two-sided ε-expander for some $\varepsilon > 0$ depending only on $c, \alpha, C, \beta, k, K$.

3.2. An algebraic description of quasirandomness

As defined above, quasirandomness is a property of representations. However, one can reformulate this property (at a qualitative level, at least) in a more algebraic fashion, by means of *Jordan's theorem*:

Theorem 3.2.1 (Jordan's theorem [**Jo1878**])**.** *Let G be a finite subgroup of $U_d(\mathbf{C})$ for some $d \geq 1$. Then G contains a normal abelian subgroup of index at most $K(d)$, where $K(d)$ depends only on d.*

A proof of this theorem (giving a rather poor value of $K(d)$) may be found in [**Ta2013**, §1.1]. The optimal value of $K(d)$ is known for almost all d, thanks to the classification of finite simple groups; for instance, it is a result of Collins [**Co2007**] that $K(d) = (d+1)!$ for $d \geq 71$ (which is attained with the example of the symmetric group S_{d+1} which acts on the space $\mathbf{C}_0^{d+1}$ of $d+1$-dimensional complex vectors whose coefficients sum to zero).

Jordan's theorem can be used to give a qualitative description of quasirandomness, providing a converse to Exercises 3.0.3 and 3.0.4:

Exercise 3.2.1. Let $D > 1$ be an integer. Let G be a perfect finite group, with the property that all proper normal subgroups of G have index greater than $K(D-1)$, where $K(D-1)$ is the quantity in Jordan's theorem. Show that G is D-quasirandom.

Conclude in particular that any finite simple nonabelian group of cardinality greater than $K(D-1)$ is D-quasirandom.

By using the classification of finite simple groups more carefully, Nikolov and Pyber [**NiPy2011**] were able to replace $K(D-1)$ here by $10^{10}D^2$. Using related arguments, they also showed that if G was not D-quasirandom, then there was a subset A of G with cardinality $\gg |G|/D$ such that $A^3 \neq G$, thus giving a reasonably tight converse to Exercise 3.1.1.

3.3. A weak form of Selberg's 3/16 theorem

Remark 3.3.1. This section presumes some familiarity with Riemannian geometry, as well as the functional analysis of Sobolev spaces and distributions. See for instance [**Ta2009b**, §2.1] for a very brief introduction to Riemannian geometry, and [**Ta2010b**, §1.13-1.14] for an introduction to distributions and Sobolev spaces. On the other hand, the material here is not directly used in later chapters.

We now give an application of quasirandomness to establish the following result, first observed explicitly in [**LuPhSa1988**] as a corollary of a famous theorem of Selberg [**Se1965**]:

Theorem 3.3.2 (Selberg's expander construction). *If S is a symmetric set of generators of $\mathrm{SL}_2(\mathbf{Z})$ that does not contain the identity, then the Cayley graphs $\mathrm{Cay}(\mathrm{SL}_2(\mathbf{F}_p), \pi_p(S))$ form a one-sided expander family, where $\pi_p \colon \mathrm{SL}_2(\mathbf{Z}) \to \mathrm{SL}_2(\mathbf{F}_p)$ is the obvious projection homomorphism, and p ranges over primes.*

This is the $d = 2$ analogue of Margulis's expander construction (Corollary 2.2.9), except that the modulus n has been restricted here to be prime. This restriction can be removed with some additional effort, but we will not discuss this issue here. The condition that S generates the entire group $\mathrm{SL}_2(\mathbf{Z})$ can be substantially relaxed; we will discuss this point in later chapters.

In the property (T) approach to expansion, one passed from discrete groups such as $\mathrm{SL}_d(\mathbf{Z})$ to continuous groups such as $\mathrm{SL}_d(\mathbf{R})$, in order to take advantage of tools from analysis (such as limits). Similarly, to prove Theorem 3.3.2, we will pass from $\mathrm{SL}_2(\mathbf{Z})$ to $\mathrm{SL}_2(\mathbf{R})$. Actually, it will be convenient to work with the quotient space $\mathbf{H} := \mathrm{SL}_2(\mathbf{R})/\mathrm{SO}_2(\mathbf{R})$, better known as the *hyperbolic plane*. We will endow this plane with the structure of a *Riemannian manifold*, in order to access the *Laplace-Beltrami operator* on that plane, which is a continuous analogue (after some renormalisation) of the adjacency operator of a Cayley graph, which enjoys some nice exact identities which are difficult to discern in the discrete world.

We now therefore digress from the topic of expansion to recall the geometry of the hyperbolic plane. It will be convenient to switch between a number of different coordinatisations of this plane. Our primary model will be the half-plane model:

Definition 3.3.3 (Poincaré half-plane model). The *Poincaré half-plane* is the upper half-plane $\mathbf{H} := \{x + iy \in \mathbf{C} : y > 0\}$ with the Riemannian metric $ds^2 := \frac{dx^2 + dy^2}{y^2}$. The (left) action of $\mathrm{SL}_2(\mathbf{R})$ on this half-plane is given by the formula

$$\begin{pmatrix} a & b \\ c & d \end{pmatrix} z := \frac{az + b}{cz + d}.$$

Exercise 3.3.1. Verify that $\mathrm{SL}_2(\mathbf{R})$ acts isometrically and transitively on $\mathbf{H}$, with stabiliser group conjugate to $\mathrm{SO}_2(\mathbf{R})$; thus $\mathbf{H}$ is isomorphic (as an $\mathrm{SL}_2(\mathbf{R})$-homogeneous space) to $\mathrm{SL}_2(\mathbf{R})/\mathrm{SO}_2(\mathbf{R})$.

Note that in some of the literature, a right action is used instead of a left action, leading to some reversals in the notational conventions used below, but this does not lead to any essential changes in the arguments or results.

Exercise 3.3.2. Show that the distance $d(z,w)$ between two points $z, w \in \mathbf{H}$ is given by the formula

$$\cosh d(z,w) = 1 + 2\frac{|z-w|^2}{4\mathrm{Im}(z)\mathrm{Im}(w)}.$$

We can also use the model of the *Poincaré disk* $\mathbf{D} := \{a + ib \in \mathbf{C} : a^2 + b^2 < 1\}$ with the Riemannian metric $ds^2 := 4\frac{da^2+db^2}{(1-a^2-b^2)^2}$.

Exercise 3.3.3. Show that the *Cayley transform* $z \mapsto \frac{z-i}{z+i}$ is an isometric isomorphism from the half-plane $\mathbf{H}$ to the disk $\mathbf{D}$.

Expressing an element $a + ib$ of the Poincaré disk in exponential polar coordinates as $\tanh(\rho/2)e^{i\theta}$, we can also model the Poincaré disk (in slightly singular coordinates) as the half-cylinder $\{(\rho,\theta) : \rho \in [0,+\infty); \theta \in \mathbf{R}/2\pi\mathbf{Z}\}$ with metric $ds^2 = d\rho^2 + \sinh^2\rho\, d\theta^2$. (Compare with the Euclidean plane in polar coordinates, which is similar but with the $\sinh^2\rho$ factor replaced by ρ^2, or the sphere in Euler coordinates, which is also similar but with ρ restricted to $[0,\pi]$ and $\sinh^2\rho$ replaced by $\sin^2\rho$. This similarity reflects the fact that these three Riemannian surfaces have constant curvature $-1, 0, +1$ respectively.)

The action of $\mathrm{SL}_2(\mathbf{R})$ can of course be described explicitly in the disk or half-plane models, but we will not need these explicit formulae here.

A Riemannian metric on a manifold always generates a measure $d\mu$ on that manifold. For the Poincaré half-plane, the measure is $d\mu = \frac{dxdy}{y^2}$. For the Poincaré disk, it is $d\mu = 4\frac{dadb}{(1-a^2-b^2)^2}$. For the half-cylinder, it is $\sinh\rho\, d\rho d\theta$. In all cases, the action of $\mathrm{SL}_2(\mathbf{R})$ will preserve the measure, because it preserves the metric, thus one can view $d\mu$ as a Haar measure on the hyperbolic plane.

A Riemannian metric also generates a Laplace-Beltrami operator Δ. In the Poincaré half-plane model, it is

$$\Delta = y^2\left(\frac{\partial^2}{\partial x^2} + \frac{\partial^2}{\partial y^2}\right);$$

in the Poincaré disk model, it is

$$\Delta = \frac{(1-a^2-b^2)^2}{4}\left(\frac{\partial^2}{\partial a^2} + \frac{\partial^2}{\partial b^2}\right);$$

and in the half-cylinder model, it is

$$\Delta = \frac{\partial^2}{\partial \rho^2} + \frac{1}{\sinh\rho}\frac{\partial}{\partial \rho} + \frac{1}{\sinh^2\rho}\frac{\partial^2}{\partial \theta^2}.$$

Again, in all cases, the Laplacian will commute with the action of $\mathrm{SL}_2(\mathbf{R})$, because this action preserves the metric.

The discrete group $\mathrm{SL}_2(\mathbf{Z})$ acts on the hyperbolic plane, giving rise to a quotient $X(1) := \mathrm{SL}_2(\mathbf{Z})\backslash\mathbf{H}$, known as the *principal modular curve* of level 1. This quotient can also be viewed by taking the (closure of a) *fundamental domain*

$$\Omega := \{z \in \mathbf{H} : |\mathrm{Re}(z)| \leq 1/2; |z| \geq 1\}$$

and then identifying $-1/2 + it$ with $1/2 + it$ on the left and right sides of this domain, and also identifying z with $-1/z$ on the lower boundary of this domain. The quotient $X(1)$ is not compact, but it does have finite measure with respect to μ; indeed, outside of a compact set, $X(1)$ behaves like the *cusp*

(3.2) $$\{x + iy : -1/2 \leq x \leq 1/2; y > C\}$$

for any constant $C > 1$, again identifying the $x = -1/2$ boundary with the $x = 1/2$ boundary, and this cusp has measure

$$\int_C^\infty \int_{-1/2}^{1/2} \frac{dxdy}{y^2} < \infty.$$

Thus μ descends to a finite Haar measure on $X(1)$.

Remark 3.3.4. One can interpret the modular curve geometrically as follows. As in Chapter 2, one can think of $\mathrm{SL}_2(\mathbf{R})$ as the space of all unimodular lattices in $\mathbf{R}^2$ (or equivalently, in $\mathbf{C}$) with two marked generators $z_1, z_2 \in \mathbf{C}$ with $\mathrm{Im}(\overline{z_1}z_2) = 1$. We can then map $\mathrm{SL}_2(\mathbf{R})$ to the Poincaré half-plane $\mathbf{H}$ by sending such a lattice with generators z_1, z_2 to the point z_2/z_1; note that the fibers of this map correspond to rotations of the lattice and marked generators, thus identifying $\mathbf{H}$ with $\mathrm{SL}_2(\mathbf{R})/\mathrm{SO}_2(\mathbf{R})$. The action of $\mathrm{SL}_2(\mathbf{Z})$ on $\mathbf{H}$ corresponds to moving the generators around while keeping the lattice (or more precisely, the lattice modulo rotations) fixed. The (interior of the) fundamental domain Ω then corresponds to the selection of generators given by setting z_1 to be the nonzero lattice element of smallest norm, and z_2 to be the generator whose z_1 component lies between $-z_1/2$ and $+z_1/2$.

The quotient $X(1)$ is not quite a smooth Riemannian manifold, due to the presence of partially fixed points of the $\mathrm{SL}_2(\mathbf{Z})$ action at $+i$ and $\pm 1/2 + \sqrt{3}/2i$ (of order 2 and order 3 respectively), and is thus technically an *orbifold* rather than a manifold. However, this distinction does not significantly affect the analysis and will be glossed over here.

The Laplace-Beltrami operator Δ is defined on smooth[3] compactly supported functions $f \in C_c^\infty(\mathbf{H})$ on $\mathbf{H}$, and then descends to an operator on

---

[3] Here, we use the smooth structure on $X(1)$ inherited from $\mathbf{H}$, thus a function is smooth at a point in $X(1)$ if it lifts to a function smooth at the preimage of that point.

3.3. A weak form of Selberg's 3/16 theorem

smooth compactly supported functions $f \in C_c^\infty(X(1))$ on $X(1)$. On $\mathbf{H}$, we have the integration by parts formula

$$\int_{\mathbf{H}} (-\Delta f) g \, d\mu = \int_{\mathbf{H}} \langle \nabla f, \nabla g \rangle \, d\mu$$

where ∇ is the gradient with respect to the Riemannian metric, and $\langle,\rangle$ the inner product; in the half-plane coordinates, we have

$$\langle \nabla f, \nabla g \rangle = y^2 \left(\frac{\partial f}{\partial x} \frac{\partial g}{\partial x} + \frac{\partial f}{\partial y} \frac{\partial g}{\partial y} \right),$$

in the disk model, it is

$$\langle \nabla f, \nabla g \rangle = \frac{(1-a^2-b^2)^2}{4} \left(\frac{\partial f}{\partial a} \frac{\partial g}{\partial a} + \frac{\partial f}{\partial b} \frac{\partial g}{\partial b} \right)$$

and in the half-cylinder model, it is

$$\langle \nabla f, \nabla g \rangle = \frac{\partial f}{\partial \rho} \frac{\partial g}{\partial \rho} + \frac{1}{\tanh^2 \rho} \frac{\partial f}{\partial \theta} \frac{\partial g}{\partial \theta}.$$

In particular, we have the positive definiteness property

$$\langle -\Delta f, f \rangle_{L^2(\mathbf{H},\mu)} = \int_{\mathbf{H}} |\nabla f|^2 \, d\mu \geq 0$$

for all $f \in C_c^\infty(\mathbf{H})$. This descends to $X(1)$:

$$\langle -\Delta f, f \rangle_{L^2(X(1),\mu)} = \int_{X(1)} |\nabla f|^2 \, d\mu \geq 0.$$

Thus $-\Delta$ is a symmetric positive-definite densely-defined operator on $L^2(X(1), \mu)$. One can in fact show (by solving some PDE, such as the wave equation or the resolvent equation, and exploiting at some point the fact that the Riemannian manifold $X(1)$ is *complete*) that $-\Delta$ is *essentially self-adjoint* and is thus subject to the *spectral theorem* (see Section 10), but we will avoid using the full force of spectral theory here.

Since $X(1)$ has finite measure and $\Delta 1 = 0$, we see that $1 \in L^2(X(1), \mu)$ is an eigenfunction of Δ (or $-\Delta$) with eigenvalue zero. We eliminate this eigenfunction by working in the space $L^2(X(1))_0$ (or $C_c^\infty(X(1))_0$) of functions in $L^2(X(1))$ (or $C_c^\infty(X(1))$) of mean zero. Let us now define the *spectral gap* $\lambda_1(X(1))$ to be the quantity

$$\lambda_1(X(1)) := \inf \left\{ \int_{X(1)} |\nabla f|^2 \, d\mu : f \in C_c^\infty(X(1))_0; \|f\|_{L^2(X(1))} = 1 \right\}.$$

Then $\lambda_1(X(1)) \geq 0$. Using the spectral theorem, one can interpret the spectral gap as the infimum of the spectrum $\sigma(-\Delta)$ of the (negative) Laplacian on $L^2(X(1))_0$. Note also that one can take f to be either real- or complex-valued, as this will not affect the value of the spectral gap. Also by a truncation and mollification argument we may allow f to range in $L^2(X(1))_0$ instead of $C_c^\infty(X(1))_0$ here if desired.

We have the following bounds:

Proposition 3.3.5 (Spectral gap of $X(1)$). *We have $0 < \lambda_1(X(1)) \leq \frac{1}{4}$.*

Proof. We first establish the upper bound $\lambda_1(X(1)) \leq \frac{1}{4}$. It will suffice to find nonzero functions $f \in C_c^\infty(X(1))_0$ whose *Rayleigh quotient*

$$\frac{\int_{X(1)} |\nabla f|^2 \, d\mu}{\int_{X(1)} |f|^2 \, d\mu}$$

is arbitrarily close to $1/4$.

We will restrict attention to smooth compactly supported functions f supported on the cusp (3.2) for a fixed C (e.g., one can take $C = 2$). In coordinates, the Raleigh quotient becomes

$$\frac{\int_C^\infty \int_{-1/2}^{1/2} |f_x|^2 + |f_y|^2 \, dxdy}{\int_C^\infty \int_{-1/2}^{1/2} \frac{|f|^2}{y^2} \, dxdy}$$

where we use subscripts to denote partial differentiation, while the mean zero condition becomes

$$\int_C^\infty \int_{-1/2}^{1/2} \frac{f}{y^2} \, dxdy = 0.$$

To build such functions, we select a large parameter $R \gg C$, and choose a function $f(x,y) = f_R(y)$ that depends only on the y variable, is supported on the region $\{C < y < 2R\}$, and equals $y^{1/2}$ in the region $\{2C \leq y \leq R\}$ and is smoothly truncated in the intermediate region (assigning enough negative mass in the region $\{C < y \leq 2C\}$ to obtain the mean zero condition). A brief calculation shows that

$$\int_C^\infty \int_{-1/2}^{1/2} \frac{|f|^2}{y^2} \, dxdy = \log R + O(1)$$

and

$$\int_C^\infty \int_{-1/2}^{1/2} |f_x|^2 + |f_y|^2 \, dxdy = \frac{1}{4} \log R + O(1)$$

(where the implied constant can depend on C but not on R), and so the claim follows by sending $R \to \infty$.

Now we show the lower bound $\lambda_1(X(1)) > 0$. Suppose this claim failed; then we may find a sequence of functions $f_n \in C_c^\infty(X(1))_0$ with $\|f_n\|_{L^2(X(1))} = 1$ such that

$$\int_{X(1)} |\nabla f_n|^2 \, d\mu = o(1)$$

3.3. A weak form of Selberg's 3/16 theorem

where $o(1)$ denotes a quantity that goes to zero as $n \to \infty$. We can take the f_n to be real-valued.

To deal with the noncompact portion of $X(1)$ (i.e., the cusp (3.2)) we now use *Hardy's inequality*. Observe that if f is smooth, real-valued, and compactly supported on a cusp (3.2) for some C (we can take $C = 2$ as before), then by integration by parts

$$\int_C^\infty \int_{-1/2}^{1/2} \frac{f f_y}{y} \, dxdy = \frac{1}{2} \int_C^\infty \int_{-1/2}^{1/2} \frac{f^2}{y^2} \, dxdy$$

and hence by Cauchy-Schwarz

$$\int_C^\infty \int_{-1/2}^{1/2} \frac{|f|^2}{y^2} \, dxdy \leq 4 \int_C^\infty \int_{-1/2}^{1/2} |f_y|^2 \, dxdy.$$

Applying this to a truncated version $f(x,y) = \chi(y/R) f_n(x,y)$ of f_n for some $R > C$ and some smooth cutoff $\chi \colon \mathbf{R}^+ \to [0,1]$ supported on $[1, +\infty)$ that equals one on $[2, +\infty)$, we conclude that

$$\int_R^\infty \int_{-1/2}^{1/2} \frac{|f_n|^2}{y^2} \, dxdy \leq 4 \int_R^\infty \int_{-1/2}^{1/2} |(f_n)_y|^2 \, dxdy + O\left(\int_R^{2R} \frac{|f_n|^2}{y^2} \, dxdy \right).$$

For any $\varepsilon > 0$, one can use the pigeonhole principle to find an $R = O_\varepsilon(1)$ (depending on n) such that

$$\int_R^{2R} \frac{|f_n|^2}{y^2} \, dxdy \leq \varepsilon$$

and thus we see that

$$\int_{R_\varepsilon}^\infty \int_{-1/2}^{1/2} \frac{|f_n|^2}{y^2} \, dxdy \ll \varepsilon + o(1)$$

for some R_ε depending only on ε. Thus, the probability measures $|f_n|^2 \, d\mu$ form a *tight sequence of measures*[4] in $X(1)$. As the f_n are also locally uniformly bounded in the Sobolev space $H^1(X(1))$, we conclude from the *Rellich compactness theorem* (or the *Poincaré inequality*) that after passing to a subsequence, the f_n converge strongly in $L^2(X(1))$ to a limit f, which then has $L^2(X(1))$ norm one, mean zero, and $\nabla f = 0$ in a distributional sense. But then by the Poincaré inequality, f is constant, which is absurd. □

Remark 3.3.6. One can in fact establish after some calculation using the theory of modular forms that $\lambda_1(X(1))$ is exactly $1/4$, but we will not do so here. By modifying the above arguments, one can in fact show that $-\Delta$ on $X(1)$ has absolutely continuous spectrum on $[1/4, +\infty)$.

---

[4] A sequence of measures is *tight* if, for every $\varepsilon > 0$, there is a compact set K_ε outside of which all but finitely many of the measures only have mass at most ε.

Now we move back towards the task of establishing expansion for the Cayley graphs $\mathrm{Cay}(\mathrm{SL}_2(\mathbf{F}_p), \pi_p(S))$. Let $\Gamma(p)$ denote the kernel of the projection map π_p; this is the group of matrices in $\mathrm{SL}_2(\mathbf{Z})$ that are equal to 1 mod p, and is known as a *principal congruence subgroup* of $\mathrm{SL}_2(\mathbf{Z})$. It is a finite index normal subgroup of $\mathrm{SL}_2(\mathbf{Z})$, and the quotient $\mathrm{SL}_2(\mathbf{Z})/\Gamma(p)$ can easily be seen to be isomorphic to $\mathrm{SL}_2(\mathbf{F}_p)$. In analogy with what we did for $X(1)$, we can then define the principal modular curve $X(p) := \Gamma(p)\backslash \mathbf{H}$, and then define the Laplacian Δ on this curve and the spectral gap $\lambda_1(X(p))$. At a qualitative level, the geometry of $X(p)$ is similar to that of $X(1)$, except that instead of having just one cusp (3.2), there are now multiple cusps (which do not necessarily go to infinity as in (3.2), but may instead go to some other point on the boundary $\mathbf{R} \cup \{\infty\}$ of the hyperbolic plane).

Remark 3.3.7. One may think of $X(p)$ as being formed by cutting up a finite number of of $X(1)$'s and then (pseudo-)randomly sowing them together to create a tangled orbifold that is a continuous analogue of an expander graph; see [**Sa2004**] for more discussion of this perspective. Indeed, one can view $X(p)$ as a continuous analogue of the Cayley graph $\mathrm{Cay}(P\mathrm{SL}_2(\mathbf{F}_p), \pi'_p(S))$, where

$$S := \left\{ \begin{pmatrix} 1 & 1 \\ 0 & 1 \end{pmatrix}, \begin{pmatrix} 1 & -1 \\ 0 & 1 \end{pmatrix}, \begin{pmatrix} 1 & 1 \\ 0 & 1 \end{pmatrix}, \begin{pmatrix} 0 & -1 \\ 1 & 0 \end{pmatrix} \right\},$$

and π'_p is the projection onto $P\mathrm{SL}_2(\mathbf{F}_p)$, with each vertex of the Cayley graph being replaced by a copy of the fundamental domain Ω of $X(1)$, with these domains then being glued together along their edges as prescribed by the edges of the Cayley graph.

By a routine modification of Proposition 3.3.5, one can show that

$$0 < \lambda_1(X(p)) \leq \frac{1}{4}.$$

(Note also that as $X(p)$ is a finite isometric cover of $X(1)$, we have the trivial bound $\lambda_1(X(p)) \leq \lambda_1(X(1))$.) However, these arguments do not keep $\lambda_1(X(p))$ *uniformly* bounded away from zero. Much more is conjectured to be true:

Conjecture 3.3.8 (Selberg's conjecture). *One has $\lambda_1(X(p)) = \frac{1}{4}$ for all p (not necessarily prime).*

This conjecture remains open (though it has been verified numerically for small values of p, in particular for all $p \leq 857$ [**BoSt2007**]. On the other hand, we have the following celebrated result of Selberg:

Theorem 3.3.9 (Selberg's 3/16 theorem [**Se1965**]). *One has $\lambda_1(X(p)) \geq \frac{3}{16}$ for all p (not necessarily prime).*

3.3. A weak form of Selberg's 3/16 theorem

Selberg's argument uses a serious amount of number-theoretic machinery (in particular, bounds for *Kloosterman sums*) and will not be reproduced here. The $\frac{3}{16}$ bound has since been improved; the best bound currently known is $\frac{975}{4096}$, due to Kim and Sarnak [**Ki2003**] and involving even more number-theoretic machinery (related to the *Langlands conjectures*); this argument will also not be discussed further here.

In [**SaXu1991**], [**Ga2002**], an argument based primarily on quasirandomness that used only very elementary number theory was introduced to obtain the following result:

Theorem 3.3.10 (Weak Selberg theorem). *One has*

$$\lambda_1(X(p)) \geq \min(\lambda_1(X(1)), \frac{5}{36} - o(1))$$

for all primes p, where $o(1)$ goes to zero as $p \to \infty$.

In particular, one has a uniform lower bound $\lambda_1(X(p)) \geq c$ for some absolute constant $c > 0$ (and, since one can compute that $\lambda_1(X(1)) = \frac{1}{4}$, one can in fact take $c = \frac{5}{36}$). Despite giving a weaker result than Theorem 3.3.9, the argument is more flexible and can be applied to other arithmetic surfaces than $X(p)$, for which the method of Selberg does not seem to apply; see [**SaXu1991**], [**Ga2002**] for further discussion.

We will not quite prove Theorem 3.3.10 here, but instead establish the following even weaker version which uses the same ideas, but in a slightly less computation-intensive fashion (at the cost of some efficiency in the argument):

Theorem 3.3.11 (Even weaker Selberg theorem). *One has $\lambda_1(X(p)) \geq \min(\lambda_1(X(1)), \frac{1}{12} - o(1))$ for all primes p.*

Of course, this result is still strong enough to supply a uniform lower bound on $\lambda_1(X(p))$.

Before we prove Theorem 3.3.11 (a spectral gap in the continuous world), let us show how it can be transferred to deduce Theorem 3.3.2 (a spectral gap in the discrete world). Suppose for contradiction that Theorem 3.3.2 failed. Then we can find a finite symmetric generating set S for $\mathrm{SL}_2(\mathbf{Z})$ (not containing the identity) and a sequence of primes p_n going to infinity such that the one-sided expansion constant of $\mathrm{Cay}(\mathrm{SL}_2(F_{p_n}), \pi_{p_n}(S))$ goes to zero. Write $G_n := \mathrm{SL}_2(F_{p_n})$. Applying the weak discrete Cheeger inequality (Exercise 2.0.3), we conclude that we can find nonempty subsets $E_n \subset G_n$ of size $|E_n| \leq \frac{1}{2}|G_n|$ which are almost $\pi_{p_n}(S)$-invariant in the sense that $|E_n \pi_{p_n}(S)| = 1 + o(1)|E_n|$. Since S generates $\mathrm{SL}_2(\mathbf{Z})$, we conclude in particular that

$$(3.3) \qquad |E_n \pi_{p_n}(s) \Delta E_n| = o(|E_n|)$$

for any $s \in \mathrm{SL}_2(\mathbf{Z})$ independent of n.

The idea now is to pass from this nearly-invariant discrete set E_n to a nearly-invariant continuous analogue f_n to which the uniform bound on the spectral gap can be applied to obtain a contradiction. (This argument is similar in spirit to Proposition 2.2.7.)

We turn to the details. Let $R \geq 1$ be a large parameter (independent of n) to be chosen later, and let z_0 be a point on $X(p_n)$ (avoiding fixed points of the G_n action); for the sake of concreteness we can take z_0 to be the projection of $2i \in \mathbf{H}$ to $X(p_n)$. Note that G_n acts on $X(p_n)$. We consider the function $f_n \colon X(p_n) \to [0,1]$ defined by the formula

$$f_n(z) := \max\left(\min\left(2 - \frac{\mathrm{dist}(z, E_n z_0)}{R}, 1\right), 0\right).$$

This function equals 1 when z is within R (in the hyperbolic metric) of a point in the orbit $E_n z_0$, and equals 0 when z is further than $2R$ of this orbit; in particular, it is compactly supported. The function is also Lipschitz with constant $O(1/R)$, so $|\nabla f_n| \leq 1/R$ (using a weak derivative).

The curve $X(p_n)$ is a $|G_n|$-fold cover of $X(1)$ and thus has volume $|G_n|\mu(X(1))$. Observe that f_n equals 1 on the R-neighbourhood of any point in $E_n z_0$. As these points are separated from each other by a bounded distance (independent of n and R), we conclude that

$$\mu(\{x \in X(p_n) : f_n(x) = 1\}) \gg |E_n|,$$

where the bound is uniform in R. Conversely, if $\gamma \in G_n$ is not of the form $\gamma = \gamma_1 \pi_{p_n}(\gamma_2)$ for some $\gamma_1 \in E_n$ and some $\gamma_2 \in \mathrm{SL}_2(\mathbf{Z})$ within distance $3R$ from the identity, we have f_n equal to 0 on the R-neighbourhood of γz_0. There are only $O_R(1)$ possible choices for γ_2; since R is independent of n, we conclude from (3.3) that all but $(1+o(1))|E_n|$ γ in G_n are of the form described above. Since $|E_n| \leq |G_n|/2$, we conclude that

$$\mu(\{x \in X(p_n) : f_n(x) = 0\}) \gg |G_n|$$

if n is sufficiently large depending on R. As a consequence, if we let

$$\tilde{f}_n := f_n - \frac{1}{\mu(X(p_n))} \int_{X(p_n)} f_n \, d\mu$$

be the mean-free component of f_n, we have the lower bound

(3.4) $$\|\tilde{f}_n\|_{L^2(X(p_n), \mu)} \gg |E_n|^{1/2},$$

for n sufficiently large depending on R.

On the other hand, $\nabla \tilde{f}_n = \nabla f_n$ is nonzero only at points which are at distance between R and $2R$ of $E_n z_0$. Call the set of such points A. To estimate the volume A, we partition $X(p_n)$ into $|G_n|$ sets of the form $\gamma \Omega$, where Ω is a fundamental domain of $X(1)$ (projected onto $X(p_n)$) and γ

3.3. A weak form of Selberg's 3/16 theorem

ranges over G_n. Because the ball of radius $2R$ centred at z_0 is precompact and thus meets only $O_R(1)$ of the translated domains $\gamma\Omega$, we see that the only γ for which $\gamma\Omega$ meets A are of the form $\gamma_1 \pi_{p_n}(\gamma_2)$, where γ_1 lies in E_n and γ_2 lies in a subset of $\mathrm{SL}_2(\mathbf{Z})$ of size $O_R(1)$ that is independent of n. From (3.3) we conclude that all but at most $o(|E_n|)$ of these γ thus lie in E_n, and so
$$\mu(A) \leq |E_n| + o(|E_n|).$$
Since $\nabla \tilde{f}_n = \nabla f_n = O(1/R)$, we conclude that
$$\|\nabla \tilde{f}_n\|_{L^2(X(p_n),\mu)} \ll \frac{1}{R}|E_n|^{1/2},$$
for n sufficiently large depending on R. But this and (3.4) contradict the uniform lower bound on the spectral gap $\lambda_1(X(p_n))$ (after regularising $\tilde{f}_n$ in a standard fashion to make it smooth rather than merely Lipschitz), giving the desired contradiction.

We now turn to the proof of Theorem 3.3.11. The first step is to show that the only source of spectrum below $1/4$ is provided by eigenfunctions.

Proposition 3.3.12 (Discrete spectrum below $1/4$). *Suppose that $\lambda_1(X(p)) < 1/4$. Then there exists a nonzero $\phi \in L^2(X(p))$ such that $-\Delta\phi = \lambda_1(X(p))\phi$ (in the distributional sense).*

Note that while ϕ is only initially in $L^2(X(p))$, it is a routine application of elliptic regularity (which we omit here) to show that ϕ is necessarily smooth.

Proof. For notational simplicity, we will just prove the claim in the $p = 1$ case, though the general case is similar. Write $\lambda := \lambda_1(X(p))$, so that $\lambda < 1/4$. The argument will be similar in spirit to the proof of the lower bound $\lambda > 0$. Indeed, by definition of λ, we can find a sequence of functions $f_n \in C_c^\infty(X(1))_0$ with $\|f_n\|_{L^2(X(1))} = 1$ such that

$$(3.5) \qquad \int_{X(1)} |\nabla f_n|^2 \, d\mu = \lambda + o(1).$$

As before, we can take the f_n to be real-valued.

Using Hardy's inequality as in the proof of Proposition 3.3.5, we see that
(3.6)
$$\int_R^\infty \int_{-1/2}^{1/2} \frac{|f_n|^2}{y^2} \, dx dy \leq 4 \int_R^\infty \int_{-1/2}^{1/2} |(f_n)_y|^2 \, dx dy + O\left(\int_R^{2R} \frac{|f_n|^2}{y^2} \, dx dy\right)$$
for any $R > C$. For any $\varepsilon > 0$, one can use the pigeonhole principle to find an $R = O_\varepsilon(1)$ (depending on n) such that
$$\int_R^{2R} \frac{|f_n|^2}{y^2} \, dx dy \leq \varepsilon$$

and thus we see that
$$\int_{R_\varepsilon}^\infty \int_{-1/2}^{1/2} \frac{|f_n|^2}{y^2}\, dxdy \ll \varepsilon + 4\lambda + o(1)$$
for some R_ε depending only on ε. If ε is small enough, $\varepsilon + 4\lambda < 1 = \int_{X(1)} |f_n|^2\, d\mu$, and thus
$$\int_{y \leq R_\varepsilon} |f_n|^2\, d\mu \gg 1$$
for all sufficiently large n. By (3.5), f_n is also uniformly bounded in H^1 norm. Thus by the Rellich compactness theorem, we may pass to a subsequence and assume that f_n converges weakly in $L^2(X(1))$ and strongly in L^2_{loc} to a limit ϕ, which is then nonzero. Also, from (3.6) we see that for each ε, R_0 and n there is an $R_0 \leq R = O_{R_0,\varepsilon}(1)$ such that
$$\lambda \int_{y>R} |f_n|^2\, d\mu \leq \int_{y>R} |\nabla f_n|^2\, d\mu + O(\varepsilon)$$
and thus
$$\lambda \int_{y\leq R} |f_n|^2\, d\mu \geq \int_{y\leq R} |\nabla f_n|^2\, d\mu + O(\varepsilon) + o(1).$$
Taking limits in weak L^2 (and strong L^2_{loc}), we conclude that for some $R = O_{\varepsilon,R_0}(1)$ larger than R_0 that
$$\lambda \int_{y\leq R} |\phi|^2\, d\mu \geq \int_{y\leq R} |\nabla \phi|^2\, d\mu + O(\varepsilon).$$
Sending $R_0 \to \infty$ using monotone convergence, we conclude that
$$\lambda \int_{X(1)} |\phi|^2\, d\mu \geq \int_{X(1)} |\nabla \phi|^2\, d\mu + O(\varepsilon)$$
for any ε; by definition of λ, we must then have
$$\lambda \int_{X(1)} |\phi|^2\, d\mu = \int_{X(1)} |\nabla \phi|^2\, d\mu.$$
Perturbing ϕ in some test function direction $g \in C_c^\infty(X_1)$ of mean zero, and using the definition of λ, we conclude that
$$\lambda \int_{X(1)} \langle \phi, g \rangle\, d\mu = \int_{X(1)} \langle \nabla \phi, \nabla g \rangle\, d\mu$$
for all such g. The mean zero condition on g can be removed since both sides of this equation vanish when g is constant. By duality we thus see that $-\Delta \phi = \lambda \phi$ in the sense of distributions, as required. $\square$

Exercise 3.3.4. Establish the above proposition for general p.

3.3. A weak form of Selberg's 3/16 theorem

Exercise 3.3.5. Show that for any $\lambda < 1/4$, the spectrum of $-\Delta$ in $[0, \lambda]$ on $X(p)$ is finite (and in particular consists only of eigenvalues), with each eigenvalue having finite multiplicity. (For this exercise you may use, without proof, the fact that $-\Delta$ is essentially self-adjoint; see Section 10.)

We will also need a variant of the above proposition:

Lemma 3.3.13 (Eigenfunctions do not concentrate in cusps). *Let $\lambda_0 < 1/4$. Then there is a compact subset F of $X(1)$, such that for any p and any eigenfunction $-\Delta \phi = \lambda \phi$ on $X(p)$ with some $\lambda < \lambda_0$, one has*

$$\int_{\eta_p^{-1}(F)} |\phi(x)|^2 \, d\mu(x) \gg_{\lambda_0} \int_{X(p)} |\phi(x)|^2 \, d\mu(x)$$

where the implied constant is independent of p, λ, and ϕ, and $\eta_p \colon X(p) \to X(1)$ is the covering map.

The lemma is basically proved by applying Hardy's inequality to each cusp of $X(p)$; see the paper of Gamburd for details.

Now we can start using quasirandomness. Let $V \subset L^2(X(p))_0$ be the space of all eigenfunctions of $-\Delta$ of eigenvalue λ:

$$V := \{\phi \in L^2(X(p))_0 : -\Delta \phi = \lambda \phi\}.$$

By the above proposition, this is a nontrivial Hilbert space. From Exercise 3.3.5, V is finite-dimensional (though we do not really need to know this fact yet in the argument that follows, as it will be a consequence of the computations). Since $\mathrm{SL}_2(\mathbf{F}_p)$ acts isometrically on $X(p)$, it also acts on V. If ϕ is a $\mathrm{SL}_2(\mathbf{F}_p)$-invariant vector in V, then it descends to a function on $L^2(X(1))_0$, which is impossible if $\lambda < \lambda_1(X(1))$. Applying the Frobenius lemma (Lemma 3.0.3), we conclude

Lemma 3.3.14 (Quasirandomness). *If $\lambda < \lambda_1(X(1))$, then V has dimension at least $\frac{p-1}{2}$.*

To complement this quasirandomness to get expansion, we need a flattening property, as in Proposition 3.1.3. In the discrete world, we applied a flattening property to the distribution μ^{*m} of a long discrete random walk. The direct analogue of such a distribution would be a *heat kernel* $e^{t\Delta}$ of the Laplacian Δ, and this is what we shall use here. (It turns out that the heat kernel is not quite the most efficient object to analyse here; see Remark 3.3.18 below.)

We first recall the formula for the heat kernel on the hyperbolic plane **H** (which can be found in many places, such as [**Ch1984**] or [**Te1985**]):

Exercise 3.3.6. Show that the heat operator $e^{t\Delta}$ on test functions f in $\mathbf{H}$ is given by the formula

$$e^{t\Delta} f(x) = \int_{\mathbf{H}} K_t(d(x,y)) f(y) \, d\mu(y)$$

where K_t is the kernel

$$K_t(\rho) := \frac{\sqrt{2}}{(4\pi t)^{3/2}} e^{-t/4} \int_\rho^\infty \frac{s e^{-s^2/4t}}{(\cosh s - \cosh \rho)^{1/2}} \, ds.$$

(*Hint:* There are two main computations. One is to show that $K_t(\rho)$ obeys the heat equation, which in half-cylindrical coordinates means that one has to verify that

(3.7) $$\frac{\partial}{\partial t} K_t(\rho) = \left(\frac{\partial^2}{\partial \rho^2} + \frac{1}{\sinh \rho} \frac{\partial}{\partial \rho} \right) K_t(\rho).$$

The other is to show that $K_t(\rho)$ resembles the Euclidean heat kernel $\frac{1}{4\pi t} e^{-\rho^2/4t}$ for small t. There are several other ways to derive this formula in terms of formulae for other operators (e.g., the wave propagator); see for instance [**Te1985**] for some discussion.)

For our purposes, we only need a crude upper bound on the heat kernel:

Exercise 3.3.7. With the notation of the preceding exercise, show that

$$K_t(\rho) \ll (t+\rho)^{O(1)} e^{-t/4} e^{-\rho/2} e^{-\rho^2/4t}$$

when $t \geq 1$ and $\rho \geq 0$.

In our applications, the polynomial factors $(t+\rho)^{O(1)}$ will be negligible; only the exponential factors will be of importance. Note that if one integrates the above estimate against the measure $d\mu = \sinh \rho \, d\rho \, d\theta$, one sees that

(3.8) $$\int_{\mathbf{H}} K_t \, d\mu \ll \int_0^\infty (t+\rho)^{O(1)} e^{-t/4} e^{+\rho/2} e^{-\rho^2/4t} d\rho.$$

The right-hand side evaluates to $O(t^{O(1)})$. On the other hand, as the heat kernel is a probability measure, one has $\int_{\mathbf{H}} K_t \, d\mu = 1$. Thus, up to polynomial factors, the above estimate is quite tight.

Remark 3.3.15. From Exercise 3.3.7, we see that the probability measure $K_t(\rho) \sinh \rho \, d\rho \, d\theta$ concentrates around the region $\rho = t + O(\sqrt{t})$; thus on the hyperbolic plane, Brownian motion moves "ballistically" away from its starting point at a unit speed, in contrast to the situation in Euclidean geometry, where after time t a Brownian motion is only expected to move by a distance $O(\sqrt{t})$. One can see this phenomenon also from the heat

3.3. A weak form of Selberg's 3/16 theorem

equation (3.7), which when expressed in terms of the probability density $u(\rho) := K_t(\rho) \sinh \rho$ becomes a *Fokker-Planck equation*

$$\frac{\partial}{\partial t} u(\rho) = \frac{\partial^2}{\partial \rho^2} u(\rho) - \frac{\partial}{\partial \rho} \left(\frac{1}{\tanh \rho} u \right)(\rho)$$

with unit diffusion and drift speed $\frac{1}{\tanh \rho}$. Since $\frac{1}{\tanh \rho}$ rapidly approaches 1 when ρ becomes large, we thus expect u to concentrate in the region $\rho = t + O(\sqrt{t})$, as is indeed the case.

We let $t \geq 1$ be a parameter to optimise in later. The heat operator $e^{t\Delta}$ on $\mathbf{H}$ descends to a heat operator on the quotient $X(p)$, defined by the formula

$$e^{t\Delta} f(x) = \int_{X(p)} \sum_{z \in \Gamma(p) y} K_t(d(x,z)) f(y) \, d\mu(y)$$

for $f \in C_c(X(p))$; note that the sum $\sum_{z \in \Gamma(p)y} K_t(d(x,z))$ is $\Gamma(p)$-invariant, and so makes sense for $x \in X(p)$ and not just for $x \in \mathbf{H}$. When applied to an eigenfunction $\phi \in V$, one has $e^{t\Delta}\phi = e^{-t\lambda}\phi$; in particular, $e^{t\Delta}$ preserves V, and thus also preserves the orthogonal complement of V. As $e^{t\Delta}$ is positive semidefinite, it therefore splits as the sum of $e^{-t\lambda} P_V$ and another positive semidefinite operator, where P_V is the orthogonal projection to V. Note that $e^{-t\lambda} P_V$ is an integral operator with kernel[5]

$$e^{-t\lambda} \sum_{i=1}^{\dim(V)} \phi_i(x) \overline{\phi_i(y)}$$

where $\phi_1, \ldots, \phi_{\dim(V)}$ is an orthonormal basis of V. Since positive semidefinite integral operators (with continuous kernel) are nonnegative on the diagonal, we conclude the pointwise inequality

$$e^{-t\lambda} \sum_{i=1}^{\dim(V)} |\phi_i(x)|^2 \leq \sum_{\gamma \in \Gamma(p)} K_t(d(x, \gamma x))$$

for all $x \in X(p)$.

This will be our starting point to get a lower bound on λ. But first we must deal with the other quantities ϕ_i, K_t in this expression. A simple way to proceed here is to integrate out in $X(p)$ to exploit the L^2 normalisation of the ϕ_i:

$$e^{-t\lambda} \dim(V) \leq \int_{X(p)} \sum_{\gamma \in \Gamma(p)} K_t(d(x, \gamma x)) \, d\mu(x).$$

However, this turns out to be a little unfavourable because the integrand on the right-hand side does not behave well enough at cusps. However, if one

---

[5] Here we use the fact that V is finite-dimensional, but if one does not want to use this fact yet, one can work instead with a finite-dimensional subspace of V in the argument that follows.

uses Lemma 3.3.13 first (assuming that $\lambda \leq 1/12$), and integrates over the resulting region $\eta_p^{-1}(F)$, we can avoid the cusps:

$$e^{-t\lambda}\dim(V) \ll \int_{\eta_p^{-1}(F)} \sum_{\gamma \in \Gamma(p)} K_t(d(x,\gamma x))\, d\mu(x).$$

Because the sum here is $\mathrm{SL}_2(\mathbf{F}_p)$-invariant, we can descend from $X(p)$ to $X(1)$:

$$e^{-t\lambda}\dim(V) \ll |\mathrm{SL}_2(\mathbf{F}_p)| \int_F \sum_{\gamma \in \Gamma(p)} K_t(d(x,\gamma x))\, d\mu(x).$$

We now insert the bound in Exercise 3.3.7, as well as the bound $|\mathrm{SL}_2(\mathbf{F}_p)| \ll p^3$:

$$e^{-t\lambda}\dim(V) \ll p^3 \int_F \sum_{\gamma \in \Gamma(p)} (t+d(x,\gamma x))^{O(1)} e^{-t/4} e^{-d(x,\gamma x)/2} e^{-d(x,\gamma x)^2/4t}\, d\mu(x).$$

Because F is compact, we can get a good bound on $d(x,\gamma x)$:

Exercise 3.3.8.

(i) For any $\gamma \in \mathrm{SL}_2(\mathbf{R})$, show that $d(i,\gamma i) = 2\log\|\gamma\| + O(1)$, where $\|\gamma\| := (a^2+b^2+c^2+d^2)^{1/2}$ is the Frobenius norm of the matrix $\gamma =: \begin{pmatrix} a & b \\ c & d \end{pmatrix}$.

(ii) More generally, if F is a compact subset of $X(1)$, show that $d(x,\gamma x) \leq C_F \log\|\gamma\| + C_F$ for some constant C_F depending only on F.

Inserting these bounds, we obtain

$$e^{-t\lambda}\dim(V) \ll p^3 \sum_{\gamma \in \Gamma(p)} (t+\log\|\gamma\|)^{O(1)} e^{-t/4} e^{-\log\|\gamma\|} e^{-(\log\|\gamma\|+O(1))^2/t}.$$

Decomposing according to the integer part R of $\log\gamma + O(1)$, we thus have

$$(3.9) \quad e^{-t\lambda}\dim(V) \ll p^3 \sum_{R=1}^{\infty} (t+R)^{O(1)} e^{-t/4} e^{-R} e^{-R^2/t} N_p(e^{R+O(1)})$$

where $N_p(T)$ is the counting function

$$N_p(T) := |\{\gamma \in \Gamma(p) : \|\gamma\| \leq T\}|.$$

So one is left with the purely number-theoretic task of estimating $N_p(T)$. This is basically the number of points of $\Gamma(p)i$ in the ball of radius $2\log T$ in $\mathbf{H}$. From the half-cylinder model, we see that the measure of this ball is $O(e^{2\log T}) = O(T^2)$. On the other hand, $\Gamma(p)$ has index $|\mathrm{SL}_2(\mathbf{F}_p)| \sim p^3$ in $\Gamma(1)$, which has bounded covolume in $\mathbf{H}$. We thus heuristically expect $N_p(T)$ to be $O(T^2/p^3)$. If this were the truth, then the right-hand side of (3.9) would be $O(t^{O(1)})$ (cf. the evaluation of (3.8)), which when combined

3.3. A weak form of Selberg's 3/16 theorem

with quasirandomness (Lemma 3.3.14) would give a lower bound of λ, that would be particularly strong when t was small.

The key is then the following "flattening lemma", that shows that $N_p(T)$ is indeed roughly of the order of $O(T^2/p^3)$ when T is large, and is the main number-theoretic input to the argument:

Lemma 3.3.16 (Flattening lemma). *For any $\varepsilon > 0$, one has*

$$N_p(T) \ll_\varepsilon \frac{T^{2+\varepsilon}}{p^3} + \frac{T^{1+\varepsilon}}{p} + T^\varepsilon.$$

Proof. Using the definition of $\Gamma(p)$, we are basically counting the number of integer solutions $(a,b,c,d) \in \mathbf{Z}^4$ to the equation

$$ad - bc = 1$$

subject to the congruences

$$a = d = 1 \bmod p; \quad b = c = 0 \bmod p$$

and the bounds

$$a,b,c,d = O(T).$$

Since b, c are both divisible by p, we see also that $ad = 1 \bmod p^2$. Similarly, as $a-1$ and $d-1$ are divisible by p, we have $(a-1)(d-1) = 0 \bmod p^2$. Subtracting, we conclude that

$$a + d = 2 \bmod p^2.$$

Now we proceed as follows. The number of integers $a = 1 \bmod p$ with $a = O(T)$ is $O(\frac{T}{p}+1)$. For each such a, the number of d with $a+d = 2 \bmod p^2$ and $d = O(T)$ is $O(\frac{T}{p^2} + 1)$. For each fixed a and d, the expression $ad - 1$ is of size $O(T^2)$; by the *divisor bound* (see [**Ta2009**, §1.6]), there are thus $O_\varepsilon(T^\varepsilon)$ ways to factor $ad - 1$ into bc. Thus, we obtain a final bound of

$$N_p(T) \ll_\varepsilon \left(\frac{T}{p}+1\right)\left(\frac{T}{p^2}+1\right) T^\varepsilon,$$

giving the claim. □

Remark 3.3.17. One can obtain improved bounds to $N_p(T)$ for some ranges of T (particularly when T ranges between p and p^2) by using more advanced tools, such as bounds on *Kloosterman sums*. Unfortunately, such improvements do not actually improve the final constants in this argument. (Kloosterman sums do however play a key role in the proof of Theorem 3.3.9, which proceeds by a different, and more highly arithmetic, argument.)

A routine calculation then finishes off the proof of Theorem 3.3.11:

Exercise 3.3.9. Using the above lemma, show that the right-hand side of (3.9) is

(3.10) $$\ll_\varepsilon e^{\varepsilon t}(1 + p^3 e^{-t/4})$$

for any $\varepsilon > 0$. Optimising this in t and using Lemma 3.3.14, establish a contradiction whenever $\lambda < \min(\frac{1}{12} - \varepsilon, \lambda_1(X(1)))$ and p is sufficiently large depending on ε, thus giving Theorem 3.3.11.

Remark 3.3.18. An inspection of the above argument shows that the $p^3 e^{-t/4}$ term in (3.10) is the main obstacle to improving the $\frac{1}{12}$ constant. This term ultimately can be "blamed" for the relatively large value of the heat kernel $K_t(\rho)$ at the origin. To improve this, one can observe that the main features of the heat kernel $K_t(\rho)$ that were needed for the argument were that it was positive definite, and had an explicit effect on eigenfunctions ϕ. It turns out that there are several other kernels with these properties, and by selecting a kernel with less concentration at the identity, one can obtain a better result. In particular, an efficient choice of kernel turns out to be the convolution of a ball of radius t with itself. By performing some additional calculations in hyperbolic geometry (in particular, using the Selberg/Harish-Chandra theory of *spherical functions*) one can use this kernel to improve the 1/12 bound given here to 5/36; see [**Ga2002**] for details. Unfortunately, the fraction 5/36 here appears to be the limit of this particular method.

Chapter 4

The Balog-Szemerédi-Gowers lemma, and the Bourgain-Gamburd expansion machine

We have now seen two ways to construct expander Cayley graphs Cay(G, S). The first, discussed in Chapter 2, is to use Cayley graphs that are projections of an infinite Cayley graph on a group with Kazhdan's property (T). The second, discussed in Chapter 3, is to combine a quasirandomness property of the group G with a flattening hypothesis for the random walk.

We now pursue the second approach more thoroughly. The main difficulty here is to figure out how to ensure flattening of the random walk, as it is then an easy matter to use quasirandomness to show that the random walk becomes mixing soon after it becomes flat. In the case of Selberg's theorem, we achieved this through an explicit formula for the heat kernel on the hyperbolic plane (which is a proxy for the random walk). However, in most situations such an explicit formula is not available, and one must develop some other tool for forcing flattening, and specifically an estimate of the form

$$(4.1) \qquad \|\mu^{*n}\|_{\ell^2(G)} \ll |G|^{-1/2+\varepsilon}$$

for some $n = O(\log |G|)$, where μ is the uniform probability measure on the generating set S.

In [**BoGa2008**] Bourgain and Gamburd introduced a general method for achieving this goal. The intuition here is that the main obstruction that prevents a random walk from spreading out to become flat over the entire group G is if the random walk gets *trapped* in some proper subgroup H of G (or perhaps in some coset xH of such a subgroup), so that $\mu^{*n}(xH)$ remains large for some moderately large n. Note that

$$\mu^{*2n}(H) \geq \mu^{*n}(Hx^{-1})\mu^{*n}(xH) = \mu^{*n}(xH)^2,$$

since $\mu^{*2n} = \mu^{*n} * \mu^{*n}$, $H = (Hx^{-1}) \cdot (xH)$, and μ^{*n} is symmetric. By iterating this observation, we see that if $\mu^{*n}(xH)$ is too large (e.g., of size $|G|^{-o(1)}$ for some n comparable to $\log|G|$), then it is not possible for the random walk μ^{*n} to converge to the uniform distribution in time $O(\log|G|)$, and so expansion does not occur.

A potentially more general obstruction of this type would be if the random walk gets trapped in (a coset of) an *approximate* group H. For any $K \geq 1$, we define a *K-approximate group* to be a subset H of a group G which is symmetric, contains the identity, and is such that $H \cdot H$ can be covered by at most K left-translates (or equivalently, right-translates) of H. Such approximate groups were studied extensively in last quarter's course. A similar argument to the one given previously shows (roughly speaking) that expansion cannot occur if $\mu^{*n}(xH)$ is too large for some coset xH of an approximate group.

It turns out that this latter observation has a converse: if a measure does not concentrate in cosets of approximate groups, then some flattening occurs. More precisely, one has the following combinatorial lemma:

Lemma 4.0.1 (Weighted Balog-Szemerédi-Gowers lemma). *Let G be a group, let ν be a finitely supported probability measure on G which is symmetric (thus $\nu(g) = \nu(g^{-1})$ for all $g \in G$), and let $K \geq 1$. Then one of the following statements hold:*

  (i) *(Flattening) One has $\|\nu * \nu\|_{\ell^2(G)} \leq \frac{1}{K}\|\nu\|_{\ell^2(G)}$.*

  (ii) *(Concentration in an approximate group) There exists an $O(K^{O(1)})$-approximate group H in G with $|H| \ll K^{O(1)}/\|\nu\|_{\ell^2(G)}^2$ and an element $x \in G$ such that $\nu(xH) \gg K^{-O(1)}$.*

This lemma is a variant of the more well-known *Balog-Szemerédi-Gowers lemma* [**BaSz1994**], [**Go1998**] in additive combinatorics; the version in [**Go1998**] roughly speaking corresponds to the case when μ is the uniform distribution on some set A, and is a polynomially quantitative version of an earlier lemma of Balog and Szemerédi [**BaSz1994**]. We will prove it below the fold.

The lemma is particularly useful when the group G in question enjoys a *product theorem* which, roughly speaking, says that the only medium-sized approximate subgroups of G are trapped inside genuine proper subgroups of G (or, contrapositively, medium-sized sets that generate the entire group G cannot be approximate groups). The fact that some finite groups (and specifically, the bounded rank finite simple groups of Lie type) enjoy product theorems is a nontrivial fact, and will be discussed in later chapters. For now, we simply observe that the presence of the product theorem, together with quasirandomness and a nonconcentration hypothesis, can be used to demonstrate expansion:

Theorem 4.0.2 (Bourgain-Gamburd expansion machine)**.** *Suppose that G is a finite group, that $S \subseteq G$ is a symmetric set of k generators, and that there are constants $0 < \kappa < 1 < \Lambda$ with the following properties.*

(1) *(Quasirandomness) The smallest dimension of a nontrivial representation $\rho \colon G \to \mathrm{GL}_d(\mathbf{C})$ of G is at least $|G|^\kappa$.*

(2) *(Product theorem) For all $\delta > 0$ there is some $\delta' = \delta'(\delta) > 0$ such that the following is true. If $H \subseteq G$ is a $|G|^{\delta'}$-approximate subgroup with $|G|^\delta \leq |H| \leq |G|^{1-\delta}$ then H generates a proper subgroup of G.*

(3) *(Nonconcentration estimate) There is some even number $n \leq \Lambda \log |G|$ such that*

$$\sup_{H < G} \mu^{*n}(H) < |G|^{-\kappa},$$

where the supremum is over all proper subgroups $H < G$.

Then $\mathrm{Cay}(G, S)$ is a two-sided ε-expander for some $\varepsilon > 0$ depending only on k, κ, Λ, and the function $\delta'(\cdot)$ (and this constant ε is, in principle, computable in terms of these constants).

This criterion for expansion is implicitly contained in [**BoGa2008**], where it was used to establish the expansion of various Cayley graphs in $\mathrm{SL}_2(\mathbf{F}_p)$ for prime p. This criterion has since been applied (or modified) to obtain expansion results in many other groups, as will be discussed in later chapters.

4.1. The Balog-Szemerédi-Gowers lemma

The Balog-Szemerédi-Gowers lemma (Lemma 4.0.1) is ostensibly a statement about group structure, but the main tool in its proof is a remarkable graph-theoretic lemma (also known as the Balog-Szemerédi-Gowers lemma) that allows one to upgrade a "statistical" structure (a structure which is only valid a small fraction of the time, say 1% of the time) to a "complete" structure (one which is valid 100% of the time), by shrinking the size of the structure slightly (and in particular, with losses of polynomial type, as opposed to exponential or worse). This is in contrast to other structure-improving results (such as *Ramsey's theorem* [**Ra1930**], *Szemerédi's theorem* [**Sz1975**], or *Freiman's theorem* [**Fr1973**]), which are qualitatively similar in spirit, but have much worse quantitative bounds (though there is some hope in the case of Freiman's theorem to only lose polynomial bounds with some improvement of existing arguments; see [**Sa2013**]).

As we shall see later, the property of $\|\nu*\nu\|_{\ell^2(G)}$ being large is a statistical assertion about ν (it asserts that $\nu * \nu$ collides with itself somewhat often), whereas approximate groups H represent a more complete sort of structure (*all* products of $H \cdot H$ are trapped in a small set, whereas only *many* of the products in $\nu * \nu$ are so constrained). The graph-theoretic Balog-Szemerédi lemma is the key to moving from the former type of structure to the latter with only polynomial losses.

We need some notation. Define a *bipartite graph* $G = G(A, B, E)$ to be a graph whose vertex set $V := A \cup B$ is partitioned into two nonempty sets A, B, and the edge set E consists only of edges between A and B. If a finite bipartite graph $G = G(A, B, E)$ is *dense* in the sense that its edge density $|E|/|A||B|$ is large, then for many vertices $a \in A$ and $b \in B$, a and b are connected by a path of length one (i.e., an edge). It is thus intuitive that many pairs of vertices $a \in A$ and $a' \in A$ will be connected by many paths of length two. Perhaps surprisingly, one can upgrade "many pairs" here to "almost all pairs", provided that one is willing to shrink the set A slightly. More precisely, one has

Lemma 4.1.1 (Balog-Szemerédi-Gowers lemma: paths of length two). *Let $G(A, B, E)$ be a finite bipartite graph with $|E| \geq |A||B|/K$. Let $\varepsilon > 0$. Then there exists a subset A' of A with $|A'| \geq \frac{|A|}{\sqrt{2K}}$ such that at least $(1-\varepsilon)|A'|^2$ of the pairs $(a, a') \in A' \times A'$ are such that a, a' are connected by at least $\frac{\varepsilon}{2K^2}|B|$ paths of length two (i.e., there exists at least $\frac{\varepsilon}{2K^2}|B|$ vertices $b \in B$ such that $\{a, b\}, \{a', b\}$ both lie in E).*

Remark 4.1.2. It is not possible to remove the ε entirely from this lemma; see [**TaVu2006**, Exercise 6.4.2] for a counterexample (involving Hamming balls).

Proof. The idea here is to use a probabilistic construction, picking A' to be a neighbourhood of a randomly selected element b of B. The rationale here is that if a pair a, a' of vertices in A are not connected by many paths of length two, then they are unlikely to lie in the same neighbourhood, and so are unlikely to "wreck" the construction.

We turn to the details. Let $b \in B$ be chosen uniformly at random, and let $A' := \{a \in A : (a, b) \in E\}$ be the neighbourhood of b. Observe that the

expected size of A' is
$$\mathbf{E}|A'| = \frac{1}{|B|}|E| \geq \frac{|A|}{K}.$$

By Cauchy-Schwarz, we conclude in particular that

(4.2) $$\mathbf{E}|A'|^2 \geq \frac{|A|^2}{K^2}.$$

Now, call a pair (a, a') *bad* if it is connected by fewer than $\frac{\varepsilon|B|}{2K^2}$ paths of length two, and let N be the number of bad pairs (a, a') in $A' \times A'$. We consider the quantity $\mathbf{E}N$. Observe that if (a, a') is a bad pair in $A \times A$, then there are at most $\frac{\varepsilon|B|}{2K^2}$ values of b for which a and a' will both lie in A', and so this bad pair contributes at most $\frac{\varepsilon}{2K^2}$ to the expectation. Since there are at most $|A|^2$ bad pairs, we conclude that

$$\mathbf{E}N \leq \frac{\varepsilon|A|^2}{2K^2}.$$

Combining this with (4.2), we see that

$$\mathbf{E}\left(|A'|^2 - \frac{N}{\varepsilon} - \frac{|A|^2}{2K^2}\right) \geq 0.$$

In particular, there exists a choice of b for which the expression on the left-hand side is nonnegative. This implies that

$$N \leq \varepsilon|A'|^2$$

and

$$|A'|^2 \geq \frac{|A|^2}{2K^2}$$

and the claim follows. □

Given that almost all pairs a, a' in A' are joined by many paths of length two, it is then plausible that almost all pairs $a \in A'$, $b \in B'$ are joined by many paths of length three, for some large subset B' of B. Remarkably, one can now upgrade "almost all" pairs here to *all* pairs:

Lemma 4.1.3 (Balog-Szemerédi-Gowers lemma: paths of length three). *Let $G(A, B, E)$ be a finite bipartite graph with $|E| \geq |A||B|/K$. Then there exists subsets A', B' of A, B respectively with $|A'| \gg K^{-O(1)}|A|$ and $|B'| \gg K^{-O(1)}|B|$, such that for every $a \in A$ and $b \in B$, a and b are joined by $\gg K^{-O(1)}|A||B|$ paths of length three.*

Remark 4.1.4. A lemma similar to this was first established in [**BaSz1994**], as a consequence of the *Szemerédi regularity lemma* [**Sz1978**]. However, as a consequence of using that lemma, the polynomial bounds $K^{-O(1)}$ in the above lemma had to be replaced by much worse bounds (of tower-exponential type in K), which turns out to be far too weak for the purposes of establishing expansion.

Proof. The idea is to first prune a few "unpopular" vertices from A and B and then apply the preceding lemma.

Let A_1 be the vertices in A of degree at least $|B|/2K$, and let E_1 be the edges connecting A_1 and B. Note that the vertices in $A \backslash A_1$ are connected to a total of at most $|A||B|/2K$ edges, and so $|E_1| \geq |A||B|/2K \geq |A_1||B|/2K$. Since $|E_1| \leq |A_1||B|$, we conclude in particular that $|A_1| \geq |A|/2K$.

Let $\varepsilon > 0$ be a sufficiently small quantity (depending on K) to be chosen later. Applying Lemma 4.1.1, one can find a subset A_2 of A_1 of cardinality $|A_2| \gg |A_1|/K \gg |A|/K^2$ such that at most $\varepsilon |A_2|^2$ of the pairs $(a, a') \in A_2 \times A_2$ are *bad* in the sense that they are not connected by $\gg \frac{\varepsilon}{K^2}|B|$ paths of length two.

Let A' be those vertices a in A_2 for which there are at most $\sqrt{\varepsilon}|A_2|$ elements a' of A_2 for which (a, a') is bad. By Markov's inequality, A' consists of all but at most $\sqrt{\varepsilon}|A_2|$ elements of A_2.

Let E_2 be the edges connecting A_2 with B. Since each vertex in A_2 has degree at least $|B|/2K$, one has

$$|E_2| \geq |A_2||B|/2K \gg |A||B|/K^3.$$

We may thus find a subset B' of B of cardinality $|B'| \gg |B|/K^3$ such that each $b \in B'$ is adjacent to $\gg |A|/K^3$ elements of A_2.

Now let $a \in A'$ and $b \in B'$. We know that b is adjacent to $\gg |A|/K^3$ elements a' of A_2, and that at most $\sqrt{\varepsilon}|A_2|$ of these elements are such that (a, a') is bad. If we choose ε to be a sufficiently small multiple of $1/K^6$, we conclude that there are $\gg |A|/K^3$ elements a' which are adjacent to b and for which (a, a') is not bad. One thus has $\gg (|A|/K^3)(\varepsilon/K^2)|B| \gg |A||B|/K^{11}$ paths of length three connecting a to b, and the claim follows. □

The exponents in K here can be improved slightly, but we will not attempt to obtain the optimal numerology here.

Remark 4.1.5. The above results are analogous to a phenomenon in additive combinatorics, namely that a "1%-structured" set (such as a small density subset of a group) can often be upgraded to a "99%-structured" set (such as the complement of a small density subset of a group) by applying a single "convolution" or "sumset" operation, and then upgraded further to

4.1. The Balog-Szemerédi-Gowers lemma

a "100%-structured" set (such as a genuine group) by applying a further[1] convolution or sumset operation.

Exercise 4.1.1 (Weighted Balog-Szemerédi-Gowers theorem). Let (X, μ) and (Y, ν) be probability spaces, and let $E \subset X \times Y$ have measure $\mu \times \nu(E) \geq 1/K$ for some $K \geq 1$.

(i) Show that for any $\varepsilon > 0$, there exists a subset X' of X of measure $\mu(X') \geq \frac{1}{\sqrt{2}K}$ such that

$$\mu \times \mu\left(\left\{(x, x') \in X' \times X' : \int_Y 1_E(x, y) 1_E(x', y) \, d\nu(y) < \frac{\varepsilon}{2K^2}\right\}\right) \leq \varepsilon \mu(X')^2.$$

(ii) Show that there exists subsets X', Y' of X, Y of measure $\mu(X') \gg K^{-O(1)}$ and $\nu(Y') \gg K^{-O(1)}$ such that

$$\int_X \int_Y 1_E(x, y') 1_E(x', y') 1_E(x', y) \, d\mu(x') d\nu(y') \gg K^{-O(1)}$$

for all $x \in X'$ and $y \in Y'$.

Exercise 4.1.2 (99% Balog-Szemerédi theorem). Let $G(A, B, E)$ be a finite bipartite graph such that $|E| \geq (1-\varepsilon)|A||B|$.

(i) Show that there exists a subset A' of A of size $|A'| \geq (1-O(\sqrt{\varepsilon}))|A|$ such that for every $a, a' \in A'$, a and a' are connected by at least $(1 - O(\sqrt{\varepsilon}))|B|$ paths of length 2. (*Hint:* Select A' to be those vertices in A' that are connected to "almost all" the vertices in B.)

(ii) Show that there also exists a subset B' of B of size $|B'| \geq (1 - O(\sqrt{\varepsilon}))|B|$ such that for every $a \in A'$ and $b \in B'$, a and b are connected by at least $(1 - O(\sqrt{\varepsilon}))|A||B|$ paths of length 3.

We now apply the graph-theoretic lemma to the group context. The main idea here is to show that various sets (e.g., product sets $A \cdot B$) are small by showing that they are in the high-multiplicity region of some convolution (e.g., $1_{A_1} * \ldots * 1_{A_k}$), or equivalently that elements g of such sets have many representations as a product $g = a_1 \ldots a_k$ with $a_1 \in A_1, \ldots, a_k \in A_k$. One can then use Markov's inequality and the trivial identity $\|1_{A_1} * \ldots * 1_{A_k}\|_{\ell^1(G)} = |A_1| \ldots |A_k|$ to get usable size bounds on such sets.

---

[1]This is basically why, for instance, it is known that almost all even natural numbers are the sum of two primes, and all but finitely many odd natural numbers are the sum of three primes; but it is not known whether all but finitely many even natural numbers are the sum of two primes.

Corollary 4.1.6 (Balog-Szemerédi-Gowers lemma, product set form). *let A, B be finite nonempty subsets of a group $G = (G, \cdot)$, and suppose that*

$$\|1_A * 1_B\|_{\ell^2(G)} \geq |A|^{3/4}|B|^{3/4}/K$$

for some $K \geq 1$. (This hypothesis should be compared with the upper bound

$$\|1_A * 1_B\|_{\ell^2(G)} \leq \|1_A\|_{\ell^{4/3}(G)}\|1_B\|_{\ell^{4/3}(G)} = |A|^{3/4}|B|^{3/4}$$

arising from Young's inequality.) Then there exists subsets A', B' of A, B respectively with $|A'| \gg K^{-O(1)}|A|$ and $|B'| \gg K^{-O(1)}|B|$ with $|A' \cdot B'| \ll K^{O(1)}|A|^{1/2}|B|^{1/2}$ and $|A' \cdot (A')^{-1}| \ll K^{O(1)}|A|$.

The quantity $\|1_A * 1_B\|_{\ell^2(G)}^2$ (or equivalently, the number of solutions to the equation $ab = a'b'$ with $a, a' \in A$ and $b, b' \in B$) is also known as the *multiplicative energy* of A and B, and is sometimes denoted $E(A, B)$ in the literature.

Proof. By hypothesis, we have

$$\sum_{(a,b) \in A \times B} 1_A * 1_B(ab) = \|1_A * 1_B\|_{\ell^2(G)}^2 \geq |A|^{3/2}|B|^{3/2}/K^2.$$

Since

$$\sum_{(a,b) \in A \times B : 1_A * 1_B(ab) \leq |A|^{1/2}|B|^{1/2}/2K^2} 1_A * 1_B(ab) \leq |A|^{3/2}|B|^{3/2}/2K^2,$$

we conclude that

$$\sum_{(a,b) \in A \times B : 1_A * 1_B(ab) > |A|^{1/2}|B|^{1/2}/2K^2} 1_A * 1_B(ab) \geq |A|^{3/2}|B|^{3/2}/2K^2.$$

Since, by Cauchy-Schwarz (or Young's inequality), we have $1_A * 1_B(ab) \leq |A|^{1/2}|B|^{1/2}$, we conclude that there is a set $E \subset A \times B$ with $|E| \geq |A||B|/2K^2$ such that

$$1_A * 1_B(ab) > |A|^{1/2}|B|^{1/2}/2K^2$$

for all $(a, b) \in E$.

By a slight abuse of notation (arising from the fact that A, B are not necessarily disjoint, and that E is a set of ordered pairs rather than unordered pairs), we can view the triplet (A, B, E) as a bipartite graph. Applying Lemma 4.1.3, we can find subsets A', B' of A, B respectively with $|A'| \gg K^{-O(1)}|A|$ and $|B'| \gg K^{-O(1)}|B|$ such that for all $a \in A'$ and $b \in B'$, one can find $\gg K^{-O(1)}|A||B|$ elements $a' \in A, b' \in B$ such that $(a, b'), (a', b'), (a', b) \in E$. In particular, we see that

$$(4.3) \quad \sum_{a' \in G} \sum_{b' \in G} 1_A * 1_B(ab') 1_A * 1_B(a'b') 1_A * 1_B(a'b) \gg K^{-O(1)}|A|^{5/2}|B|^{5/2}.$$

Observe that $1_A * 1_B(a'b') = 1_{B^{-1}} * 1_{A^{-1}}((b')^{-1}(a')^{-1})$. Using the identity $(ab')((b')^{-1}(a')^{-1})(a'b) = ab$, we note that triples $(ab', (b')^{-1}(a')^{-1}, a'b)$ for $a', b' \in G$ are precisely those triples $(g_1, g_2, g_3) \in G \times G$ with $g_1 g_2 g_3 = ab$. Thus the left-hand side of (4.3) is equal to $F(ab)$, where

$$F := 1_A * 1_B * 1_{B^{-1}} * 1_{A^{-1}} * 1_A * 1_B.$$

But since
$$\|F\|_{\ell^1} = |A||B||B^{-1}||A^{-1}||A||B| = |A|^3|B|^3,$$
we see from Markov's inequality that there are at most $O(K^{O(1)}|A|^{1/2}|B|^{1/2})$ possible values for ab, which gives the bound $|A' \cdot B'| \ll K^{O(1)}|A|^{1/2}|B|^{1/2}$.

The second bound $|A' \cdot (A')^{-1}| \ll K^{O(1)}|A|$ can be proven similarly to the first (noting that any $a, a' \in A'$ are connected by $\gg K^{-O(1)}|A|^2|B|^2$ paths of length six), but can also be from the former bound as follows. Observe that any element $a(a')^{-1} \in A' \cdot (A')^{-1}$ has at least $|B'|$ representations of the form $a(a')^{-1} = (ab)(a'b)^{-1}$ with $b \in B'$, and hence $ab, a'b \in A' \cdot B'$, thus

$$1_{A'B'} * 1_{(A'B')^{-1}} \geq |B'| \gg K^{-O(1)}|B|$$

on $A'(A')^{-1}$. On the other hand, the left-hand side has an $\ell^1(G)$ norm of $|A'B'||(A'B')^{-1}| \ll K^{O(1)}|A||B|$, and the bound $|A' \cdot (A')^{-1}| \ll K^{O(1)}|A|$ then follows from Markov's inequality. □

Exercise 4.1.3. In the converse direction, show that if A, B are nonempty finite subsets of G with $|AB| \leq K|A|^{1/2}|B|^{1/2}$, then $\|1_A * 1_B\|_{\ell^2(G)} \geq |A|^{3/4}|B|^{3/4}/K^{1/2}$.

Exercise 4.1.4. If A, B, C are three nonempty finite subsets of G, establish the *Ruzsa triangle inequality* $|A \cdot C^{-1}| \leq \frac{|A \cdot B^{-1}||B \cdot C^{-1}|}{|B|}$. (*Hint:* Mimic the final part of the proof of Corollary 4.1.6.)

We now give a variant of this corollary involving approximate groups.

Lemma 4.1.7 (Balog-Szemerédi-Gowers lemma, approximate group form). *Let A be a finite symmetric subset of a group $G = (G, \cdot)$, and suppose that*

$$\|1_A * 1_A\|_{\ell^2(G)} \geq |A|^{3/2}/K$$

for some $K \geq 1$. Then there exists a $K^{O(1)}$-approximate group H with $|H| \ll K^{O(1)}|A|$ such that $|A \cap gH| \gg K^{-O(1)}|A|$ for some $g \in H$.

Proof. By Corollary 4.1.6, we may find a subset $A' \subset A$ with $|A'| \gg K^{-O(1)}|A|$ such that

(4.4) $$|A'(A')^{-1}| \ll K^{O(1)}|A|.$$

By Exercise 4.1.3, this implies that

$$\|1_{A'} * 1_{(A')^{-1}}\|_{\ell^2(G)}^2 \gg K^{-O(1)}|A|^3.$$

Observe that the left-hand side is equal to

$$1_{A'} * 1_{(A')^{-1}} * 1_{A'} * 1_{(A')^{-1}}(1)$$
$$= 1_{(A')^{-1}} * 1_{A'} * 1_{(A')^{-1}} * 1_{A'}(1)$$
$$= \|1_{(A')^{-1}} * 1_{A'}\|_{\ell^2(G)}^2.$$

We conclude that

$$\sum_{s \in G}(1_{(A')^{-1}} * 1_{A'}(s))^2 \gg K^{-O(1)}|A|^3.$$

On the other hand, we have

$$\sum_{s \in G} 1_{(A')^{-1}} * 1_{A'}(s) = |A'||A'| \leq |A|^2.$$

As a consequence, we see that if we set

$$S := \{s \in G : 1_{(A')^{-1}} * 1_{A'}(s) \geq C^{-1}K^{-C}|A|\}$$

for some sufficiently large absolute constant C, then

$$\sum_{s \in G \setminus S}(1_{(A')^{-1}} * 1_{A'}(s))^2 \leq C^{-1}K^{-C}|A|^3,$$

and thus (for C large enough)

$$\sum_{s \in S}(1_{(A')^{-1}} * 1_{A'}(s))^2 \gg K^{-O(1)}|A|^3.$$

Since $1_{(A')^{-1}} * 1_{A'}(s) \leq |A'| \leq |A|$, we conclude that

$$|S| \gg K^{-O(1)}|A|.$$

Also, S is clearly symmetric and contains the origin.

Now let us consider an element $g = a_0 s_1 \ldots s_5 b_6^{-1}$ of the product $(A')S^5(A')^{-1}$. By construction of S, we can write each s_i as a product $b_i^{-1} a_i$ with $a_i, b_i \in A'$ in at least $C^{-1}K^{-C}|A|$ ways. Doing so for each $i = 1, \ldots, 5$ gives rise to a factorisation,

$$g = g_1 \ldots g_6,$$

where $g_i := a_{i-1} b_i^{-1} \in A'(A')^{-1}$; as the $g_1, \ldots, g_6$ uniquely determine the a_i, b_i (for fixed $a_0, s_1, \ldots, s_5, b_6$), we conclude that each element g of $(A')S^5(A')^{-1}$ has at least $\gg K^{-O(1)}|A|^5$ such factorisations. But by (4.4), there are at most $O(K^{O(1)}|A|^6)$ such tuples $g_1, \ldots, g_6$, and so there are at most $O(K^{O(1)}|A|)$ possible values for g, thus

(4.5) $$|(A')S^5(A')^{-1}| \ll K^{O(1)}|A|.$$

In particular,

$$|S^5| \ll K^{O(1)}|S|.$$

4.1. The Balog-Szemerédi-Gowers lemma

By the Ruzsa covering lemma (see Exercise 4.1.5), this implies that S^4 is covered by $O(K^{O(1)})$ left-translates of S^2, and so $H := S^2$ is a $K^{O(1)}$-approximate group. Finally, from (4.5) one has

$$|A'H| \ll K^{O(1)}|A|$$

and thus by Exercise 4.1.3

$$\|1_{A'} * 1_H\|_{\ell^2(G)} \gg K^{-O(1)}|A|^{3/2}.$$

In particular, since the support of $1_{A'} * 1_H$ has size $O(K^{O(1)}|A|)$, one has

$$1_{A'} * 1_H(g) \gg K^{-O(1)}|A|$$

for some $g \in G$, or equivalently that

$$|A' \cap Hg| \gg K^{-O(1)}|A|.$$

Increasing A' to A and taking inverses, we conclude that $|gH \cap A| \ll K^{-O(1)}|A|$, and the claim follows. $\square$

Exercise 4.1.5 (Ruzsa covering lemma). Let A, B be finite nonempty subsets of a group G. Show that A can be covered by at most $\frac{|AB|}{|B|}$ left-translates of BB^{-1}. (*Hint:* Consider a maximal disjoint collection of translates aB of B with $a \in A$.)

Exercise 4.1.6 (Converse to Balog-Szemerédi-Gowers). Let A be a finite symmetric subset of a group $G = (G, \cdot)$, and suppose there exists a K-approximate group H with $|H| \leq K|A|$ such that $|A \cap gH| \geq |A|/K$ for some $g \in H$. Show that

$$\|1_A * 1_A\|_{\ell^2(G)} \geq K^{-3}|A|^{3/2}.$$

Exercise 4.1.7. Let A, B be finite nonempty subsets of a group G, and suppose that $\|1_A * 1_B\|_{\ell^2(G)}^2 \geq |A|^{3/2}|B|^{3/2}/K$. Show that there exists a $O(K^{O(1)})$-approximate group H with

$$K^{-O(1)}|A|^{1/2}|B|^{1/2} \leq |H| \leq K^{O(1)}|A|^{1/2}|B|^{1/2}$$

and elements $g, h \in G$ such that $|A \cap gH| \gg K^{-O(1)}|H|$ and $|B \cap Hh| \gg K^{-O(1)}|H|$.

Finally, we can prove Lemma 4.0.1. Fix G, ν, K. We may assume that

(4.6) $$\|\nu * \nu\|_{\ell^2(G)} > \frac{1}{K}\|\nu\|_{\ell^2(G)}$$

and we need to use this to locate an $O(K^{O(1)})$-approximate group H in G with $|H| \ll K^{O(1)}/\|\nu\|_{\ell^2(G)}^2$ and an element $x \in G$ such that $\nu(xH) \gg K^{-O(1)}$.

Let us write $M := 1/\|\nu\|_{\ell^2(G)}^2$. Intuitively, M represents the "width" of the probability measure ν, as can be seen by considering the model example $\nu = \frac{1}{M}1_A$ where A is a symmetric set of cardinality M (i.e., ν is the uniform probability measure on A). If we were actually in this model case, we could apply Lemma 4.1.7 immediately and be done. Of course, in general, ν need not be a uniform measure on a set of size M. However, it turns out that one can use (4.6) to conclude that the "bulk" of ν is basically of this form.

More precisely, let us split $\nu = \nu_< + \nu_> + \nu_=$, where

$$\nu_< := \nu 1_{\nu \leq \frac{1}{100K^2M}},$$
$$\nu_> := \nu 1_{\nu \geq \frac{10K}{M}},$$
$$\nu_= := \nu - \nu_< - \nu_>.$$

Observe that

$$\|\nu_<\|_{\ell^2(G)}^2 \leq \frac{1}{100K^2M}\|\nu\|_{\ell^1(G)} = \frac{1}{100K^2M}$$

and so by Young's inequality

$$\|\nu_< * \nu\|_{\ell^2(G)}, \|\nu * \nu_<\|_{\ell^2(G)} \leq \frac{1}{10KM^{1/2}}.$$

In a similar vein, we have

$$\|\nu_>\|_{\ell^1(G)} \leq \frac{M}{10K}\|\nu\|_{\ell^2(G)}^2 = \frac{1}{10K}$$

and thus by Young's inequality (and the normalisation $\|\nu\|_{\ell^2(G)} = 1/M^{1/2}$)

$$\|\nu_> * \nu\|_{\ell^2(G)}, \|\nu * \nu_>\|_{\ell^2(G)} \leq \frac{1}{10KM^{1/2}}.$$

Finally, from (4.6) one has

$$\|\nu * \nu\|_{\ell^2(G)} \geq \frac{1}{KM^{1/2}}.$$

Subtracting using the triangle inequality (ignoring some slight double-counting), we conclude that

$$\|\nu_= * \nu_=\|_{\ell^2(G)} \gg \frac{1}{KM^{1/2}}.$$

If we then set $A := \{g \in G : \nu(g) > \frac{1}{100K^2M}\}$, we conclude in particular that

$$\|1_A * 1_A\|_{\ell^2(G)} \gg K^{-O(1)}M^{3/2}.$$

On the other hand, from Markov's inequality one has $|A| \ll K^2M$. Applying Lemma 4.1.7, we conclude the existence of a $O(K^{O(1)})$-approximate group H with $|H| \ll K^{O(1)}M$ such that $|A \cap gH| \gg K^{-O(1)}M$ for some $g \in G$, which by definition of A implies that $\nu(gH) \gg K^{-O(1)}$, and the claim follows.

4.2. The Bourgain-Gamburd expansion machine

We can now prove Theorem 4.0.2. We can assume that $|G|$ is sufficiently large depending on the parameters $k, \kappa, \Lambda, \delta'$, since the claim is trivial for bounded G (note that as S generates G, the Cayley graph $\mathrm{Cay}(G, S)$ will be an ε-expander for some $\varepsilon > 0$). Henceforth, we allow all implied constants in the asymptotic notation to depend on $k, \kappa, \Lambda, \delta'$.

To show expansion, it suffices from the quasirandomness hypothesis (and Proposition 3.1.3), it will suffice to show that

$$\|\mu^{*n}\|_{\ell^2(G)} \leq |G|^{-1/2+\kappa/2} \tag{4.7}$$

for some $n = O(\log |G|)$.

From Young's inequality, $\|\mu^{*n}\|_{\ell^2(G)}$ is decreasing in n, and is initially equal to 1 when $n = 0$. We need to "flatten" the $\ell^2(G)$ norm of μ^{*n} as n increases. We first use the nonconcentration hypothesis to obtain an initial amount of flattening:

Proposition 4.2.1. *For any $n \geq \frac{1}{2}\Lambda \log |G|$, one has*

$$\|\mu^{*n}\|_{\ell^2(G)} \leq |G|^{-\kappa/4}. \tag{4.8}$$

Furthermore, we have

$$\mu^{*n}(gH) \leq |G|^{-\kappa/2} \tag{4.9}$$

for all proper subgroups H of G and all $g \in G$.

Proof. By the nonconcentration hypothesis, we can find $n_0 \leq \frac{1}{2}\Lambda \log |G|$ such that

$$\mu^{*2n_0}(H) \leq |G|^{-\kappa}$$

for all proper subgroups H of G. If we write $\mu^{*2n_0}(H)$ as $\mu^{*n_0} * \mu^{*n_0}(Hgg^{-1}H)$, we see that

$$\mu^{*2n_0}(H) \geq \mu^{*n_0}(Hg)\mu^{*n_0}(g^{-1}H)$$

for all $g \in G$. By symmetry, $\mu^{*n_0}(g^{-1}H) = \mu^{*n_0}(Hg)$, and thus

$$\sup_{g \in G} \mu^{*n_0}(gH) \leq |G|^{-\kappa/2}.$$

If $n \geq \frac{1}{2}\Lambda \log |G|$, then we may write μ^{*n} as the convolution of a probability measure $\mu^{*(n-n_0)}$ and μ^{*n_0}. From this, we see that

$$\mu^{*n}(g'H) \leq \sup_{g \in G} \mu^{*n_0}(gH) \leq |G|^{-\kappa/2}$$

for all $g' \in G$, giving the claim (4.9). Specialising this to the case when H is the trivial group, one has

$$\|\mu^{*n}\|_{\ell^\infty(G)} \leq |G|^{-\kappa/2}.$$

Since we also have
$$\|\mu^{*n}\|_{\ell^1(G)} = 1,$$
the claim (4.8) then follows from Hölder's inequality. □

Now we obtain additional flattening using the product theorem hypothesis:

Lemma 4.2.2 (Flattening lemma). *Suppose $n \geq \frac{1}{2}\Lambda \log|G|$ is such that*

(4.10) $$\|\mu^{*n}\|_{\ell^2(G)} \geq |G|^{-1/2+\kappa/2}.$$

Then one has
$$\|\mu^{*n} * \mu^{*n}\|_{\ell^2(G)} \leq |G|^{-\varepsilon} \|\mu^{*n}\|_{\ell^2(G)}$$
for some $\varepsilon > 0$ depending only on κ and δ'.

Proof. Suppose the claim fails for some ε to be chosen later, thus
$$\|\mu^{*n} * \mu^{*n}\|_{\ell^2(G)} > |G|^{-\varepsilon} \|\mu^{*n}\|_{\ell^2(G)}.$$

Applying Lemma 4.0.1, we may thus find an $O(|G|^{O(\varepsilon)})$-approximate group H with
$$|H| \ll |G|^{O(\varepsilon)}/\|\mu^{*n}\|_{\ell^2(G)}^2$$
and $g \in G$ such that
$$\mu^{*n}(gH) \gg |G|^{-O(\varepsilon)}.$$
Since $\mu^{*n}\|_{\ell^\infty(G)} \leq |G|^{-\kappa/2}$ by (4.9), we see that
$$|H| \gg |G|^{\kappa/2 - O(\varepsilon)}.$$
Meanwhile, from (4.10) one has
$$|H| \ll |G|^{1-\kappa+O(\varepsilon)}.$$
Applying the product hypothesis (assuming ε sufficiently small depending on κ and δ), we conclude that H generates a proper subgroup K of G, and thus
$$\mu^{*n}(gK) \gg |G|^{-O(\varepsilon)}.$$
But this contradicts (4.9) (again if ε is sufficiently small). □

Iterating the above lemma $O(1)$ times we obtain (4.7) for some $n = O(\log|G|)$, as desired.

Remark 4.2.3. Roughly speaking, the three hypotheses in Theorem 4.0.2 govern three separate stages of the life cycle of the random walk and its distributions μ^{*n}. In the early stage $n = o(\log|G|)$, the nonconcentration hypotheses creates some initial spreading of this random walk, in particular, ensuring that the walk "escapes" from cosets of proper subgroups. In the middle stage $n \sim \log|G|$, the product theorem steadily flattens the distribution of the random walk, until it is very roughly comparable to the uniform distribution. Finally, in the late stage $n \gg \log|G|$, the quasirandomness property can smooth out the random walk almost completely to obtain the mixing necessary for expansion.

Chapter 5

Product theorems, pivot arguments, and the Larsen-Pink nonconcentration inequality

In Chapter 4, we saw that one could derive expansion of Cayley graphs from three ingredients: nonconcentration, product theorems, and quasirandomness. Quasirandomness was discussed in Chapter 3. In this chapter we will discuss product theorems. Roughly speaking, these theorems assert that in certain circumstances, a finite subset A of a group G either exhibits expansion (in the sense that A^3, say, is significantly larger than A), or is somehow "close to" or "trapped" by a genuine group. A typical result is the following.

Theorem 5.0.1 (Product theorem in $\mathrm{SL}_d(k)$)**.** *Let $d \geq 2$, let k be a finite field, and let A be a finite subset of $G := \mathrm{SL}_d(k)$. Let $\varepsilon > 0$ be sufficiently small depending on d. Then at least one of the following statements holds:*

(i) *(Expansion) One has $|A^3| \geq |A|^{1+\varepsilon}$.*

(ii) *(Close to G) One has $|A| \geq |G|^{1-O_d(\varepsilon)}$.*

(iii) *(Trapping) A is contained in a proper subgroup of G.*

We will prove this theorem (which was proven first in the $d = 2, 3$ cases for fields F of prime order in [**He2008**], [**He2011**], and then for $d = 2$ and general F in [**Di2010**], and finally, for general d and F independently in [**PySz2010**], [**BrGrTa2011**]) later in this chapter. A more qualitative version of this proposition was also previously obtained in [**Hr2012**]. There are also generalisations of the product theorem of importance to number theory, in which the field k is replaced by a cyclic ring $\mathbf{Z}/q\mathbf{Z}$ (with q not necessarily prime); this was achieved first for $d = 2$ and q square-free in [**BoGaSa2010**], in [**Va2012**] for general d and q square-free, and finally in [**BoVa2012**] for arbitrary d and q.

Exercise 5.0.1 (Girth bound). Assuming Theorem 5.0.1, show that whenever S is a symmetric set of generators of $\mathrm{SL}_d(k)$ for some finite field k and some $d \geq 2$, then any element of $\mathrm{SL}_d(k)$ can be expressed as the product[1] of $O_d(\log^{O_d(1)} |k|)$ elements from S. This is a special case of a conjecture of Babai and Seress [**BaSe1992**], who conjectured that the bound should hold uniformly for all finite simple groups (in particular, the implied constants here should not actually depend on d. The methods used to handle the SL_d case can handle other finite groups of Lie type of bounded rank, but at present we do not have bounds that are independent of the rank. On the other hand, a recent paper of Helfgott and Seress [**HeSe2011**] has almost resolved the conjecture for the permutation groups A_n.

A key tool to establish product theorems is an argument which is sometimes referred to as the *pivot argument*. To illustrate this argument, let us first discuss a much simpler (and older) theorem, essentially due to Freiman [**Fr1973b**], which has a much weaker conclusion but is valid in any group G:

Theorem 5.0.2 (Baby product theorem). *Let G be a group, and let A be a finite nonempty subset of G. Then one of the following statements hold:*

(i) *(Expansion) One has $|A^{-1}A| \geq \frac{3}{2}|A|$.*

(ii) *(Close to a subgroup) A is contained in a left-coset of a group H with $|H| < \frac{3}{2}|A|$.*

To prove this theorem, we suppose that the first conclusion does not hold, thus $|A^{-1}A| < \frac{3}{2}|A|$. Our task is then to place A inside the left-coset of a fairly small group H.

To do this, we take a group element $g \in G$, and consider the intersection $A \cap gA$. A priori, the size of this set could range anywhere from 0 to $|A|$. However, we can use the hypothesis $|A^{-1}A| < \frac{3}{2}|A|$ to obtain an important dichotomy, reminiscent of the classical fact that two cosets gH, hH of a subgroup H of G are either identical or disjoint:

---

[1]Equivalently, if we add the identity element to S, then $S^m = \mathrm{SL}_d(k)$ for some $m = O_d(\log^{O_d(1)} |k|)$.

5. Products, pivots, and Larsen-Pink

Proposition 5.0.3 (Dichotomy). *If $g \in G$, then exactly one of the following occurs:*

(i) *(Noninvolved case) $A \cap gA$ is empty.*

(ii) *(Involved case) $|A \cap gA| > \frac{|A|}{2}$.*

Proof. Suppose we are not in the pivot case, so that $A \cap gA$ is nonempty. Let a be an element of $A \cap gA$, then a and $g^{-1}a$ both lie in A. The sets $A^{-1}a$ and $A^{-1}g^{-1}a$ then both lie in $A^{-1}A$. As these sets have cardinality $|A|$ and lie in $A^{-1}A$, which has cardinality less than $\frac{3}{2}|A|$, we conclude from the inclusion-exclusion formula that

$$|A^{-1}a \cap A^{-1}g^{-1}a| > \frac{|A|}{2}.$$

But the left-hand side is equal to $|A \cap gA|$, and the claim follows. $\square$

The above proposition provides a clear separation between two types of elements $g \in G$: the "noninvolved" elements, which have nothing to do with A (in the sense that $A \cap gA = \emptyset$, and the "involved" elements, which have a lot to do with A (in the sense that $|A \cap gA| > |A|/2$. The key point is that there is a significant "gap" between the noninvolved and involved elements; there are no elements that are only "slightly involved", in that A and gA intersect a little but not a lot. It is this gap that will allow us to upgrade approximate structure to exact structure. Namely:

Proposition 5.0.4. *The set H of involved elements is a finite group, and is equal to AA^{-1}.*

Proof. It is clear that the identity element 1 is involved, and that if g is involved then so is g^{-1} (since $A \cap g^{-1}A = g^{-1}(A \cap gA)$). Now suppose that g, h are both involved. Then $A \cap gA$ and $A \cap hA$ have cardinality greater than $|A|/2$ and are both subsets of A, and so have nonempty intersection. In particular, $gA \cap hA$ is nonempty, and so $A \cap g^{-1}hA$ is nonempty. By Proposition 5.0.3, this makes $g^{-1}h$ involved. It is then clear that H is a group.

If $g \in AA^{-1}$, then $A \cap gA$ is nonempty, and so from Proposition 5.0.3 g is involved. Conversely, if g is involved, then $g \in AA^{-1}$. Thus we have $H = AA^{-1}$ as claimed. In particular, H is finite. $\square$

Now we can quickly wrap up the proof of Theorem 5.0.2. By construction, $|A \cap gA| > |A|/2$ for all $g \in H$, which by double counting shows that $|H| < 2|A|$. As $H = AA^{-1}$, we see that A is contained in a right coset Hg of H; setting $H' := g^{-1}Hg$, we conclude that A is contained in a left coset gH' of H'. H' is a conjugate of H, so $|H'| < 2|A|$. If $h \in H'$, then A and Ah

both lie in H' and have cardinality $|A|$, so must overlap, and so $h \in AA^{-1}$. Thus $AA^{-1} = H'$, and so $|H'| < \frac{3}{2}|A|$, and Theorem 5.0.2 follows.

Exercise 5.0.2. Show that the constant $3/2$ in Theorem 5.0.2 cannot be replaced by any larger constant.

Exercise 5.0.3. Let $A \subset G$ be a finite nonempty set such that $|A^2| < 2|A|$. Show that $AA^{-1} = A^{-1}A$. (*Hint:* If $ab^{-1} \in AA^{-1}$, show that $ab^{-1} = c^{-1}d$ for some $c, d \in A$.)

Exercise 5.0.4. Let $A \subset G$ be a finite nonempty set such that $|A^2| < \frac{3}{2}|A|$. Show that there is a finite group H with $|H| < \frac{3}{2}|A|$ and a group element $g \in G$ such that $A \subset Hg \cap gH$ and $H = AA^{-1}$.

Exercise 5.0.5 ([**BrGrTa2013**]). Let $k \geq 1$ be a natural number. Show that if $\varepsilon = \varepsilon_k > 0$ is sufficiently small depending on k, then given any Cayley graph $\text{Cay}(G, S)$ of a symmetric set S of order k in a finite set which is a one-sided ε-expander, one either has that $\text{Cay}(G, S)$ is a two-sided ε'-expander for some $\varepsilon' > 0$ depending only on ε and k, or else there is an index two subgroup H of G such that S is disjoint from H. (*Hint:* If $\text{Cay}(G, S)$ fails to be a two-sided expander, find a subset A of G of density close to $1/2$ with the property that sA is close to the complement of A for all $s \in A$, and conclude that for all $g \in G$, the set $A_g := A \cap Ag$ has the property that SA_g has approximately the same cardinality as A_g for all $s \in S$. Combine this with one-sided expansion to obtain a dichotomy concerning the size of A_g which one can then use to run a pivot argument.)

We now turn to some further examples of the pivot argument in other group-like situations, including Theorem 5.0.2 and also the "sum-product theorem" from [**BoKaTa2004**], [**BoGlKo2006**].

5.1. The sum-product theorem

Consider a finite nonempty subset A of a field k. Then we may form the sumset
$$A + A := \{a + b : a, b \in A\}$$
and the product set
$$A \cdot A := \{ab : ab \in A\}.$$
The minimal sizes of such sets are well understood:

Exercise 5.1.1. Let A be a finite nonempty subset of a field k.
- (i) Show that $|A + A| \geq |A|$, with equality occurring if and only if A is an additive coset $A = x + H$ of an finite additive subgroup H of k with some $x \in k$.

5.1. The sum-product theorem

(ii) Show that $|A \cdot A| \geq |A|$, with equality occurring if and only if A is either equal to a multiplicative coset $A = gH$ of a finite multiplicative subgroup H of $k^\times := k \backslash \{0\}$ with some $g \in k^\times$, or the set $\{0\}$, or the set $\{0\} \cup gH$ where gH is a multiplicative coset.

(iii) Show that $\max(|A+A|, |A \cdot A|) \geq |A|$, with equality occurring if and only if A is either equal to a multiplicative dilate $A = cF$ of a finite subfield F of k with $c \in k^\times$, or a singleton set.

The *sum-product phenomenon* is a robust version of the above observation, asserting that one of $A + A$ or $A \cdot A$ must be significantly larger than A if A is not somehow "close" to a genuine subfield of k. Here is one formulation of this phenomenon:

Theorem 5.1.1 (Sum-product theorem)**.** *Let $\varepsilon > 0$ be a sufficiently small number. Then for any field k and any finite nonempty subset A, one of the following statements hold:*

(i) *(Expansion)* $\max(|A+A|, |A \cdot A|) \geq |A|^{1+\varepsilon}$.

(ii) *(Close to a subfield) There is a dilate cF of a subfield F of k with $|F| \ll |A|^{1+O(\varepsilon)}$ and $c \neq 0$ which contains all but $O(|A|^{O(\varepsilon)})$ elements of A.*

(iii) *(Smallness) A is an additive subgroup of order 2.*

If k has characteristic zero, then the second option here cannot occur, and we conclude that $\max(|A+A|, |A \cdot A|) \geq |A|^{1+\varepsilon}$ for some absolute constant $\varepsilon > 0$ as soon as A contains at least two nonzero elements, a claim first established in **R** by Erdős and Szemerédi [**ErSz1983**]. When k is a finite field of prime order, the second option can only occur when $F = k$, and we conclude that $\max(|A+A|, |A \cdot A|) \geq |A|^{1+\varepsilon}$ as soon as $|A| \leq |k|^{1-C\varepsilon}$ whenever ε is sufficiently small, C is an absolute constant, and A has at least two nonzero elements. A preliminary version of this result (which required more size assumptions on A, in particular a bound of the shape $|A| \geq |k|^\delta$) was obtained in [**BoKaTa2004**], with the version stated above first obtained in [**BoGlKo2006**]. The proof given here is drawn from the one in [**TaVu2006**], and was originally inspired by the arguments in [**BoKo2003**].

Remark 5.1.2. There has been a substantial amount of literature on trying to optimise the exponent ε in the sum-product theorem. In **R**, the best result currently in this direction is by Solymosi [**So2009**], who established that one can take ε arbitrarily close to $1/3$; for **C**, the best result currently is by Rudnev [**Ru2011**], who shows that one can take ε arbitrarily close to $19/69$. For fields of prime order, one can take ε arbitrarily close to $1/11$ [**Ru2012**];

an extension to arbitrary finite fields was then obtained in [**LiRo2011**]. For sum-product theorems in other rings, see [**Ta2009c**].

We now start proving Theorem 5.1.1. As with Theorem 5.0.2, the engine of the proof is a dichotomy similar to that of Proposition 5.0.3. Whilst the former proposition was modeled on the basic group-theoretic assertion that cosets gH of a subgroup where either identical or disjoint, this proposition is modeled on the basic linear algebra fact that if F is a subfield of k and $\xi \in k$, then $F + \xi F$ is either of size $|F|^2$, or of size $|F|$.

Lemma 5.1.3 (Dichotomy). *Let k be a field, let A be a finite nonempty subset of k, and let $\xi \in k$. Then at least one of the following statements hold:*

(i) *(Noninvolved case)* $|A + \xi A| = |A|^2$.

(ii) *(Involved case)* $|A + \xi A| \leq |(A - A)A + (A - A)A|$.

Proof. Suppose that we are not in the noninvolved case, thus $|A + \xi A| \neq |A|^2$. Then the map $(a, b) \mapsto a + \xi b$ from $A \times A$ to k is not injective, and so there exists $a, b, c, d \in A$ with $(a, b) \neq (c, d)$ and

$$a + \xi b = c + \xi d.$$

In particular, $b \neq d$. We then have $\xi = (a - c)/(d - b)$ and so

$$|A + \xi A| = |(d - b)A + (a - c)A| \leq |(A - A)A + (A - A)A|. \qquad \square$$

Remark 5.1.4. One can view $A + \xi A$ as measuring the extent to which the dilate ξA of A is "transverse" to A. As the "slope" ξ varies, ξA "pivots" around the origin, encountering both the (relatively rare) involved slopes, and the (generic) noninvolved slopes. It is this geometric picture which led to the term "pivot argument", as used in particular by Helfgott [**He2011**] (who labeled the noninvolved slopes as "pivots").

This dichotomy becomes useful if there is a significant gap between $|(A - A)A + (A - A)A|$ and $|A|^2$. Let us see how. To prove Theorem 5.1.1, we may assume that $|A|$ is larger than some large absolute constant C, as the claim follows from Exercise 5.1.1 otherwise (making ε small enough depending on C). By deleting 0 from A, and tweaking ε, noting that we may then assume that A does not contain 0. We suppose that

$$|A + A|, |A \cdot A| \leq K|A|$$

for some $K \leq |A|^{\varepsilon_0}$ and some sufficiently small absolute constant ε_0. In particular, we see that $|A|$ will exceed any quantity of the form $O(K^{O(1)})$ if we make ε_0 small enough and C large enough.

5.1. The sum-product theorem

We would like to boost this control of sums and products to more complex combinations of A. We will need some basic tools from additive combinatorics.

Lemma 5.1.5 (Ruzsa triangle inequality). *If A, B, C are nonempty finite subsets of k, then $|A - C| \leq \frac{|A-B||B-C|}{|B|}$.*

Proof. This is the additive version of Exercise 4.1.4. □

Lemma 5.1.6 (Ruzsa covering lemma). *If A, B are nonempty finite subsets of k, then A can be covered by at most $\frac{|A+B|}{|B|}$ translates of $B - B$.*

Proof. This is the additive version of Exercise 4.1.5. □

Exercise 5.1.2 (Sumset estimates). If A, B are nonempty finite subsets of k such that $|A + B| \leq K|A|^{1/2}|B|^{1/2}$, show that A and B can both be covered by $O(K^{O(1)})$ translates of the same $O(K^{O(1)})$-approximate group H, with $|H| \ll K^{O(1)}|A|$. Conclude that

$$|n_1 A - n_2 A + n_3 B - n_4 B| \ll_{n_1, n_2, n_3, n_4} K^{O(|n_1|+|n_2|+|n_3|+|n_4|)}|A|$$

for any natural numbers n_1, n_2, n_3, n_4, where $nA := A + \ldots + A$ denotes the sumset of n copies of A. (*Hint:* Use the additive form of Exercise 4.1.7 and the preceding lemmas.)

These lemmas allow us to improve the sum-product properties of A by passing to a large subset B (cf. Corollary 4.1.6):

Lemma 5.1.7 (Katz-Tao lemma). *Let A be as above. Then there is a subset B of A with $|B| \geq |A|/2K$ such that $|B^2 - B^2| \ll K^{O(1)}|B|$.*

Proof. The dilates aA of A with $a \in A$ all lie in a set A^2 of cardinality at most $K|A|$. Intuitively, this should force a lot of collision between the aA, which we will exploit using the sumset estimates. More precisely, observe that

$$\|\sum_{a \in A} 1_{aA}\|_{\ell^1} = |A|^2$$

and hence by Cauchy-Schwarz

$$\|\sum_{a \in A} 1_{aA}\|_{\ell^2}^2 \geq |A|^3/K.$$

The left-hand side can be written as

$$\sum_{b \in A} \sum_{a \in A} |aA \cap bA|$$

and so by the pigeonhole principle we can find $b_0 \in A$ such that
$$\sum_{a \in A} |aA \cap b_0 A| \geq |A|^2/K.$$
We apply a dilation to set $b_0 = 1$ (recall that A does not contain 0). If we set $B := \{a \in A : |aA \cap A| \geq |A|/2K\}$, we conclude that
$$\sum_{a \in B} |aA \cap A| \geq |A|^2/2K$$
which implies in particular that
$$|B| \geq |A|/2K.$$
If $a \in B$, then
$$|aA \cap A| \geq |A|/2K.$$
Since $|aA + aA| \leq K|A|$ we also have
$$|aA + (aA \cap A)| \leq K|A|$$
and similarly
$$|A + (aA \cap A)| \leq K|A|$$
and thus by the Ruzsa triangle inequality
$$|aA - A| \ll K^{O(1)}|A|$$
whenever $a \in B$. Informally, let us call a nonzero element a of k *good* if $|aA - A| \ll K^{O(1)}|A|$ (but note that this notion of "good" is a bit fuzzy, as it depends on the choice of implied constants in the $O()$ notation). Observe that if a, a' are good, then
$$|aa'A - aA|, |aA - A| \ll K^{O(1)}|A|$$
and thus by the Ruzsa triangle inequality
$$|aa'A - A| \ll K^{O(1)}|A|;$$
thus the product of two good elements are good (with somewhat worse implied constants). Similarly, from the Ruzsa covering lemma we see that aA and $a'A$ are both covered by $O(K^{O(1)})$ translates of $A - A$, and from this and sumset estimates we see that
$$|(a + a')A - A| \ll K^{O(1)}|A|$$
and so the sum of two good elements is again good. Similarly the difference of good elements is good. Applying all these facts, we conclude that all the elements of $B^2 - B^2$ are good, thus $|gA - A| \ll K^{O(1)}|A|$ for all $g \in B^2 - B^2$. In particular, since $|A|$ exceeds $K^{O(1)}$, we see from the Cauchy-Schwarz inequality that for each $g \in B^2 - B^2$, there are $\gg |A|^3/K^{O(1)}$ solutions to the equation $ga_1 - a_2 = ga_3 - a_4$ with $a_1 \neq a_3$ and $a_1, a_2, a_3, a_4 \in A$. However, there are only $|A|^4$ possible choices for a_1, a_2, a_3, a_4, and each such

choice uniquely determines g, so there are at most $O(K^{O(1)}|A|)$ possible choices for g, and the claim follows. □

Note that one could replace $B^2 - B^2$ in the above lemma by any other homogeneous polynomial combination of B.

By applying a dilation, we may assume that B contains 1. Applying Lemma 5.1.3 to this set B (and using sumset estimates), we arrive at the following dichotomy: every field element $\xi \in k$ is either "noninvolved" in the sense that $|B + \xi B| = |B|^2$, or is "involved" in the sense that $|B + \xi B| \leq C_1 K^{C_1} |B|$ for some fixed absolute constant C_1. By sumset estimates we have $|B + BB| \ll K^{O(1)} |B|$; as we can assume that $|A|$, and hence $|B|$, is larger than any quantity of the form $O(K^{O(1)})$, this forces all elements of B to be involved.

To exploit this, observe (by repeating the proof of Lemma 5.1.3) that if ξ_1, ξ_2 are involved, then the quantities $\xi = \xi_1 \xi_2, \xi_1 + \xi_2, \xi_1 - \xi_2$ are somewhat involved in the sense that

$$|B + \xi B| \ll K^{O(1)} |B|$$

for those choices of ξ (where the implied constants depend on C_1). But as we can assume that $|A|$, and hence $|B|$, is larger than any quantity of the form $O(K^{O(1)})$, we see from Lemma 5.1.3 that this forces ξ to be involved as well (this is the crucial step at which approximate structure is improved to exact structure). We thus see that the set F of all involved elements is closed under multiplication, addition, and subtraction; as it also contains 0, it is a subring of k. Arguing as in the proof of Lemma 5.1.3, we have that F is finite with $|F| \ll K^{O(1)} |A|$; in particular, F must now be a finite subfield of k.

Now we enter the "endgame", in which we use this F to control A. By previous discussion, F contains B, and thus $|A \cap F| \gg K^{-O(1)} |A|$. By the Ruzsa triangle inequality applied to $A, A \cap F, F$, this implies that $|A + F| \ll K^{O(1)} |F|$, and so A can be covered by $O(K^{O(1)})$ translates of F. A similar argument applied multiplicatively shows that A can be covered by $O(K^{O(1)})$ dilates of F. Since a nontrivial translate of F and a nontrivial dilate of F intersect in at most one point, we conclude that A has at most $O(K^{O(1)})$ elements outside of F, and the claim follows.

Remark 5.1.8. One can abstract this argument by replacing the multiplicative structure here by an abelian group action; see [**He2011**] for details. The argument can also extend to noncommutative settings, such as division algebras or more generally to arbitrary rings (though in the latter case, the presence of nontrivial zero divisors becomes a very significant issue); see [**Ta2009c**] for details.

5.2. Finite subgroups of SL_2

We will shortly establish Theorem 5.0.1, which can be viewed as a way to describe approximate subgroups of $SL_d(k)$. Before we do so, let us first warm up and digress slightly by studying *genuine* finite subgroups A of $SL_d(k)$, in the model case $d = 2$, which is simplified by the availability of some additional *ad hoc* explicit calculations. In order to make the algebraic geometry of the situation cleaner, it is convenient to embed the field k in its algebraic closure $\overline{k}$, and similarly embed $SL_2(k)$ in $SL_2(\overline{k})$. This is a group which is also an algebraic variety (identifying the space of 2×2 matrices with coefficients in $\overline{k}$ with $\overline{k}^4$), whose group operations are algebraic (in fact, polynomial) maps; in other words, $SL_2(\overline{k})$ is an *algebraic group*. We now consider the question of what finite subgroups of $SL_2(\overline{k})$ can look like. This is a classical question, with a complete classification obtained by Dickson [**Di1901**]. The precise classification is somewhat complicated; to give just a taste of this complexity, we observe that the symmetry group of the isocahedron is a finite subgroup of $SO_3(\mathbf{R})$, which can be lifted to the spin group $Spin_3(\mathbf{R})$ (giving what is known as the *binary isocahedral group*, a group of order 120), which is a subgroup of $Spin_3(\mathbf{C})$, which can be identified with $SL_2(\mathbf{C})$. Because of this, it is possible for some choices of finite field k to embed the binary isocahedral group into $SL_2(\overline{k})$ or $SL_2(k)$. Similar considerations obtain for the symmetry group of other Platonic solids. However, if one is willing to settle for a "rough" classification, in which one ignores groups of bounded size (and more generally, is willing just to describe a bounded index subgroup of the group A), the situation becomes much simpler. In the characteristic zero case $k = \mathbf{C}$, for instance, we have *Jordan's theorem* (Theorem 3.2.1), which asserts that given a finite subgroup A of $SL_d(\mathbf{C})$ for some $d = O(1)$, a bounded index subgroup of A is abelian. The finite characteristic case is inherently more complicated though (due in large part to the proliferation of finite subfields), with a satisfactory rough classification only becoming available for general d with the work of Larsen and Pink [**LaPi2011**] (published in 2011, but which first appeared as a preprint in 1998). However, the $d = 2$ case is significantly simpler and can be treated by somewhat *ad hoc* methods, as we shall now do. The discussion here is loosely based on [**Ko2012**].

We pause to recall some basic structural facts about $SL_2(\overline{k})$. Elements of this group are 2×2 matrices with determinant one, and thus have two (algebraic, possibly repeated) eigenvalues t, t^{-1} for some $t \in \overline{k}$ (note here that we are using the algebraically closed nature of $\overline{k}$). This allows us to classify elements of $SL_2(\overline{k})$ into three classes:

5.2. Finite subgroups of SL_2

(i) the *central* elements ± 1;

(ii) the *regular unipotent elements* and their negations, which are non-central elements with a double eigenvalue at $+1$ (or a double eigenvalue at -1);

(iii) the *regular semisimple elements*, which have two distinct eigenvalues.

We collectively refer to regular unipotent elements and their negations as *regular projectively unipotent elements*.

Remark 5.2.1. The presence of the nonidentity central element -1 leads to some slight technical annoyances (for instance, it means that SL_2 is merely an *almost simple* algebraic group rather than a simple one, in the sense that the only normal algebraic subgroups are finite). One can eliminate this element by working instead with the *projective special linear group* $P SL_2 := SL_2 /\{\pm 1\}$, but we will not do so here. We remark that if one works in SL_d for $d > 2$ then the classification of elements becomes significantly more complicated; for instance, there exist elements which are semisimple (i.e., diagonalisable) but neither regular nor central, because some but not all of the eigenvalues may be repeated.

One can distinguish the unipotent elements from the semisimple ones using the trace: unipotent elements have trace $+2$, their negations have trace -2, and the semisimple elements have traces distinct from ± 2. The ability to classify elements purely from the trace is a very special fact concerning SL_2 which breaks down completely for higher rank matrix groups, but we will not hesitate to take advantage of this fact here.

Associated to the above classification are some natural algebraic subgroups of $SL_2(\overline{k})$, including the *standard maximal torus*

$$T(\overline{k}) := \left\{ \begin{pmatrix} t & 0 \\ 0 & t^{-1} \end{pmatrix} : t \in \overline{k}^\times \right\},$$

the one-dimensional *standard unipotent group*

$$U(\overline{k}) := \left\{ \begin{pmatrix} 1 & x \\ 0 & 1 \end{pmatrix} : x \in \overline{k} \right\},$$

and the two-dimensional *standard Borel subgroup*

$$B(\overline{k}) := \left\{ \begin{pmatrix} t & x \\ 0 & t^{-1} \end{pmatrix} : x \in \overline{k}; t \in \overline{k}^\times \right\}.$$

More generally, we define[2] a maximal torus of $\mathrm{SL}_2(\overline{k})$ to be a conjugate (in $\mathrm{SL}_2(\overline{k})$) of the standard maximal torus, a unipotent group to be a conjugate of the standard unipotent group, and a Borel subgroup to be a conjugate of the standard Borel subgroup. Note that one can also think of a Borel subgroup as the stabiliser of a one-dimensional subspace of $\overline{k}^2$ (using the obvious action of $\mathrm{SL}_2(\overline{k})$ on $\overline{k}^2$). Using the *Jordan normal form* (again taking advantage of the algebraically closed nature of $\overline{k}$), we can see how these groups interact with group elements:

(i) The central elements lie in every maximal torus and every Borel subgroup. The identity $+1$ lies in every unipotent group, but -1 lies in none of them.

(ii) Every regular unipotent element lies in exactly one unipotent group, which in turn lies in exactly one Borel subgroup (the normaliser of the unipotent group). Conversely, a unipotent group consists entirely of regular unipotent elements and the identity $+1$.

(iii) Every regular semisimple element lies in exactly one maximal torus, which in turn lies in exactly two Borel subgroups (the stabiliser of one of the eigenspaces of a regular semisimple element in the torus). Conversely, a maximal torus consists entirely of regular semisimple elements and the central elements ± 1.

Remark 5.2.2. If one was working in a nonalgebraically closed field F instead of in $\overline{k}$, one could subdivide the regular semisimple elements into two classes, the *split* case when the elements can be diagonalised inside F, and the *nonsplit* case when they can only be diagonalised in a quadratic extension of F. This similarly subdivides maximal tori into two families, the split tori and the nonsplit tori. In the case when one is working over the field $\mathbf{R}$, the unipotent, split semisimple, and nonsplit semisimple elements are referred to as parabolic, hyperbolic, and elliptic elements of $\mathrm{SL}_2(\mathbf{R})$ respectively. Fortunately, in our applications we can work in algebraically closed fields and avoid these sorts of finer distinctions.

Ignoring the exceptional small examples of subgroups of $\mathrm{SL}_2(\overline{k})$, such as the binary isocahedral group mentioned earlier, there are two obvious ways to generate subgroups of $\mathrm{SL}_2(\overline{k})$. One is to pass from $\overline{k}$ to a subfield F, creating "arithmetic" subgroups of the form $\mathrm{SL}_2(F)$ (or conjugates thereof). If F is not closed under square roots, one can also create a slightly larger subgroup $\widetilde{\mathrm{SL}_2(F)}$, consisting of matrices in $\mathrm{SL}_2(\overline{k})$ of the form cM, where M is a 2×2 matrix with entries in F and c is a scalar with $c^2 \in F$; one can

---

[2]This is not really the "right" way to define these groups, for the purpose of generalisation to other algebraic groups, but will suffice as long as we are only working with SL_2. See Section 12.2 for a discussion of the more general setting of a Chevalley group over an arbitrary field.

check that this is a group that contains $\mathrm{SL}_2(F)$ as a subgroup of index at most two.

The other way to generate subgroups of $\mathrm{SL}_2(\overline{k})$ is to replace SL_2 with an algebraic subgroup of the three-dimensional group SL_2, such as[3] the maximal tori, unipotent groups, and Borel subgroups mentioned earlier.

Observe that if $A = \mathrm{SL}_2(F)$ is an arithmetic subgroup, then its intersections $T(F) := A \cap T(\overline{k})$, $U(F) := A \cap U(\overline{k})$, $B(F) := A \cap B(\overline{k})$ capture a portion of A proportionate to the dimensions involved, or more precisely that

$$|A \cap T(\overline{k})| \ll |A|^{1/3}; |A \cap U(\overline{k})| \ll |A|^{1/3}; \quad |A \cap B(\overline{k})| \ll |A|^{2/3}.$$

Indeed, it is easy to see that $|A| = |\mathrm{SL}_2(F)| \sim |F|^3$, $|A \cap T(\overline{k})| = |T(F)| \sim |F|$, and so forth. An important and general observation of Larsen and Pink [**LaPi2011**] is that this sort of behaviour is shared by all other finite subgroups of algebraic groups such as $\mathrm{SL}_2(\overline{k})$, as long as these groups are not (mostly) trapped in a proper algebraic subgroup. We first illustrate this phenomenon for the torus groups:

Proposition 5.2.3 (Larsen-Pink inequality, special case). *Let A be a finite subgroup of $\mathrm{SL}_2(\overline{k})$. Then one of the following statements hold:*

  (i) *(Nonconcentration) For any maximal torus T, one has $|A \cap T| \ll |A|^{1/3}$.*

  (ii) *(Trapping) There is a Borel subgroup B such that $|A \cap B| \gg |A|$.*

Proof. Suppose that the trapping hypothesis fails, thus $|A \cap B| = o(|A|)$ for all Borel subgroups B, where we interpret[4] $o(|A|)$ here to mean "less than $\varepsilon |A|$ for an arbitrarily small constant $\varepsilon > 0$ which we are at liberty to choose". In particular, we see that any coset of B occupies a fraction $o(1)$ at most of A. Thus, for instance, if we select an element $\begin{pmatrix} a & b \\ c & d \end{pmatrix}$ from A uniformly at random, then with probability $1 - o(1)$, b is nonzero, and similarly for a, c, d. To put it more informally, the matrix entries of an element of A are "generically" nonzero; similarly, if we first conjugate A by a fixed group element.

We need to show that $|A \cap T| \ll |A|^{1/3}$ for any maximal torus T. By conjugation we may take T to be the standard maximal torus $T = T(\overline{k})$.

---

[3]Actually, these are the only (connected) proper algebraic subgroups of SL_2, as can be seen by consideration of the associated Lie algebras.

[4]If one is uncomfortable with this type of definition, one can instead consider a sequence of potential counterexamples $A = A_n$ to the above proposition in various groups $\mathrm{SL}_2(\overline{k_n})$, in which $\sup_B |A \cap B| = o_{n \to \infty}(|A|)$. Alternatively, one can also rephrase this argument if desired in the language of nonstandard analysis.

Set $A' := A \cap T(\overline{k})$, then A' is a subgroup of A of the form
$$A' := \left\{ \begin{pmatrix} t & 0 \\ 0 & t^{-1} \end{pmatrix} : t \in H \right\}$$
for some finite multiplicative subgroup H of $\overline{k}^\times$. We may assume that $|H|$ is larger than any given absolute constant, as the claim is trivial otherwise. Our task is to show that $|H|^3 \ll |A|$.

Let $g = \begin{pmatrix} a & b \\ c & d \end{pmatrix}$ be a typical element of A. By the preceding discussion, we may assume that a, b, c, d are all nonzero. Since A' is a subgroup of A, we have $A'gA'gA' \subset A$, thus
$$\begin{pmatrix} t_1 & 0 \\ 0 & t_1^{-1} \end{pmatrix} \begin{pmatrix} a & b \\ c & d \end{pmatrix} \begin{pmatrix} t_2 & 0 \\ 0 & t_2^{-1} \end{pmatrix} \begin{pmatrix} a & b \\ c & d \end{pmatrix} \begin{pmatrix} t_3 & 0 \\ 0 & t_3^{-1} \end{pmatrix} \in A$$
for all $t_1, t_2, t_3 \in H$. We evaluate the inner matrix products to obtain that
$$\begin{pmatrix} t_1 & 0 \\ 0 & t_1^{-1} \end{pmatrix} \begin{pmatrix} a^2 t_2 + bc t_2^{-1} & act_2 + bdt_2^{-1} \\ act_2 + cdt_2^{-1} & bct_2 + d^2 t_2^{-1} \end{pmatrix} \begin{pmatrix} t_3 & 0 \\ 0 & t_3^{-1} \end{pmatrix} \in A$$
for $t_1, t_2, t_3 \in H$.

Because a, b, c, d are nonzero, we see that for all but $O(1)$ values of t_2, all four entries of the middle matrix here are nonzero. As a consequence, if one fixes t_2 and lets t_1, t_3 vary, all of the triple products given above are distinct. Note that if one takes the above triple product and multiplies the diagonal entries together, the t_1, t_3 terms cancel and one obtains $(a^2 t_2 + bct_2^{-1})(bct_2 + d^2 t_2^{-1})$. This rational map (as a function of t_2) is at most four-to-one; each value of this map is associated to at most four values of t_2. Putting all this together, we conclude that there are $\gg |H|^3$ different triple products one can form here as $t_1, t_2, t_3 \in H$ vary, and the claim follows. $\square$

Exercise 5.2.1. Establish a variant of Proposition 5.2.3 in which the maximal tori are replaced by unipotent groups.

Given a group element $g \in \mathrm{SL}_2(\overline{k})$, let $\mathrm{Conj}(g) := \{hgh^{-1} : h \in \mathrm{SL}_2(\overline{k})\}$ be the conjugacy class of g. The behaviour of this class depends on the nature of g:

Exercise 5.2.2. Let g be an element of $\mathrm{SL}_2(\overline{k})$.

(i) If g is central, show that $\mathrm{Conj}(g) = \{g\}$.

(ii) If g is regular unipotent, show that $\mathrm{Conj}(g)$ is the space of all regular unipotent elements.

(iii) If g is negative of a regular unipotent element, show that $\mathrm{Conj}(g)$ is the space of all negatives of regular unipotent elements.

(iv) If g is regular semisimple, show that $\mathrm{Conj}(g) := \{g' \in \mathrm{SL}_2(\overline{k}) : \mathrm{tr}(g) = \mathrm{tr}(g')\}$.

5.2. Finite subgroups of SL_2

We can "dualise" the upper bound on maximal tori in Proposition 5.2.3 into a lower bound on conjugacy classes:

Proposition 5.2.4 (Large conjugacy classes). *Let A be a finite subgroup of $SL_2(\bar{k})$. Then one of the following statements hold:*

  (i) *(Large conjugacy classes) For any regular semisimple or regular projectively unipotent $g \in A$, one has $|A \cap \mathrm{Conj}(g)| \gg |A|^{2/3}$.*

  (ii) *(Trapping) There is a Borel subgroup B such that $|A \cap B| \gg |A|$.*

Proof. As before we may assume that $|A \cap B| = o(|A|)$ for all Borel subgroups B. Let $g \in A$ be regular semisimple or regular projectively unipotent, and consider the map $\phi : h \mapsto hgh^{-1}$ from A to $A \cap \mathrm{Conj}(g)$. For each $g' \in A \cap \mathrm{Conj}(g)$, the preimage of g' by ϕ is contained in a coset of the centraliser $C(g') := \{h \in SL_2(\bar{k}) : hg' = g'h\}$ of g'. As g (and hence g') is regular semisimple or regular projectively unipotent, this centraliser is a maximal torus or (two copies of) a unipotent group (this can be seen by placing g in Jordan normal form). By Proposition 5.2.3 or Exercise 5.2.1, we conclude that each preimage of ϕ has cardinality $O(|A|^{1/3})$, which forces the range to have cardinality $\gg |A|^{2/3}$ as claimed. $\square$

We remark that this gives a dichotomy analogous to Lemma 5.0.3 or Lemma 5.1.3 in the case $|A \cap B| = o(|A|)$. Namely, for any $g \in SL_2(\bar{k})$, either $A \cap \mathrm{Conj}(g)$ is empty, or $|A \cap \mathrm{Conj}(g)| \gg |A|^{2/3}$. We will take advantage of a dichotomy similar to this (but for tori instead of conjugacy classes) in the next section.

We can match the lower bound in Proposition 5.2.4 with an upper bound:

Proposition 5.2.5 (Larsen-Pink inequality, another special case). *Let A be a finite subgroup of $SL_2(\bar{k})$. Then one of the following statements hold:*

  (i) *(Nonconcentration) For any regular semisimple $g \in SL_2(\bar{k})$, one has $|A \cap \mathrm{Conj}(g)| \ll |A|^{2/3}$.*

  (ii) *(Trapping) There is a Borel subgroup B such that $|A \cap B| \gg |A|$.*

Proof. Again, we may assume that $|A \cap B| = o(|A|)$ for all Borel subgroups B. In particular, we may take $|A|$ larger than any given absolute constant. Let g be regular semisimple, and let $S := A \cap \mathrm{Conj}(g) = \{s \in A : \mathrm{tr}(s) = \mathrm{tr}(g)\}$; our task is to show that $|S| \ll |A|^{2/3}$. Note from Exercise 5.2.2 that g is conjugate to g^{-1}, and so S is symmetric: $S = S^{-1}$. Also, $Sa = aS$ for all $a \in A$.

Observe that whenever $a, b \in A$ and $s \in S \cap a^{-1}S \cap b^{-1}S$, then the triple (s, as, bs) lies in S^3; conversely, every triple in S^3 arises in this manner.

Thus we have the identity
$$|S|^3 = \sum_{a,b \in A} |S \cap a^{-1}S \cap b^{-1}S|.$$

We will show that
(5.1) $$\sum_{a,b \in A} |S \cap a^{-1}S \cap b^{-1}S| \ll |A|^2 + |A|^{4/3}|S|,$$
which will give $|S| \ll |A|^{2/3}$ as required.

We now establish (5.1). We divide into several contributions. First suppose that $a = \pm 1$. Then we bound the summand by $|S|$; there are $O(|A|)$ summands here, leading to a total contribution of $O(|A||S|)$, which is acceptable. Similarly if $b = \pm 1$, or $a = \pm b$, so we may restrict to the remaining cases when $\pm 1, \pm a, \pm b$ are distinct. In particular, a, b are now either regular unipotent or regular semisimple.

We now consider the case in which $1, a, b$ are linearly dependent (in the space $M_2(\overline{k})$ of 2×2 matrices). For fixed a, this constrains b to either a maximal torus or a unipotent group (depending on whether a is regular semisimple or regular projectively unipotent); this is easiest to see by placing a in Jordan canonical form. By the preceding results, we see that there are $O(|A|^{1/3})$ choices of b for each $|A|$, leading to a contribution of $O(|A|^{4/3}|S|)$ in this case, which is acceptable. So we may now take $1, a, b$ to be linearly independent.

The set $S \cap a^{-1}S \cap b^{-1}S$ is the intersection of A with the affine line
$$\ell := \{s \in M_2(\overline{k}); \operatorname{tr}(s) = \operatorname{tr}(as) = \operatorname{tr}(bs) = \operatorname{tr}(g)\};$$
this is indeed a line when $1, a, b$ are linearly independent. In most cases, this line ℓ will intersect $\operatorname{SL}_2(\overline{k})$ (which we can view as a quadric surface in $M_2(\overline{k})$) in at most two points, leading to a contribution of $O(|A|^2)$ for this case, which is acceptable. The only cases left to treat are when the line ℓ is incident to $\operatorname{SL}_2(\overline{k})$. This only occurs when the line ℓ takes the form hU for some $h \in \operatorname{SL}_2(\overline{k})$ and unipotent group U; this is easiest to see by multiplying ℓ on the left so that it contains the identity, and then placing another element of the line in Jordan normal form. In that case, we have
$$\operatorname{tr}(hu) = \operatorname{tr}(ahu) = \operatorname{tr}(bhu) = \operatorname{tr}(g)$$
for all $u \in U$. This forces h, ah, bh to all lie in the Borel subgroup B associated to U (this is easiest to see by first conjugating U into the standard unipotent group $U(\overline{k})$). In particular, a, b both lie in B. Furthermore, if we write $\operatorname{tr}(g) = t + t^{-1}$, then the diagonal entries of h, ah, bh are t, t^{-1} or t^{-1}, t, and so the diagonal entries of a, b are either $1, 1$ or t^{-2}, t^2 or t^2, t^{-2}. In particular, U is the stabiliser of one of the eigenvectors of a; so for fixed a, there are at most two choices for U (recall that a was regular). Furthermore,

5.2. Finite subgroups of SL_2

for fixed a and U, b is constrained to lie in at most three cosets of U. As such, there are only $O(|A|^{1/3})$ choices of b here for each a, giving another contribution of $O(|A|^{4/3}|S|)$, and the claim follows. □

Exercise 5.2.3. Let A be a finite subgroup of $SL_2(\overline{k})$, such that $|A \cap B| = o(|A|)$ for all Borel subgroups B. Show that at most $O(|A|^{2/3})$ of the elements of A are unipotent.

We can use the upper bound on conjugacy classes to obtain a lower bound on tori:

Proposition 5.2.6 (Large tori). *Let A be a finite subgroup of $SL_2(\overline{k})$. Then one of the following statements hold:*

(i) *(Large torus) For any regular semisimple $g \in A$, one has $|A \cap T| \gg |A|^{1/3}$, where T is the unique maximal torus containing g.*

(ii) *(Trapping) There is a Borel subgroup B such that $|A \cap B| \gg |A|$.*

Proof. We can of course assume that the trapping case does not occur. We consider the map $\phi : a \mapsto aga^{-1}$ from A to $A \cap \mathrm{Conj}(g)$. By Proposition 5.2.5, the range of ϕ has cardinality $O(|A|^{2/3})$, so by the pigeonhole principle, there is a preimage of $A \cap \mathrm{Conj}(g)$ of cardinality $\gg |A|^{1/3}$. But all preimages are conjugate to each other, so the preimage of g has cardinality $\gg |A|^{1/3}$. But this preimage is the intersection of A with the centraliser of g, which is T, and so $|A \cap T| \gg |A|^{1/3}$ as required. □

Exercise 5.2.4. Establish a variant of Proposition 5.2.6 in which g is regular unipotent instead of regular semisimple, and T is replaced with the unique unipotent group containing g.

This gives a second (and particularly useful) dichotomy: assuming A is not trapped by a Borel subgroup, for a maximal torus T, $|A \cap T|$ is either at most two, or is comparable to $|A|^{1/3}$.

To exploit this, we use the following counting argument of Larsen and Pink [**LaPi2011**] (which is also reminiscent of an old argument of Jordan [**Jo1878**], used to prove Theorem 3.2.1), followed by some *ad hoc* arguments specific to SL_2. We continue to assume that A is not trapped by a Borel subgroup. Let $Z := A \cap \{+1, -1\}$ denote the central elements of A, thus $|Z|$ is either 1 or 2. Observe that every element in $A \backslash Z$ is either regular projectively unipotent or regular semisimple; in the latter case, the element lies in a unique maximal torus, which also contains Z. We conclude the *class equation*

$$|A| - |Z| = u + \sum_T (|A \cap T| - |Z|)$$

where T ranges over all the maximal tori that intersect A, and u is the number of regular projective unipotents in A.

If we conjugate a maximal torus T by an element of A, we get another maximal torus, or the same maximal torus if the element used to conjugate T in was in the normaliser $N_A(T) := \{a \in A : aT = Ta\}$ of T. Thus, by the *orbit-stabilizer theorem*, there are exactly $|A|/|N_A(T)|$ tori conjugate to T in A. We thus see that

$$|A| - |Z| = u + \sum_{T \in \mathcal{T}} \frac{|A|}{|N_A(T)|}(|A \cap T| - |Z|)$$

where $\mathcal{T}$ is a collection of representatives of conjugacy classes of maximal tori intersecting A in a regular semisimple element. We rearrange this as

$$1 = \frac{u + |Z|}{|A|} + \sum_{T \in \mathcal{T}} \frac{1}{[N_A(T) : A \cap T]}\left(1 - \frac{|Z|}{|A \cap T|}\right).$$

Note that if T is a maximal torus, the normaliser of T in $\mathrm{SL}_2(\bar{k})$ has index 2. As such, $A \cap T$ has index at most two in $N_A(T)$, and so $\frac{1}{[N_A(T):A\cap T]}$ is either equal to 1 or $1/2$ for each T. From the preceding bounds on tori and unipotent elements, we also have $\frac{|Z|}{|A\cap T|} \sim |A|^{-1/3}$ and $\frac{u+|Z|}{|A|} = O(|A|^{-1/3})$. As we are assuming $|A|$ to be large, the above equation is only consistent when $\mathcal{T}$ has cardinality 1 or 2, and $\frac{u+|Z|}{|A|}$ is comparable to $|A|^{-1/3}$, or equivalently that u is comparable to $|A|^{2/3}$. Thus, A has plenty of regular projective unipotents (matching the upper bound from Exercise 5.2.3); in particular, there is at least one regular unipotent.

Applying a conjugation, we may assume that A contains $e := \begin{pmatrix} 1 & 1 \\ 0 & 1 \end{pmatrix}$, thus

$$A \cap U(\bar{k}) = \left\{ \begin{pmatrix} 1 & t \\ 0 & 1 \end{pmatrix} : t \in E \right\}$$

for some additive group $E \subset \bar{k}$ containing 1. By Exercise 5.2.1, $|E| \ll |A|^{1/3}$; by Exercise 5.2.4, we have $|E| \gg |A|^{1/3}$ also.

The map $a \mapsto a(A \cap U(\bar{k}))a^{-1}$ maps A to unipotent groups that intersect A in $\sim |A|^{1/3}$ regular unipotents. As there are $\sim |A|^{2/3}$ regular unipotent elements in A, we see that there are only $O(|A|^{1/3})$ such unipotent groups available. From the pigeonhole principle and conjugation, we conclude that the preimage of $U(\bar{k})$ in this map has cardinality $\gg |A|^{2/3}$. But this preimage is simply $A \cap B(\bar{k})$. In particular, the quotient $(A \cap B(\bar{k}))/(A \cap U(\bar{k}))$ has cardinality $\gg |A|^{1/3}$. Observe that each element of this quotient acts on $A \cap U(\bar{k})$ by conjugation, and the corresponding action on E is multiplicative (by the square of a diagonal entry of an element in the quotient). As such, if we set

$$F := \{\xi \in \bar{k} : \xi E \subset E\}$$

to be the "multiplicative symmetry set" of E, then we have $|F| \gg |A|^{1/3}$. As E is a finite additive group, F is a field of size at most $|E| \ll |A|^{1/3}$, thus F is a finite field of cardinality $|F| \sim |A|^{1/3}$. Also, E is a vector space over F; as E contains 1 and has cardinality $O(|A|^{1/3})$, we see that $E = F$. Thus we have

$$(5.2) \qquad A \cap U(\overline{k}) = \begin{pmatrix} 1 & F \\ 0 & 1 \end{pmatrix}.$$

Also, as $(A \cap B(\overline{k}))/(A \cap U(\overline{k}))$ has to stabilise F, we see that all elements of $A \cap B(\overline{k})$ have diagonal elements whose square lies in F. Combining this with (5.2), we see that $A \cap B(\overline{k})$ takes the form

$$(5.3) \qquad A \cap B(\overline{k}) = \left\{ \begin{pmatrix} t & f(t) + tx \\ 0 & t^{-1} \end{pmatrix} : t \in H, x \in F \right\}$$

for some multiplicative group H of elements whose square lies in $F^\times$ (in particular, $H \cap F^\times$ has index at most 2 in H), and some function $f \colon H \to \overline{k}$; since $A \cap B(\overline{k})$ has cardinality $\gg |A|^{2/3} \sim |F|^2$, H must have cardinality $\sim |F|$. By taking the commutators of two matrices in (5.3), we see that

$$(5.4) \qquad f(t)(s - s^{-1}) - f(s)(t - t^{-1}) \in F$$

for all $s, t \in H$.

If we select $t_0 \in H \cap F^\times$ such that $t_0 - t_0^{-1}$ is nonzero, then by conjugating A by a suitable element of $U(\overline{k})$ (which does not affect any of the previous control established on A) we may normalise $f(t_0)$ to be zero. From (5.4) this makes $f(t) \in tF$ for all $t \in H$. In particular, $A \cap B(\overline{k})$ is almost completely contained in $B(F)$:

$$(5.5) \qquad |A \cap B(\overline{k}) \cap B(F)| \geq |A \cap B(\overline{k})|/2.$$

Now for any $g \in A$, the subgroups $A \cap B(\overline{k})$ and $g^{-1}(A \cap B(\overline{k}))g$ of A have index $O(|A|^{1/3})$, so their intersection must have cardinality $\gg |A|^{1/3} \gg |F|$, thus

$$|A \cap B(F) \cap gB(F)g^{-1}| \gg |F|.$$

In particular, there must exist either a regular unipotent or a regular semisimple element $h \in A$ of $B(F)$ such that ghg^{-1} also lies in $B(F)$. If h is regular semisimple in $B(F)$, it has an eigenbasis in F^2, and so g must map such an eigenbasis to another eigenbasis, and is thus either diagonal or antidiagonal in this basis; in either case we can check that g lies in $\widetilde{SL_2(F)}$, which we recall is the group of matrices in $SL_2(\overline{k})$ of the form cM where M is a 2×2 matrix with entries in F and $c^2 \in F$.

If instead h is regular unipotent, it has the line $\{0\} \times \overline{k}$ as the unique (geometric) eigenspace; g must preserve this eigenspace and thus lies in $B(\overline{k})$; from (5.3) and the fact that $f(t) \in tF$, this implies that $g \in \widetilde{SL_2(F)}$.

Combining the cases, we conclude that $A \subset \widetilde{\mathrm{SL}_2(F)}$. We may therefore summarise our discussion as follows:

Theorem 5.2.7 (Rough description of finite subgroups of SL_2). *Let A be a finite subgroup of $\mathrm{SL}_2(\overline{k})$. Then one of the following statements hold:*

(i) *(Arithmetic subgroup) There is a finite subfield F of $\overline{k}$ with $|F| \sim |A|^{1/3}$ such that A is contained in a conjugate of $\widetilde{\mathrm{SL}_2(F)}$ (and is thus a subgroup of that conjugate of index $O(1)$).*

(ii) *(Trapping) There is a Borel subgroup B such that $|A \cap B| \gg |A|$.*

In principle, the trapping case can be analysed further (using manipulations similar to those used to reach (5.5)) but we will not pursue this here. We remark that while these computations were somewhat lengthy (and less elementary and precise than the more classical results of [**Di1901**]), they can extend to more complicated algebraic groups, such as $\mathrm{SL}_d(\overline{k})$, or more generally to any algebraic group of bounded rank; see [**LaPi2011**] for details. In particular, in [**LaPi2011**] these methods were used to establish an important subcase of the famous *classification of finite simple groups*, namely by verifying this classification for sufficiently large subgroups of a linear group of bounded rank over a field of arbitrary characteristic. It is conceivable that these methods may be extended in the future to give an alternate proof of the full classification (for sufficiently large groups, at least).

5.3. The product theorem in $\mathrm{SL}_2(k)$

In this section we prove the $d = 2$ case of Theorem 5.0.1. This result was first established (for fields of prime order) by Helfgott [**He2008**] and then in the general case by Dinai [**Di2010**]; we will present a variant of Helfgott's original argument which was developed independently in [**BrGrTa2011**] and [**PySz2010**]. It is convenient to rephrase Helfgott's theorem as follows:

Theorem 5.3.1 (Product theorem in $\mathrm{SL}_2(k)$, alternate form). *Let k be a finite field, and let A be a K-approximate group in $G := \mathrm{SL}_2(k)$ that generates G for some $K \geq 2$. Then one of the following holds:*

(i) *(Close to trivial) One has $|A| \ll K^{O(1)}$.*

(ii) *(Close to G) One has $|A| \geq K^{-O(1)}|G|$.*

Exercise 5.3.1. Show that Theorem 5.0.1 follows from Theorem 5.3.1. (*Hint:* If $|A^3| \leq |A|^{1+\varepsilon}$, use the multiplicative form of the Rusza triangle and covering lemmas to show that $(A \cup \{1\} \cup A^{-1})^2$ is a $O(|A|^{O(\varepsilon)})$-approximate group.)

5.3. The product theorem in $\mathrm{SL}_2(k)$

The problem now concerns the behaviour of finite approximate subgroups A of $\mathrm{SL}_2(k)$. The first step will be to establish analogues[5] of the Larsen-Pink nonconcentration inequalities of the preceding section, but for approximate subgroups rather than genuine subgroups. We begin by eliminating concentration in linear subgroups.

Lemma 5.3.2 (Escape from subspaces). *Let A, K be as in Theorem 5.3.1, and let $C > 0$. Then one of the following holds:*

(i) *(Close to trivial) One has $|A| \ll_C K^{O_C(1)}$.*

(ii) *(Escape) For any $d = 0, 1, 2, 3$ and any d-dimensional subspace V of $\overline{k}^4$, such that $V \cap \mathrm{SL}_2(\overline{k})$ is a subgroup of $\mathrm{SL}_2(\overline{k})$, one has $|A^2 \cap V| \leq K^{-C}|A|$.*

In practice, we will only apply the escape conclusion for Borel subgroups of $\mathrm{SL}_2(\overline{k})$, which are intersections of $\mathrm{SL}_2(\overline{k})$ with three-dimensional subspaces; however, we need to work with the more general escape construction in the *proof* of the lemma, for inductive purposes. The claim can in fact be established for any d-dimensional subspace V, or more generally for bounded complexity d-dimensional algebraic varieties; this will be discussed in the next section.

Proof. We induct on d. For $d = 0$, the claim is trivial, since $|A^2 \cap V| = 1$ in that case. Now suppose that $d = 1, 2, 3$, and the claim has already been proven for smaller values of d.

Let V be a d-dimensional subspace of $\mathrm{SL}_2(\overline{k})$ with $V \cap \mathrm{SL}_2(\overline{k})$ a group, and suppose for contradiction that $|A^2 \cap V| > K^{-C}|A|$. As A^2 can be covered by K copies of A, we can find $a \in A$ such that

$$|aA \cap V| > K^{-C-1}|A|. \tag{5.6}$$

Suppose that there exists an element b of A such that $bVb^{-1} \neq V$, so that $bVb^{-1} \cap V$ has dimension strictly less than V. From (5.6) we have

$$|A^4 \cap bVb^{-1}| \geq |baAb^{-1} \cap bVb^{-1}| > K^{-C-1}|A|.$$

Recall that A^4 can be covered by at most K^3 left translates gA of A; we may restrict to those g for which gA intersects A^4, so $g \in A^5$. By the pigeonhole principle, we may therefore find $g \in A^5$ such that

$$|gA \cap bVb^{-1}| > K^{-C-4}|A|.$$

---

[5] The observation that these inequalities could be usefully extended to the approximate group setting is due to Hrushovski [**Hr2012**] (based on a more general model-theoretic observation in [**HrWa2008**]), although for the specific case of $\mathrm{SL}_2(k)$, the most important of these inequalities for the purposes of proving Theorem 5.3.1 were first established in [**He2008**].

Let $A_1 := aA \cap V$ and $A_2 := gA \cap bVb^{-1}$. Then $A_1 A_2^{-1}$ is contained in A^7, and so $1_{A_1} * 1_{A_2^{-1}}$ is supported on a set of cardinality at most $K^6|A|$. Since

$$\|1_{A_1} * 1_{A_2^{-1}}\|_{\ell^1} = |A_1||A_2| \geq K^{-2C-5}|A|^2$$

we thus see from the pigeonhole principle that

$$|1_{A_1} * 1_{A_2^{-1}}(x)| \geq K^{-2C-11}|A|.$$

The left-hand side is $|A_1 \cap xA_2|$, and thus

$$|(A_1 \cap xA_2)^{-1} \cdot (A_1 \cap xA_2)| \geq K^{-2C-11}|A|.$$

The set in the left-hand side is contained in both A^2 and in $V \cap (bVb^{-1})$ (here we use the group nature of $V \cap \mathrm{SL}_2(\bar{k})$), and so

$$|A^2 \cap V \cap (bVb^{-1})| \geq K^{-2C-11}|A|.$$

Applying the induction hypothesis, we conclude that $|A| \leq K^{O_C(1)}$, and the claim follows.

The only remaining case is when $bVb^{-1} = V$ for all $b \in A$. As A generates $\mathrm{SL}_2(k)$, this implies that V is normalised by $\mathrm{SL}_2(k)$. But this is impossible if V has dimension $1, 2, 3$; see Exercise 5.3.2 below. $\square$

Exercise 5.3.2 (Almost simplicity of $\mathrm{SL}_2(k)$). Let V be a subspace of $\bar{k}^4$ of dimension 1, 2, or 3. Show that the group $\{g \in \mathrm{SL}_2(k) : gVg^{-1} = V\}$ does not contain all of $\mathrm{SL}_2(k)$.

Now we can obtain an approximate version of Proposition 5.2.3:

Proposition 5.3.3 (Larsen-Pink inequality, special case). *Let A, K be as in Theorem 5.3.1. Then for any maximal torus T, one has $|A^2 \cap T| \ll K^{O(1)}|A|^{1/3}$.*

In the context of SL_2, this bound on torus concentration was first established in [**He2008**] (and extended to SL_d in [**He2011**]).

Proof. We may assume that $|A| \geq K^C$ for any given constant C, as the claim is trivial otherwise. Similarly, by Lemma 5.3.2, we may assume that $|A^2 \cap B| \leq K^{-C}|B|$ for all Borel subgroups B.

We need to show that $|A^2 \cap T| \ll K^{O(1)}|A|^{1/3}$ for any maximal torus T. By conjugation we may take T to be the standard maximal torus $T = T(\bar{k})$. (This may make A generate a conjugate of $\mathrm{SL}_2(k)$, rather than $\mathrm{SL}_2(k)$ itself, but this will not impact our argument). Set $A' := A^2 \cap T(\bar{k})$, then

$$A' := \left\{ \begin{pmatrix} t & 0 \\ 0 & t^{-1} \end{pmatrix} : t \in H \right\}$$

5.3. The product theorem in $SL_2(k)$

for some finite subset H of $\overline{k}^\times$. We may assume that $|H| \geq K^C$, as the claim is trivial otherwise. Our task is to show that $|H|^3 \ll K^{O(1)}|A|$.

As in the proof of Proposition 5.2.3, we may find an element $g = \begin{pmatrix} a & b \\ c & d \end{pmatrix}$ of A^2 with a, b, c, d all nonzero. Since $A'gA'gA' \subset A^{10}$, thus

$$\begin{pmatrix} t_1 & 0 \\ 0 & t_1^{-1} \end{pmatrix} \begin{pmatrix} a & b \\ c & d \end{pmatrix} \begin{pmatrix} t_2 & 0 \\ 0 & t_2^{-1} \end{pmatrix} \begin{pmatrix} a & b \\ c & d \end{pmatrix} \begin{pmatrix} t_3 & 0 \\ 0 & t_3^{-1} \end{pmatrix} \in A^{10}$$

for all $t_1, t_2, t_3 \in H$. Arguing as in Proposition 5.3.3, we have

$$|H|^3 \ll |A^{10}|,$$

and the claim follows. □

Exercise 5.3.3. Show that if the nonconcentration conclusion in Proposition 5.3.3 holds, then for every maximal torus T and every $m \geq 1$, one has $|A^m \cap T| \ll_m K^{O_m(1)}|A|^{1/3}$.

We can now establish variants of the other Larsen-Pink inequalities from the preceding section:

Exercise 5.3.4. Establish a variant of Proposition 5.3.3 in which the maximal tori are replaced by unipotent groups.

Exercise 5.3.5 (Large conjugacy classes). Let A, K be as in Theorem 5.3.1. Show that for any regular semisimple or regular projectively unipotent $g \in A$, one has $|A^3 \cap \text{Conj}(g)| \gg K^{-O(1)}|A|^{2/3}$.

Exercise 5.3.6 (Larsen-Pink inequality, another special case). Let A, K be as in Theorem 5.3.1. Show that for any regular semisimple $g \in SL_2(\overline{k})$ and any $m \geq 1$, one has $|A^m \cap \text{Conj}(g)| \ll_m K^{O_m(1)}|A|^{2/3}$.

Exercise 5.3.7 (Unipotent bound). Let A, K be as in Theorem 5.3.1. Show that $O(K^{O(1)}|A|^{2/3})$ of the elements of A are unipotent.

Exercise 5.3.8 (Large tori). Let A, K be as in Theorem 5.3.1. Show that for any regular semisimple $g \in A^2$, one has $|A^4 \cap T| \gg K^{-O(1)}|A|^{1/3}$, where T is the unique maximal torus containing g. In fact one has $|A^2 \cap T| \gg K^{-O(1)}|A|^{1/3}$. (For the latter claim, cover A^4 by left translates of A.)

We remark that versions of most of the results in the above exercises were first obtained in the context of SL_2 in [**He2008**] (and extended to SL_d in [**He2011**]). However, the lower bound in Exercise 5.3.8 was only obtained in those papers for some maximal tori meeting A^2 nontrivially, rather than for all such tori, leading to some additional technical complications in Helfgott's proof of Theorem 5.3.1.

We now have a dichotomy: given a maximal torus T, either $A^2 \cap T$ has no regular semisimple elements (and thus contains only central elements), or else has cardinality $\gg K^{-O(1)}|A|^{1/3}$. We exploit this dichotomy as follows. Call a maximal torus T *involved* if $A^2 \cap T$ contains a regular semisimple element.

Lemma 5.3.4 (Key lemma). *Let A, K be as in Theorem 5.3.1. Then one of the following statements hold:*

(i) *(Involved) If T is an involved torus and $a \in A$, then aTa^{-1} is an involved torus.*

(ii) *(Close to trivial) One has $|A| \ll K^{O(1)}$.*

Proof. Let T be an involved torus, then by the preceding exercise we have $|A^2 \cap T| \gg K^{-O(1)}|A|^{1/3}$, and thus $|A^4 \cap aTa^{-1}| \gg K^{-O(1)}|A|^{1/3}$. Thus, one has $|gA \cap aTa^{-1}| \gg K^{-O(1)}|A|^{1/3}$ for some $g \in G$, which implies that $|A^2 \cap aTa^{-1}| \gg K^{-O(1)}|A|^{1/3}$. In particular, if A is not close to trivial, then $A^2 \cap aTa^{-1}$ contains a regular semisimple element and so aTa^{-1} is involved, as desired. $\square$

We can now finish the proof of Theorem 5.3.1. Suppose A is not close to trivial. As there are at most $O(K^{O(1)}|A|^{2/3})$ unipotent elements and $O(1)$ central elements in A, then A has at least one regular semisimple element, so there is at least one involved torus. By the above lemma, and the fact that A generates G, we see that the set of involved tori is invariant under conjugation by G. As G has cardinality $\gg |k|^3$, and its intersection with the stabiliser of a single torus has cardinality $O(|k|)$, we conclude that there are $\gg |k|^2 \ll |G|^{2/3}$ involved tori. By Exercise 5.3.8, each of these tori contains $\gg K^{-O(1)}|A|^{1/3}$ regular semisimple elements of A^2. Since each regular semisimple element belongs to a unique maximal torus, we conclude that

$$|A^2| \gg |G|^{2/3} K^{-O(1)} |A|^{1/3};$$

as $|A^2| \leq K|A|$, we conclude that $|A| \gg K^{-O(1)}|G|$, as claimed.

Exercise 5.3.9. Let A be a finite K-approximate subgroup of $\mathrm{SL}_2(\overline{k})$ for some algebraically closed field $\overline{k}$. Show that one of the following statements hold:

(i) (Close to group) A generates a finite subgroup G of $\mathrm{SL}_2(\overline{k})$ with $|G| \ll K^{O(1)}|A|$.

(ii) (Concentrated in Borel) There is a Borel subgroup B of $\mathrm{SL}_2(\overline{k})$ with $|A \cap B| \gg K^{-O(1)}|B|$.

(*Hint:* This does not follow directly from Theorem 5.3.1, but can be established by a modification of the proof of that theorem.)

Note that the above exercise can be combined with Theorem 5.2.7 to give a more detailed description of A. The Borel group B is solvable, and by using tools from additive combinatorics, such as inverse product theorems (see e.g., [**Ta2013**, §1.8] for a discussion), one can give even more precise descriptions of A (at the cost of losing polynomial dependence of the bounds on K), but we will not discuss these topics here.

Exercise 5.3.10. Use Exercise 5.3.9 to give an alternate proof of Theorem 5.1.1. (*Hint:* There are a number of ways to embed the sum-product problem in a field k into a product problem in $\mathrm{SL}_2(k)$ (or $\mathrm{SL}_2(\bar{k})$). For instance, one considers the tripling properties of sets of the form $\{\begin{pmatrix} a & b \\ c & d \end{pmatrix} : a,b,c,d \in A\}$ in terms of sets such as $A^2 + A^2$ or $A^3 + A^3 + A^3 + A^3$, and then projects this set onto $\mathrm{SL}_2(k)$ (or $P\mathrm{SL}_2(k)$), and combines this with the Katz-Tao lemma to obtain Theorem 5.1.1. More details of this connection can be found in [**BrGrTa2011**, Section 8].) This is of course a much more complicated and inefficient way to establish the sum-product theorem, but it does illustrate the link between the two results (beyond the fact that both proofs exploit a dichotomy). Note also that the original proof of the product theorem in $\mathrm{SL}_2(\mathbf{F}_p)$ in [**He2008**] actually used the sum-product theorem in $\mathbf{F}_p$ as a key tool.

5.4. The product theorem in $\mathrm{SL}_d(k)$

We now discuss the extension of the $\mathrm{SL}_2(k)$ product theory to the more general groups $\mathrm{SL}_d(k)$. Actually, the arguments here will be valid in any almost simple connected algebraic group of bounded rank, but for the sake of concreteness we will work with $\mathrm{SL}_d(k)$. (This also has the (very) minor advantage that SL_d is an affine variety rather than a projective one, so we can work entirely in affine spaces such as $\mathbf{A}^{d^2}(\bar{k}) := \bar{k}^{d^2}$; related to this, the only regular maps we need to consider will be polynomial in nature.) There is also some recent work on product theorems in other algebraic groups than the almost simple ones; see for instance the papers of [**PySz2010**], [**BrGrTa2011**], [**GiHe2010**]; see [**PySz2012**] for a recent survey of these results.

The treatment of the $d = 2$ case relied on a number of *ad hoc* computations which were only valid in SL_2, and also on the pleasant fact that the only nonregular elements of SL_2 were the central elements ± 1, which is certainly false for higher values of d. In [**He2011**], the original $d = 2$ arguments from [**He2008**] were pushed to the $d = 3$ case, but again the arguments were somewhat *ad hoc* in nature and did not seem to extend to the general setting. However, the arguments based on the Larsen-Pink concentration estimates have proven to be quite general and, in particular, can handle the situation of $\mathrm{SL}_d(k)$. The one catch is that instead of working with very concrete and explicit subsets of SL_2, such as Borel subgroups or other intersections $SL_2 \cap V$ with linear spaces V, one has to work with more general algebraic subvarieties of SL_d. As such, a certain amount of basic algebraic geometry becomes necessary. Also, because we are seeking results with quantitative bounds, we will need to keep some track of the "complexity" of the varieties that one encounters in the course of the argument.

We now very quickly review some algebraic geometry notions, though for reasons of space we will not attempt to develop the full theory of algebraic geometry here, referring instead to standard texts such as [**Ha1995**], [**Mu1999**], [**Gr1994**]. As usual, algebraic geometry is cleanest when working over an algebraically closed field, so we will work primarily over $\overline{k}$.

Definition 5.4.1 (Variety). Let $M \geq d \geq 0$ be integers, and let $\overline{k}$ be an algebraically closed field. We write $\mathbf{A}^d(\overline{k})$ for the affine space $\overline{k}^d$.

  (i) An (affine) *variety* $V = V(\overline{k}) \subset \mathbf{A}^d(\overline{k})$ of complexity[6] at most M is a set of the form
  $$V = \{x \in \overline{k}^d : P_1(x) = \ldots = P_m(x) = 0\},$$
  where $0 \leq m \leq M$ and $P_1, \ldots, P_m : \mathbf{A}^d(\overline{k}) \to \overline{k}$ are polynomials of degree at most M. Note that the union or intersection of two varieties of complexity at most M, is another variety of complexity at most $O_M(1)$.

  (ii) A variety is *irreducible* if it cannot be expressed as the union of two proper (i.e., strict) subvarieties.

Thus, for instance, $SL_d(\overline{k})$ is a variety of complexity $O_d(1)$ in $\mathbf{A}^{d^2}(\overline{k})$ (after identifying this latter affine space with the space of $d \times d$ matrices over $\overline{k}$).

It is known that any variety can be expressed as the union of a finite number of irreducible components, and this decomposition is unique if we require that no component is contained in any other. Furthermore, to each irreducible variety V one can assign a *dimension* $\dim(V)$, defined as the maximal integer D for which there exists a chain
$$\emptyset \neq V_0 \subsetneq \ldots \subsetneq V_D = V$$
of irreducible varieties. For instance, it can be shown that $\mathbf{A}^d(\overline{k})$ has dimension d (as expected). We define the dimension of a nonirreducible variety V to be the least integer D such that V can be covered by finitely many irreducible varieties of dimension D. If a (nonempty) variety V can be cut

---

[6]Thus the complexity parameter M controls the dimension, degree, and number of polynomials needed to cut out the variety. Note that we do not assume our varieties to be irreducible, and as such what we call a variety corresponds to what is sometimes known as an *algebraic set* in the literature.

5.4. The product theorem in $\mathrm{SL}_d(k)$

out from an irreducible variety W by setting m polynomials to zero, then one has $\dim(W) - m \leq \dim(V) \leq \dim(W)$. Since $\mathrm{SL}_d(\overline{k})$ can be cut out from $\mathbf{A}^{d^2}(\overline{k})$ by a single polynomial, and is not equal to all of $\mathbf{A}^{d^2}(\overline{k})$, we conclude in particular that $\mathrm{SL}_d(\overline{k})$ has dimension $d^2 - 1$.

One can show that the image of a D-dimensional variety by a polynomial map $P\colon \mathbf{A}^{d_1} \to \mathbf{A}^{d_2}$ is contained in a variety of dimension at most D. One can thus produce upper bounds on the dimension of varieties, by covering them with polynomial images of varieties already known to be bounded by the same dimension.

An *algebraic subgroup* of $\mathrm{SL}_d(\overline{k})$ is a subvariety of $\mathrm{SL}_d(\overline{k})$ which is also a subgroup of $\mathrm{SL}_d(\overline{k})$. For instance, the *standard maximal torus* $T(\overline{k})$, consisting of all the diagonal elements of $\mathrm{SL}_d(\overline{k})$, is an algebraic subgroup; more generally, any *maximal torus*, by which we mean a conjugate of the standard maximal torus, is an algebraic subgroup.

Exercise 5.4.1. Show that every maximal torus has dimension $d - 1$ and complexity $O_d(1)$.

Dual to the maximal tori are the conjugacy classes $\mathrm{Conj}(g) := \{hgh^{-1} : h \in \mathrm{SL}_d(\overline{k})\}$ of regular semisimple elements. We call a element g of $\mathrm{SL}_d(\overline{k})$ *regular semisimple* if it has d distinct eigenvalues, and is thus diagonalisable. Observe that each regular semisimple element lies in precisely one maximal torus.

Exercise 5.4.2. Show that every conjugacy class of a regular semisimple element has dimension $d^2 - d$ and complexity $O_d(1)$.

If F is a finite subfield of $\overline{k}$, then $\mathrm{SL}_d(F)$ is a finite subgroup of $\mathrm{SL}_d(\overline{k})$, and is thus technically a 0-dimensional algebraic subgroup of $\mathrm{SL}_d(\overline{k})$. However, the complexity of this algebraic group is huge (comparable to the cardinality of $\mathrm{SL}_d(F)$). It turns out that $\mathrm{SL}_d(F)$ is "effectively Zariski-dense" in the sense that it cannot be captured in a low complexity algebraic variety:

Lemma 5.4.2 (Schwartz-Zippel lemma for $\mathrm{SL}_d(F)$). *Let V be a proper subvariety of $\mathrm{SL}_d(\overline{k})$ of complexity at most M. Let F be a finite subfield of $\overline{k}$. Then*
$$|\mathrm{SL}_d(F) \cap V| \ll_{M,d} |F|^{d^2 - 2}.$$

Proof. $\mathrm{SL}_d(\overline{k})$ is the hypersurface in $\mathbf{A}^{d^2}(\overline{k})$ cut out by the determinant polynomial. As V is a proper subvariety, we can find a polynomial $P\colon \overline{k}^{d^2} \to \overline{k}$ which is not a multiple of the determinant polynomial, but which vanishes on V; by the complexity hypothesis we may take P to have degree $O_{M,d}(1)$. Our task is then to show that
$$|\{x \in \mathrm{SL}_d(F) : P(x) = 0\}| \ll_{M,d} |F|^{d^2 - 2}.$$

Let us write the d^2 coordinates of $\mathbf{A}^{d^2}(\overline{k})$ arbitrarily as $x_1, \ldots, x_{d^2}$. In a given element of $\mathrm{SL}_d(F)$, not all of the x_i can be zero; thus by symmetry and relabeling if necessary it suffices to show that

(5.7) $$|\{x \in \mathrm{SL}_d(F) : P(x) = 0; x_{d^2} \neq 0\}| \ll_{M,d} |F|^{d^2-2}.$$

But then one can express x_{d^2} as a rational function of the other $d^2 - 1$ coordinates, and the left-hand side of (5.7) is contained in a set of the form $\{x \in F^{d^2-1} : Q(x) = 0\}$ for some polynomial Q of degree $O_{M,d}(1)$ that is not identically zero. The claim then follows from the *Schwartz-Zippel lemma*, which we give as an exercise below. □

Exercise 5.4.3 (Schwartz-Zippel lemma). Let F be a finite field, and let $Q \colon F^d \to F$ be a polynomial of degree D that is not identically zero. Show that

$$|\{x \in F^d : Q(x) = 0\}| \ll_d D|F|^{d-1}.$$

For an additional challenge, obtain the sharper bound

$$|\{x \in F^d : Q(x) = 0\}| \leq D|F|^{d-1}.$$

We contrast this with the size of $\mathrm{SL}_d(F)$ itself:

Exercise 5.4.4. Let F be a finite field. Show that $|F|^{d^2-1} \ll_d |\mathrm{SL}_d(F)| \ll_d |F|^{d^2-1}$.

The key nonconcentration inequality we will need is the following.

Proposition 5.4.3 (Larsen-Pink inequality). *Let A be a K-approximate subgroup of $\mathrm{SL}_d(\overline{k})$ for some $K \geq 2$, and let V be a subvariety of $\mathrm{SL}_d(\overline{k})$ of complexity at most M. Let $m \geq 1$. Then one of the following is true:*

(i) *(Nonconcentration) One has*

(5.8) $$|A^m \cap V| \ll_{M,d,m} K^{O_{M,d,m}(1)} |A|^{\dim(V)/\dim(\mathrm{SL}_d)}.$$

(ii) *(Trapping) A is contained in a proper algebraic subgroup H of $\mathrm{SL}_d(\overline{k})$ of complexity $O_{M,d,m}(1)$.*

This inequality subsumes results such as Proposition 5.3.3, Exercise 5.3.4, and Exercise 5.3.6. Note that if A generates $\mathrm{SL}_d(F)$ for some finite field F, then (5.8) holds unconditionally; indeed, from Lemma 5.4.2 (and Exercise 5.4.4) that the trapping option of the above proposition cannot occur with $|F|$ is sufficiently large depending on M, d, m, while the nonconcentration claim (5.8) is trivial when $|F| = O_{M,d,m}(1)$.

The proof of Proposition 5.4.3 is somewhat complicated and is deferred to the next section. We record some particular consequences of this inequality.

Exercise 5.4.5 (Consequences of the nonconcentration inequality). Let A be a K-approximate subgroup of $\mathrm{SL}_d(\overline{k})$ for some $K \geq 2$, which generates $\mathrm{SL}_d(F)$ for some finite field F.

(i) If T is a maximal torus (and thus of dimension $d-1$), show that $|A^{10} \cap T| \ll_d K^{O_d(1)} |A|^{\frac{1}{d+1}}$.

(ii) If T_0 denotes the elements of a maximal torus T which are not regular semisimple, show that $|A^{10} \cap T| \ll_d K^{O_d(1)} |A|^{\frac{d-2}{d^2-1}}$.

(iii) If $g \in \mathrm{SL}_d(\overline{k})$ is regular semisimple, show that $|A^{10} \cap \mathrm{Conj}(g)| \ll_d K^{O_d(1)} |A|^{\frac{d}{d+1}}$.

(iv) Show that at most $O_d(K^{O_d(1)} |A|^{\frac{d^2-2}{d^2-1}})$ of the elements of A are not regular semisimple.

(v) For any regular semisimple $g \in A$, show that $|A^3 \cap \mathrm{Conj}(g)| \gg_d K^{-O_d(1)} |A|^{\frac{d}{d+1}}$.

(vi) For any regular semisimple $g \in A$, show that
$$|A^2 \cap T| \gg_d K^{-O_d(1)} |A|^{\frac{1}{d+1}}.$$

Exercise 5.4.6. By repeating the arguments of the preceding section, establish Theorem 5.0.1 for general d.

Remark 5.4.4. There is an analogue of Exercise 5.3.9, in which the role of the Borel subgroups is replaced by proper algebraic subgroups of bounded complexity; see [**BrGrTa2011**, Theorem 5.5] for a more precise statement.

5.5. Proof of the Larsen-Pink inequality

We now prove Proposition 5.4.3. In order to escape the burden of having to keep track of the complexity of everything, we will use the tool of *ultraproducts* (which we will phrase in the language of *nonstandard analysis*). See [**Ta2013**, §1.7] for a discussion of ultraproducts and how they can be used to turn quantitative (or "hard") analysis tasks into qualitative (or "soft") analysis tasks. One can also use the machinery of schemes and inverse limits as a substitute for the ultraproduct formalism; this is the approach taken in [**LaPi2011**]. The paper [**BrGrTa2011**] has a slightly reduced reliance on ultraproducts, at the cost of more complexity bookkeeping, while the paper [**PySz2010**] avoids ultraproducts altogether but has perhaps the most bookkeeping of all the papers mentioned here (but, by the same token, is the only argument currently known which gives effective bounds). We will thus presume some familiarity both with ultraproducts (and nonstandard analysis) and with algebraic geometry in this section.

As in [**Ta2013**, §1.7], we select a nonprincipal ultrafilter $\alpha \in \beta\mathbf{N}\backslash\mathbf{N}$, and use it to construct ultraproducts and nonstandard objects. (To ensure the existence of such an object, we shall assume the *axiom of choice*, as we have already been doing implicitly throughout this course.) We also use the usual nonstandard asymptotic notation, thus, for instance, $O(1)$ denotes a nonstandard quantity bounded in magnitude by a standard number.

The quantitative Larsen-Pink inequality (Proposition 5.4.3) can then be deduced from the following nonstandard version, in which all references to complexity are now absent:

Proposition 5.5.1 (Larsen-Pink inequality). *Let $d \geq 2$ be standard. Let $\overline{k} = \prod_{n\to\alpha} \overline{k_n}$ be a nonstandard algebraically complete field (i.e., an ultraproduct of standard algebraically complete fields). Let $K = \lim_{n\to\alpha} K_n \geq 2$ be a nonstandard natural number, and let A be a nonstandard K-approximate subgroup of $\mathrm{SL}_d(\overline{k})$ (i.e., an ultraproduct $A = \prod_{n\to\alpha} A_n$ of standard K_n-approximate subgroups of $\mathrm{SL}_d(\overline{k_n})$), and let V be a subvariety of $\mathrm{SL}_d(\overline{k})$. Then one of the following is true:*

(i) *(Nonconcentration) One has*

(5.9) $$|A^m \cap V| \ll K^{O(1)}|A|^{\dim(V)/\dim(\mathrm{SL}_d)}$$

for all standard $m \geq 1$, where $|A| := \lim_{n\to\alpha} |A_n|$ is the nonstandard cardinality of A.

(ii) *(Trapping) A is contained in a proper algebraic subgroup H of $\mathrm{SL}_d(\overline{k})$.*

Let us see why Proposition 5.5.1 implies Proposition 5.4.3. Suppose for contradiction that Proposition 5.4.3 failed. Carefully negating all the quantifiers (and using the axiom of choice), this means that there is a sequence $\overline{k_n}$ of standard algebraically closed fields, a sequence $K_n \geq 2$ of standard numbers, a sequence A_n of K_n-approximate subgroups of $\mathrm{SL}_d(\overline{k_n})$, and a standard $M \geq 1$, a sequence V_n of subvarieties of $\mathrm{SL}_d(\overline{k_n})$ of complexity at most M, and a standard $m \geq 1$, such that for each n, A_n is not contained in a proper algebraic subgroup of $\mathrm{SL}_d(\overline{k_n})$ of complexity n or less, and one has
$$|A_n^m \cap V_n| \geq nK^n|A|^{\dim(V_n)/\dim(\mathrm{SL}_d)}.$$
Now one forms the ultralimit $K := \lim_{n\to\alpha} K_n$ and the ultraproducts $\overline{k} := \prod_{n\to\alpha} \overline{k_n}$, $A := \prod_{n\to\alpha} A_n$, $V := \prod_{n\to\alpha} V_n$. Then $\overline{k}$ is an algebraically closed field, A is a nonstandard K-approximate subgroup of $\mathrm{SL}_d(\overline{k})$, and V is an algebraic subvariety of $\mathrm{SL}_d(\overline{k})$ (here we use the uniform complexity bound). One can also show that $\dim(V) = \lim_{n\to\alpha} \dim(V_n)$; see [**Ta2012**, Lemma 2.1.8]. As such, we have
$$|A^m \cap V| \not\ll K^{O(1)}|A|^{\dim(V)/\dim(\mathrm{SL}_d)},$$

so by Proposition 5.5.1, A is contained in a proper algebraic subgroup H of $\mathrm{SL}_d(\overline{k})$. By unpacking the coefficients of all the polynomials over $\overline{k}$ used to cut out H, we see that H is itself an ultraproduct $H = \prod_{n \to \alpha} H_n$ of proper algebraic subgroups of $\mathrm{SL}_d(\overline{k_n})$, of complexity bounded uniformly in n. By *Los's theorem* (see e.g., [**BrGrTa2011b**, Theorem A.5]), one has $A_n \subset H_n$ for all n sufficiently close to α, which gives a contradiction for n large enough.

It remains to establish Proposition 5.5.1. By Łos's theorem, the ultraproduct $\overline{k}$ of algebraically closed fields is again algebraically closed, which allows us to use algebraic geometry in the nonstandard field $\overline{k}$ without difficulty.

Let $\langle A \rangle$ be the group generated by A, and consider the Zariski closure $\overline{\langle A \rangle}$ of this group, that is to say the intersection of all the varieties containing $\langle A \rangle$. This is again an algebraic variety (here we use the Noetherian property of varieties, that there does not exist any infinite descending chain of varieties), and is also a group (exercise!), and is thus an algebraic subgroup of $\mathrm{SL}_d(\overline{k})$. If this subgroup is proper then we have the trapping property, so we may assume that the closure is all of $\mathrm{SL}_d(\overline{k})$. In other words, $\langle A \rangle$ is *Zariski dense* in $\mathrm{SL}_d(\overline{k})$.

For any dimension D between 0 and $\dim(\mathrm{SL}_d)$ inclusive, and any standard real σ, let us call σ D-*admissible* if one has the bound

$$|A^m \cap V| \ll K^{O(1)} |A|^\sigma$$

whenever $m \geq 1$ is standard and V is a D-dimensional subvariety of $\mathrm{SL}_d(\overline{k})$. Our task is to show that $D/\dim(\mathrm{SL}_d)$ is admissible for all $0 \leq D \leq \dim(\mathrm{SL}_d)$. This claim is trivial at the two endpoints $D = 0$ and $D = \dim(\mathrm{SL}_d)$; the difficulty is to somehow "interpolate" between these two endpoints. We need the following combinatorial observation.

Exercise 5.5.1 (Extreme dimensions). Suppose for the sake of contradiction that $D/\dim(\mathrm{SL}_d)$ is inadmissible for some $0 < D < \dim(\mathrm{SL}_d)$. Show that we can find dimensions

$$0 < D_1 \leq D_2 < \dim(\mathrm{SL}_d)$$

and a real number $\theta \geq 1/\dim(\mathrm{SL}_d)$ such that

(i) $D_1 \theta$ is not D_1-admissible;

(ii) $D_2 \theta$ is not D_2-admissible;

(iii) $D\theta$ is D-admissible whenever $0 \leq D < D_1$ or $D_2 < D \leq \dim(\mathrm{SL}_d)$;

(iv) $(D+1)\theta$ is D-admissible for any $0 \leq D \leq \dim(\mathrm{SL}_d)$.

Let D_1, D_2, θ be as in the above exercise. By construction, we can then find subvarieties V_1, V_2 of $\mathrm{SL}_d(\overline{k})$ of dimension D_1, D_2 respectively and standard positive integers m_1, m_2 such that

(5.10) $$|A^{m_1} \cap V_1| \ll K^{O(1)} |A|^{\theta D_1}$$

and

(5.11) $$|A^{m_2} \cap V_2| \ll K^{O(1)} |A|^{\theta D_2}.$$

On the other hand, we have

(5.12) $$|A^m \cap V| \ll K^{O(1)} |A|^{\theta(\dim(V)+1)}$$

whenever V is a subvariety of $\mathrm{SL}_d(\overline{k})$, with the improvement

(5.13) $$|A^m \cap V| \ll K^{O(1)} |A|^{\theta \dim(V)}$$

whenever V has dimension strictly less than D_1, or strictly greater than D_2.

We can use (5.12), (5.13) to show that $A^{m_1} \times A^{m_2}$ is "quantitatively Zariski dense" in $V_1 \times V_2$:

Lemma 5.5.2 (Quantitative Zariski density). *For any proper subvariety W of $V_1 \times V_2$, we have*

$$|(A^{m_1} \times A^{m_2}) \cap W| \ll K^{O(1)} |A|^{\theta(D_1+D_2)}.$$

Proof. W has dimension at most $D_1 + D_2 - 1$. By standard algebraic geometry, we see that for each $0 \leq D \leq D_1$, the set of $y \in V_2$ for which the slice $\{x \in V_1 : (x, y) \in W\}$ has dimension D, has dimension at most $D_1 + D_2 - D - 1$. In particular, if $D < D_1$, then by (5.12), (5.13) the contribution of such x to $|(A^{m_1} \times A^{m_2}) \cap W|$ is at most

$$K^{O(1)} \times |A|^{\theta D} \times K^{O(1)} |A|^{\theta(D_1+D_2-D-1+1)}$$

while if $D = D_1$, then the contribution is at most[7]

$$K^{O(1)} \times |A|^{\theta(D+1)} \times K^{O(1)} |A|^{\theta(D_1+D_2-D-1)}.$$

Summing over all D we obtain the claim. $\square$

We will now use a counting argument (which is, unsurprisingly, related to the counting argument used to establish Proposition 5.3.3, or any of the other Larsen-Pink inequalities in preceding sections) to obtain a contradiction from these four estimates.

First, by decomposing V_1, V_2 into irreducible components (and using (5.12) to eliminate all lower-dimensional components) we may assume that V_1, V_2 are both irreducible.

---

[7] One may wonder about the question of uniformity in the $O()$ notation, but in nonstandard analysis one can automatically gain such uniformity through countable saturation; see [**Ta2013**, Exercise 1.7.20].

5.5. Proof of the Larsen-Pink inequality

The product $V_1 \cdot V_2$ is not necessarily a variety, but it is still a constructible set (i.e., a finite Boolean combination of varieties), and can still be assigned a dimension (by equating the dimension of a constructible set with the dimension of its Zariski closure). As it contains a translate of V_2, it has dimension at least D_2. It would be convenient if $V_1 \cdot V_2$ had dimension strictly greater than V_2. This is not necessarily the case, but it turns out that it becomes so after a generic conjugation, thanks to the almost simplicity of SL_d:

Exercise 5.5.2 (Almost simplicity). Show that the only proper normal subgroups of $\mathrm{SL}_d(\overline{k})$ are those contained in the centre of $\mathrm{SL}_d(\overline{k})$, i.e., in the identity matrix multipled by the dth roots of unity. (*Hint:* Let G be a normal subgroup of $\mathrm{SL}_d(\overline{k})$ that contains an element which is not a multiple of the identity. Place that element in Jordan normal form and divide it by one of its conjugates to make it fix a subspace of $\overline{k}^d$; iterate this procedure until one finds an element in G that is the direct sum of the identity in $\mathrm{SL}_{d-2}(\overline{k})$ and a noncentral element of $\mathrm{SL}_2(\overline{k})$. Then use this to generate all of $\mathrm{SL}_d(\overline{k})$.)

Proposition 5.5.3 (Generic skewness). *For generic $g \in \mathrm{SL}_d(\overline{k})$ (i.e., for all g in $\mathrm{SL}_d(\overline{k})$ outside of a lower-dimensional variety), the set $V_1 \cdot g \cdot V_2$ has dimension strictly greater than D_2.*

Proof. Let $g \in \mathrm{SL}_d(\overline{k})$, and assume that $V_1 \cdot g \cdot V_2$ has dimension exactly D_2. This set contains all the translates xgV_2 with $x \in V_1$, which are each D_2-dimensional irreducible varieties. By splitting up $V_1 \cdot g \cdot V_2$ into components, we conclude that there are only finitely many distinct translates xgV_2. If we denote one of these translates as W, the set $\{x \in \mathrm{SL}_d(\overline{k}) : xgV_2 = W\}$ is easily seen to be a variety (as it is the intersection of varieties $Wy^{-1}g^{-1}$ for $y \in V_2$); as a finite number of these sets cover V_1, at least one of them has to be all of V_1; thus there is a W such that $xgV_2 = W$ for all $x \in V_1$. In particular, this implies that $g^{-1}y^{-1}xgV_2 = V_2$ for all $x, y \in V_1$.

Let $S := \{h \in \mathrm{SL}_d(\overline{k}) : hV_2 = V_2\}$. Arguing as before, S is a variety, and also a group; it is thus an algebraic group, and by the preceding discussion we have $g^{-1}V_1^{-1}V_1 g \subset S$.

The set $\{g \in \mathrm{SL}_d(\overline{k}) : g^{-1}V_1^{-1}V_1 g \subset S\}$ is a variety. If it has dimension strictly less than that of SL_d, we are done, so we may assume this set is all of SL_d; thus $g^{-1}V_1^{-1}V_1 g \subset S$ for all $g \in \mathrm{SL}_d(\overline{k})$. By almost simplicity, the normal subgroup generated by $V_1^{-1}V_1$ is all of $\mathrm{SL}_d(\overline{k})$; thus S must be all of $\mathrm{SL}_d(\overline{k})$, thus $hV_2 = V_2$ for all $h \in \mathrm{SL}_d(\overline{k})$. But this forces $V_2 = \mathrm{SL}_d(\overline{k})$, a contradiction since D_2 is strictly less than $\dim(\mathrm{SL}_d)$. $\square$

Combining this proposition with the Zariski density of $\langle A \rangle$, we see that we can find $g \in A^m$ for some standard m such that $V_1 \cdot g \cdot V_2$ has dimension D strictly greater than D_2.

Fix this g. Let $\phi \colon V_1 \times V_2 \to \overline{V_1 \cdot g \cdot V_2}$ be the twisted product map $\phi(x,y) := xgy$. We have the double counting identity
$$\sum_{z \in A^{m_1+m+m_2} \cap \overline{V_1 \cdot g \cdot V_2}} |A^{m_1} \times A^{m_2} \cap \phi^{-1}(\{z\})| = |A^{m_1} \cap V_1||A^{m_2} \cap V_2|$$
and thus by (5.10), (5.11)
$$\sum_{z \in A^{m_1+m+m_2} \cap \overline{V_1 \cdot g \cdot V_2}} |A^{m_1} \times A^{m_2} \cap \phi^{-1}(\{z\})| \lll K^{O(1)} |A|^{\theta(D_1+D_2)}.$$

Now, ϕ is a map from an irreducible $D_1 + D_2$-dimensional variety to a D-dimensional variety with Zariski-dense image, and is thus a *dominant map*. Among other things, this implies that there is a subvariety S of $V_1 \times V_2$ of dimension at most $D_1 + D_2 - 1$ such that for all $x \in \overline{V_1 \cdot g \cdot V_2}$, the set $\phi^{-1}(\{x\}) \setminus S$ has dimension $D_1 + D_2 - D$. By (5.13), we then have
$$|A^{m_1} \times A^{m_2} \cap \phi^{-1}(\{z\}) \setminus S| \ll K^{O(1)} |A|^{\theta(D_1+D_2-D)}$$
for all $z \in A^{m_1+m+m_2} \cap \overline{V_1 \cdot g \cdot V_2}$; by another application of (5.13), we have
$$|A^{m_1+m+m_2} \cap \overline{V_1 \cdot g \cdot V_2}| \ll K^{O(1)} |A|^{\theta D}.$$
Combining these estimates we see that
$$\sum_{z \in A^{m_1+m+m_2} \cap \overline{V_1 \cdot g \cdot V_2}} |A^{m_1} \times A^{m_2} \cap \phi^{-1}(\{z\}) \cap S| \lll K^{O(1)} |A|^{\theta(D_1+D_2)}.$$
The left-hand side simplifies to $|A^{m_1} \times A^{m_2} \cap S|$. But this then contradicts Lemma 5.5.2.

Chapter 6

Nonconcentration in subgroups

In the last few chapters, we discussed the Bourgain-Gamburd expansion machine and two of its three ingredients, namely quasirandomness and product theorems, leaving only the nonconcentration ingredient to discuss. We can summarise the results of the last three chapters, in the case of fields of prime order, as the following theorem.

Theorem 6.0.1 (Nonconcentration implies expansion in SL_d). *Let p be a prime, let $d \geq 1$, and let S be a symmetric set of elements in $G := SL_d(\mathbf{F}_p)$ of cardinality $|S| = k$ not containing the identity. Write $\mu := \frac{1}{|S|} \sum_{s \in S} \delta_s$, and suppose that one has the nonconcentration property*

(6.1) $$\sup_{H < G} \mu^{*n}(H) < |G|^{-\kappa}$$

for some $\kappa > 0$ and some even integer $n \leq \Lambda \log |G|$. Then $\mathrm{Cay}(G, S)$ is a two-sided ε-expander for some $\varepsilon > 0$ depending only on k, d, κ, Λ.

Proof. From (6.1) we see that μ^{*n} is not supported in any proper subgroup H of G, which implies that S generates G. The claim now follows from the Bourgain-Gamburd expansion machine (Theorem 4.0.2), the product theorem (Theorem 5.0.1), and quasirandomness (Exercise 3.0.9). □

Remark 6.0.2. The above type of theorem was generalised to the setting of cyclic groups $\mathbf{Z}/q\mathbf{Z}$ with q square-free by Varju [**Va2012**], to arbitrary q by Bourgain and Varju [**BoVa2012**], and to more general algebraic groups than SL_d and square-free q by Salehi-Golsefidy and Varju [**SGVa2012**].

It thus remains to construct tools that can establish the nonconcentration property (6.1). The situation is particularly simple in $\mathrm{SL}_2(\mathbf{F}_p)$, as we have a good understanding of the subgroups of that group. Indeed, from Theorem 5.2.7, we obtain the following corollary to Theorem 6.0.1:

Corollary 6.0.3 (Nonconcentration implies expansion in SL_2)**.** *Let p be a prime, and let S be a symmetric set of elements in $G := \mathrm{SL}_2(\mathbf{F}_p)$ of cardinality $|S| = k$ not containing the identity. Write $\mu := \frac{1}{|S|} \sum_{s \in S} \delta_s$, and suppose that one has the nonconcentration property*

$$\sup_B \mu^{*n}(B) < |G|^{-\kappa} \tag{6.2}$$

for some $\kappa > 0$ and some even integer $n \leq \Lambda \log |G|$, where B ranges over all Borel subgroups of $\mathrm{SL}_2(\overline{F})$. Then, if $|G|$ is sufficiently large depending on k, κ, Λ, $\mathrm{Cay}(G, S)$ is a two-sided ε-expander for some $\varepsilon > 0$ depending only on k, κ, Λ.

It turns out (6.2) can be verified in many cases by exploiting the solvable nature of the Borel subgroups B. We give two examples of this in this chapter. The first result, due to Bourgain and Gamburd [**BoGa2008**] (with earlier partial results by Gamburd [**Ga2002**] and by Shalom [**Sh2000**]) generalises Selberg's expander construction to the case when S generates a thin subgroup of $\mathrm{SL}_2(\mathbf{Z})$:

Theorem 6.0.4 (Expansion in thin subgroups)**.** *Let S be a symmetric subset of $\mathrm{SL}_2(\mathbf{Z})$ not containing the identity, and suppose that the group $\langle S \rangle$ generated by S is not virtually solvable (i.e., it does not have a finite index subgroup which is solvable). Then as p ranges over all sufficiently large primes, the Cayley graphs $\mathrm{Cay}(\mathrm{SL}_2(\mathbf{F}_p), \pi_p(S))$ form a two-sided expander family, where $\pi_p \colon \mathrm{SL}_2(\mathbf{Z}) \to \mathrm{SL}_2(\mathbf{F}_p)$ is the usual projection.*

Remark 6.0.5. One corollary of Theorem 6.0.4 (or of the nonconcentration estimate (6.3) below) is that $\pi_p(S)$ generates $\mathrm{SL}_2(\mathbf{F}_p)$ for all sufficiently large p, if $\langle S \rangle$ is not virtually solvable. This is a special case of a much more general result, known as the *strong approximation theorem*, although this is certainly not the most direct way to prove such a theorem. Conversely, the strong approximation property is used in generalisations of this result to higher rank groups than SL_2.

Exercise 6.0.1. In the converse direction, if $\langle S \rangle$ is virtually solvable, show that for sufficiently large p, $\pi_p(S)$ fails to generate $\mathrm{SL}_2(\mathbf{F}_p)$. (*Hint:* Use Theorem 5.2.7 to prevent $\mathrm{SL}_2(\mathbf{F}_p)$ from having bounded index solvable subgroups.)

Exercise 6.0.2 (Lubotzky's 1-2-3 problem)**.** Let $S := \{\begin{pmatrix} 1 & \pm 3 \\ 0 & 1 \end{pmatrix}, \begin{pmatrix} 1 & 0 \\ \pm 3 & 1 \end{pmatrix}\}$.

(i) Show that S generates a free subgroup of $SL_2(\mathbf{Z})$. (*Hint:* Use a ping-pong argument, as in Exercise 2.3.1.)

(ii) Show that if v, w are two distinct elements of the sector $\{(x, y) \in \mathbf{R}_+^2 : x/2 < y < 2x\}$, then there is no element $g \in \langle S \rangle$ for which $gv = w$. (*Hint:* This is another ping-pong argument.) Conclude that $\langle S \rangle$ has infinite index in $SL_2(\mathbf{Z})$. (Contrast this with the situation in which the 3 coefficients in S are replaced by 1 or 2, in which case $\langle S \rangle$ is either all of $SL_2(\mathbf{Z})$, or a finite index subgroup, as demonstrated in Exercise 2.3.1).

(iii) Show that $\mathrm{Cay}(SL_2(\mathbf{F}_p), \pi_p(S))$ for sufficiently large primes p form a two-sided expander family.

Remark 6.0.6. Theorem 6.0.4 has been generalised to arbitrary linear groups, and with $\mathbf{F}_p$ replaced by $\mathbf{Z}/q\mathbf{Z}$ for square-free q; see [**SGVa2012**]. In this more general setting, the condition of virtual solvability must be replaced by the condition that the connected component of the Zariski closure of $\langle S \rangle$ is *perfect* (i.e., it is equal to its own commutator group). An effective version of Theorem 6.0.4 (with completely explicit constants) was recently obtained in [**Ko2012**].

The second example concerns Cayley graphs constructed using random elements of $SL_2(\mathbf{F}_p)$.

Theorem 6.0.7 (Random generators expand)**.** *Let p be a prime, and let x, y be two elements of $SL_2(\mathbf{F}_p)$ chosen uniformly at random. Then with probability $1 - o_{p \to \infty}(1)$, $\mathrm{Cay}(SL_2(\mathbf{F}_p), \{x, x^{-1}, y, y^{-1}\})$ is a two-sided ε-expander for some absolute constant ε.*

Remark 6.0.8. As with Theorem 6.0.4, Theorem 6.0.7 has also been extended to a number of other groups, such as the Suzuki groups (see [**BrGrTa2011c**]), and more generally to finite simple groups of Lie type of bounded rank (in [**BrGrTa2013**]). There are a number of other constructions of expanding Cayley graphs in such groups (and in other interesting groups, such as the alternating groups) beyond those discussed in this chapter; see [**Lu2012**] for further discussion. It has been conjectured by Lubotzky and Weiss [**LuWe1993**] that *any* pair x, y of (say) $SL_2(\mathbf{F}_p)$ that generates the group, is a two-sided ε-expander for an absolute constant ε: in the case of $SL_2(\mathbf{F}_p)$, this has been established for a density one set of primes in [**BrGa2010**].

6.1. Expansion in thin subgroups

We now prove Theorem 6.0.4. The first observation is that the expansion property is monotone in the group $\langle S \rangle$:

Exercise 6.1.1. Let S, S' be symmetric subsets of $\mathrm{SL}_2(\mathbf{Z})$ not containing the identity, such that $\langle S \rangle \subset \langle S' \rangle$. Suppose that $\mathrm{Cay}(\mathrm{SL}_2(\mathbf{F}_p), \pi_p(S))$ is a two-sided expander family for sufficiently large primes p. Show that $\mathrm{Cay}(\mathrm{SL}_2(\mathbf{F}_p), \pi_p(S'))$ is also a two-sided expander family.

As a consequence, Theorem 6.0.4 follows from the following two statments:

Theorem 6.1.1 (Tits alternative). *Let $\Gamma \subset \mathrm{SL}_2(\mathbf{Z})$ be a group. Then exactly one of the following statements holds:*

(i) *Γ is virtually solvable.*

(ii) *Γ contains a copy of the free group F_2 of two generators as a subgroup.*

Theorem 6.1.2 (Expansion in free groups). *Let $x, y \in \mathrm{SL}_2(\mathbf{Z})$ be generators of a free subgroup of $\mathrm{SL}_2(\mathbf{Z})$. Then as p ranges over all sufficiently large primes, the Cayley graphs $\mathrm{Cay}(\mathrm{SL}_2(\mathbf{F}_p), \pi_p(\{x, y, x^{-1}, y^{-1}\}))$ form a two-sided expander family.*

Theorem 6.1.1 is a special case of the famous *Tits alternative* [**Ti1972**], which among other things allows one to replace $\mathrm{SL}_2(\mathbf{Z})$ by $\mathrm{GL}_d(k)$ for any $d \geq 1$ and any field k of characteristic zero (and fields of positive characteristic are also allowed, if one adds the requirement that Γ be finitely generated). We will not prove the full Tits alternative here, but instead just give an *ad hoc* proof of the special case in Theorem 6.1.1 in the following exercise.

Exercise 6.1.2. Given any matrix $g \in \mathrm{SL}_2(\mathbf{Z})$, the singular values are $\|g\|_{\mathrm{op}}$ and $\|g\|_{\mathrm{op}}^{-1}$, and we can apply the singular value decomposition to decompose
$$g = u_1(g)\|g\|_{\mathrm{op}} v_1^*(g) + u_2(g)\|g\|_{\mathrm{op}}^{-1} v_2(g)^*$$
where $u_1(g), u_2(g) \in \mathbf{C}^2$ and $v_1(g), v_2(g) \in \mathbf{C}^2$ are orthonormal bases. (When $\|g\|_{\mathrm{op}} > 1$, these bases are uniquely determined up to phase rotation.) We let $\tilde{u}_1(g) \in \mathbf{CP}^1$ be the projection of $u_1(g)$ to the projective complex plane, and similarly define $\tilde{v}_2(g)$.

Let Γ be a subgroup of $\mathrm{SL}_2(\mathbf{Z})$. Call a pair $(u, v) \in \mathbf{CP}^1 \times \mathbf{CP}^1$ a *limit point* of Γ if there exists a sequence $g_n \in \Gamma$ with $\|g_n\|_{\mathrm{op}} \to \infty$ and $(\tilde{u}_1(g_n), \tilde{v}_2(g_n)) \to (u, v)$.

(i) Show that if Γ is infinite, then there is at least one limit point.

(ii) Show that if (u, v) is a limit point, then so is (v, u).

(iii) Show that if there are two limit points $(u, v), (u', v')$ with $\{u, v\} \cap \{u', v'\} = \emptyset$, then there exist $g, h \in \Gamma$ that generate a free group.

(Hint: Choose $(\tilde{u}_1(g), \tilde{v}_2(g))$ close to (u,v) and $(\tilde{u}_1(h), \tilde{v}_2(h))$ close to (u',v'), and consider the action of g and h on $\mathbf{CP}^1$, and specifically on small neighbourhoods of u,v,u',v', and set up a ping-pong type situation.)

(iv) Show that if $g \in \mathrm{SL}_2(\mathbf{Z})$ is hyperbolic (i.e., it has an eigenvalue greater than 1), with eigenvectors u, v, then the projectivisations $(\tilde{u}, \tilde{v})$ of u, v form a limit point. Similarly, if g is regular parabolic (i.e., it has an eigenvalue at 1, but is not the identity) with eigenvector u, show that $(\tilde{u}, \tilde{bu})$ is a limit point.

(v) Show that if Γ has no free subgroup of two generators, then all hyperbolic and regular parabolic elements of Γ have a common eigenvector. Conclude that all such elements lie in a solvable subgroup of Γ.

(vi) Show that if an element $g \in \mathrm{SL}_2(\mathbf{Z})$ is neither hyperbolic nor regular parabolic, and is not a multiple of the identity, then g is conjugate to a rotation by $\pi/2$ (in particular, $g^2 = -1$).

(vii) Establish Theorem 6.1.1. (Hint: Show that two square roots of -1 in $\mathrm{SL}_2(\mathbf{Z})$ cannot multiply to another square root of -1.)

Now we prove Theorem 6.1.2. Let Γ be a free subgroup of $\mathrm{SL}_2(\mathbf{Z})$ generated by two generators x, y. Let $\mu := \frac{1}{4}(\delta_x + \delta_{x^{-1}} + \delta_y + \delta_{y^{-1}})$ be the probability measure generating a random walk on $\mathrm{SL}_2(\mathbf{Z})$, thus $(\pi_p)_*\mu$ is the corresponding generator on $\mathrm{SL}_2(\mathbf{F}_p)$. By Corollary 6.0.3, it thus suffices to show that

(6.3) $$\sup_B ((\pi_p)_*\mu)^{(n)}(B) < p^{-\kappa}$$

for all sufficiently large p, some absolute constant $\kappa > 0$, and some even $n = O(\log p)$ (depending on p, of course), where B ranges over Borel subgroups.

As π_p is a homomorphism, one has $((\pi_p)_*\mu)^{(n)}(B) = (\pi_p)_*(\mu^{*n})(B) = \mu^{*n}(\pi_p^{-1}(B))$ and so it suffices to show that

$$\sup_B \mu^{*n}(\pi_p^{-1}(B)) < p^{-\kappa}.$$

To deal with the supremum here, we will use an argument of Bourgain and Gamburd [**BoGa2008**], taking advantage of the fact that all Borel groups of SL_2 obey a common group law, the point being that free groups such as Γ obey such laws only very rarely. More precisely, we use the fact that the Borel groups are solvable of derived length two; in particular, we have

(6.4) $$[[a,b],[c,d]] = 1$$

for all $a,b,c,d \in B$. Now, μ^{*n} is supported on matrices in $\mathrm{SL}_2(\mathbf{Z})$ whose coefficients have size $O(\exp(O(n)))$ (where we allow the implied constants to depend on the choice of generators x, y), and so $(\pi_p)_*(\mu^{*n})$ is supported on matrices in $\mathrm{SL}_2(\mathbf{F}_p)$ whose coefficients also have size $O(\exp(O(n)))$. If n is less than a sufficiently small multiple of $\log p$, these coefficients are then less than $p^{1/10}$ (say). As such, if $\tilde{a}, \tilde{b}, \tilde{c}, \tilde{d} \in \mathrm{SL}_2(\mathbf{Z})$ lie in the support of μ^{*n} and their projections $a = \pi_p(\tilde{a}), \ldots, d = \pi_p(\tilde{d})$ obey the word law (6.4) in $\mathrm{SL}_2(\mathbf{F}_p)$, then the original matrices $\tilde{a}, \tilde{b}, \tilde{c}, \tilde{d}$ obey[1] the word law (6.4) in $\mathrm{SL}_2(\mathbf{Z})$.

To summarise, if we let $E_{n,p,B}$ be the set of all elements of $\pi_p^{-1}(B)$ that lie in the support of μ^{*n}, then (6.4) holds for all $a, b, c, d \in E_{n,p,B}$. This severely limits the size of $E_{n,p,B}$ to only be of polynomial size, rather than exponential size:

Proposition 6.1.3. *Let E be a subset of the support of μ^{*n} (thus, E consists of words in x, y, x^{-1}, y^{-1} of length n) such that the law (6.4) holds for all $a, b, c, d \in E$. Then $|E| \ll n^2$.*

The proof of this proposition is laid out in the exercise below.

Exercise 6.1.3. Let Γ be a free group generated by two generators x, y. Let B be the set of all words of length at most n in x, y, x^{-1}, y^{-1}.

(i) Show that if $a, b \in \Gamma$ commute, then a, b lie in the same cyclic group, thus $a = c^i, b = c^j$ for some $c \in \Gamma$ and $i, j \in \mathbf{Z}$.

(ii) Show that if $a \in \Gamma$, there are at most $O(n)$ elements of B that commute with a.

(iii) Show that if $a, c \in \Gamma$, there are at most $O(n)$ elements b of B with $[a, b] = c$.

(iv) Prove Proposition 6.1.3.

Now we can conclude the proof of Theorem 6.0.4:

Exercise 6.1.4. Let Γ be a free group generated by two generators x, y.

(i) Show that $\|\mu^{*n}\|_{\ell^\infty(\Gamma)} \ll c^n$ for some absolute constant $0 < c < 1$. (For much more precise information on μ^{*n}, see [**Ke1959**].)

(ii) Conclude the proof of Theorem 6.0.4.

6.2. Random generators expand

We now prove Theorem 6.0.7. Let $\mathbf{F}_2$ be the free group on two formal generators a, b, and let $\mu := \frac{1}{4}(\delta_a + \delta_b + \delta_{a^{-1}} + \delta_{b^{-1}})$ be the generator of the random walk. For any word $w \in \mathbf{F}_2$ and any x, y in a group G, let $w(x, y) \in G$ be the element of G formed by substituting x, y for a, b respectively in

---

[1] This lifting of identities from the characteristic p setting of $\mathrm{SL}_2(\mathbf{F}_p)$ to the characteristic 0 setting of $\mathrm{SL}_2(\mathbf{Z})$ is a simple example of the "Lefschetz principle".

6.2. Random generators expand

the word w; thus w can be viewed as a map $w\colon G \times G \to G$ for any group G. Observe that if w is drawn randomly using the distribution μ^{*n}, and $x,y \in \mathrm{SL}_2(\mathbf{F}_p)$, then $w(x,y)$ is distributed according to the law $\tilde{\mu}^{*n}$, where $\tilde{\mu} := \frac{1}{4}(\delta_x + \delta_y + \delta_{x^{-1}} + \delta_{y^{-1}})$. Applying Corollary 6.0.3, it suffices to show that whenever p is a large prime and x,y are chosen uniformly and independently at random from $\mathrm{SL}_2(\mathbf{F}_p)$, that with probability $1 - o_{p\to\infty}(1)$, one has

$$\text{(6.5)} \qquad \sup_B \mathbf{P}_w(w(x,y) \in B) \leq p^{-\kappa}$$

for some absolute constant κ, where B ranges over all Borel subgroups of $\mathrm{SL}_2(\overline{\mathbf{F}_p})$ and w is drawn from the law μ^{*n} for some even natural number $n = O(\log p)$.

Let B_n denote the words in $\mathbf{F}_2$ of length at most n. We may use the law (6.4) to obtain good bound on the supremum in (6.5) assuming a certain nondegeneracy property of the word evaluations $w(x,y)$:

Exercise 6.2.1. Let n be a natural number, and suppose that $x,y \in \mathrm{SL}_2(\mathbf{F}_p)$ is such that $w(x,y) \neq 1$ for $w \in B_{100n}\setminus\{1\}$. Show that

$$\sup_B \mathbf{P}_w(w(x,y) \in B) \ll \exp(-cn)$$

for some absolute constant $c > 0$, where w is drawn from the law μ^{*n}. (*Hint:* Use (6.4) and the hypothesis to lift the problem up to $\mathbf{F}_2$, at which point one can use Proposition 6.1.3 and Exercise 6.1.4.)

In view of this exercise, it suffices to show that with probability $1 - o_{p\to\infty}(1)$, one has $w(x,y) \neq 1$ for all $w \in B_{100n}\setminus\{1\}$ for some n comparable to a small multiple of $\log p$. As B_{100n} has $\exp(O(n))$ elements, it thus suffices by the union bound to show that

$$\text{(6.6)} \qquad \mathbf{P}_{x,y}(w(x,y) = 1) \leq p^{-\gamma}$$

for some absolute constant $\gamma > 0$, and any $w \in \mathbf{F}_2\setminus\{1\}$ of length less than $c \log p$ for some sufficiently small absolute constant $c > 0$.

Let us now fix a nonidentity word w of length $|w|$ less than $c \log p$, and consider w as a function from $\mathrm{SL}_2(k) \times \mathrm{SL}_2(k)$ to $\mathrm{SL}_2(k)$ for an arbitrary field k. We can identify $\mathrm{SL}_2(k)$ with the set $\{(a,b,c,d) \in k^4 : ad - bc = 1\}$. A routine induction then shows that the expression $w((a,b,c,d),(a',b',c',d'))$ is then a polynomial in the eight variables a,b,c,d,a',b',c',d' of degree $O(|w|)$ and coefficients which are integers of size $O(\exp(O(|w|)))$. Let us then make the additional restriction to the case $a, a' \neq 0$, in which case we can write $d = \frac{bc+1}{a}$ and $d' = \frac{b'c'+1}{a'}$. Then $w((a,b,c,d),(a',b',c',d'))$ is now a rational function of a,b,c,a',b',c' whose numerator is a polynomial of degree $O(|w|)$ and coefficients of size $O(\exp(O(|w|)))$, and the denominator is a monomial of a, a' of degree $O(|w|)$.

We then specialise this rational function to the field $k = F_p$. It is conceivable that when one does so, the rational function collapses to the constant polynomial $(1, 0, 0, 1)$, thus $w((a, b, c, d), (a', b', c', d')) = 1$ for all $(a, b, c, d), (a', b', c', d') \in \mathrm{SL}_2(\mathbf{F}_p)$ with $a, a' \neq 0$. (For instance, this would be the case if $w(x, y) = x^{|\mathrm{SL}_2(\mathbf{F}_p)|}$, by *Lagrange's theorem*, if it were not for the fact that $|w|$ is far too large here.) But suppose that this rational function does not collapse to the constant rational function. Applying the Schwarz-Zippel lemma (Exercise 5.4.3), we then see that the set of pairs $(a, b, c, d), (a', b', c', d') \in \mathrm{SL}_2(\mathbf{F}_p)$ with $a, a' \neq 0$ and $w((a, b, c, d), (a', b', c', d')) = 1$ is at most $O(|w|p^5)$; adding in the $a = 0$ and $a' = 0$ cases, one still obtains a bound of $O(|w|p^5)$, which is acceptable since $|\mathrm{SL}_2(\mathbf{F}_p)|^2 \sim p^6$ and $|w| = O(\log p)$. Thus, the only remaining case to consider is when the rational function $w((a, b, c, d), (a', b', c', d'))$ is identically 1 on $\mathrm{SL}_2(\mathbf{F}_p)$ with $a, a' \neq 0$.

Now we perform another "Lefschetz principle" maneuvre to change the underlying field. Recall that the denominator of the rational function

$$w((a, b, c, d), (a', b', c', d'))$$

is monomial in a, a', and the numerator has coefficients of size $O(\exp(O(|w|)))$. If $|w|$ is less than $c \log p$ for a sufficiently small p, we conclude in particular (for p large enough) that the coefficients all have magnitude less than p. As such, the only way that this function can be identically 1 on $\mathrm{SL}_2(\mathbf{F}_p)$ is if it is identically 1 on $\mathrm{SL}_2(k)$ for all k with $a, a' \neq 0$, and hence for $a = 0$ or $a' = 0$ also by taking Zariski closures.

On the other hand, we know that for some choices of k, e.g., $k = \mathbf{R}$, $\mathrm{SL}_2(k)$ contains a copy Γ of the free group on two generators (see, e.g., Exercise 2.3.1). As such, it is not possible for any nonidentity word w to be identically trivial on $\mathrm{SL}_2(k) \times \mathrm{SL}_2(k)$. Thus this case cannot actually occur, completing the proof of (6.6) and hence of Theorem 6.0.7.

Remark 6.2.1. We see from the above argument that the existence of subgroups Γ of an algebraic group with good "independence" properties—such as that of generating a free group—can be useful in studying the expansion properties of that algebraic group, even if the field of interest in the latter is distinct from that of the former. For more complicated algebraic groups than SL_2, in which laws such as (6.4) are not always available, it turns out to be useful to place further properties on the subgroup Γ, for instance, by requiring that all nonabelian subgroups of that group be Zariski dense (a property which has been called *strong density*), as this turns out to be useful for preventing random walks from concentrating in proper algebraic subgroups. See [**BrGrTa2012**] for constructions of strongly dense free subgroups of algebraic groups and further discussion.

Chapter 7

Sieving and expanders

We now discuss how (a generalisation of) the expansion results obtained in the preceding chapters can be used for some number-theoretic applications, and in particular to locate almost primes inside orbits of thin groups, following the work of Bourgain, Gamburd, and Sarnak [**BoGaSa2010**]. We will not attempt here to obtain the sharpest or most general results in this direction, but instead focus on the simplest instances of these results which are still illustrative of the ideas involved.

One of the basic general problems in analytic number theory is to locate tuples of primes of a certain form; for instance, the famous (and still unsolved) *twin prime conjecture* asserts that there are infinitely many pairs (n_1, n_2) in the line $\{(n_1, n_2) \in \mathbf{Z}^2 : n_2 - n_1 = 2\}$ in which both entries are prime. In a similar spirit, one of the *Landau conjectures* (also still unsolved) asserts that there are infinitely many primes in the set $\{n^2 + 1 : n \in \mathbf{Z}\}$. The *Mersenne prime conjecture* (also unsolved) asserts that there are infinitely many primes in the set $\{2^n - 1 : n \in \mathbf{Z}\}$, and so forth.

More generally, given some explicit subset V in $\mathbf{R}^d$ (or $\mathbf{C}^d$, if one wishes), such as an algebraic variety, one can ask the question of whether there are infinitely many integer lattice points $(n_1, \ldots, n_d)$ in $V' := V \cap \mathbf{Z}^d$ in which all the coefficients $n_1, \ldots, n_d$ are simultaneously prime; let us refer to such points as *prime points*.

At this level of generality, this problem is extremely difficult. Indeed, even the much simpler problem of deciding whether the set V' is nonempty (let alone containing prime points) when V is a hypersurface $\{x \in \mathbf{R}^d : P(x) = 0\}$ cut out by a polynomial P is essentially *Hilbert's tenth problem*, which is known to be undecidable in general by *Matiyasevich's theorem* (see, e.g., [**Ma1993**]). So one needs to restrict attention to a more special class of sets V, in which the question of finding integer points is not so difficult. One model case is to consider *orbits* $V' = V = \Gamma b$, where $b \in \mathbf{Z}^d$ is a fixed lattice vector and Γ is some discrete group that acts on $\mathbf{Z}^d$ somehow (e.g., Γ might be embedded as a subgroup of the special linear group $\mathrm{SL}_d(\mathbf{Z})$, or on the affine group $\mathrm{SL}_d(\mathbf{Z}) \ltimes \mathbf{Z}^d$). In such a situation it is then quite easy to show that V is large; for instance, V will be infinite precisely when the stabiliser of b in Γ has infinite index in Γ.

Even in this simpler setting, the question of determining whether an orbit $V = \Gamma b$ contains infinitely prime points is still extremely difficult; indeed, the three examples given above of the twin prime conjecture, Landau conjecture, and Mersenne prime conjecture are essentially of this form (possibly after some slight modification of the underlying ring $\mathbf{Z}$, see [**BoGaSa2010**] for details), and are all unsolved (and generally considered well out of reach of current technology). Indeed, the list of nontrivial orbits $V = \Gamma b$ which are known to contain infinitely many prime points is quite slim; *Euclid's theorem* on the infinitude of primes handles the case $V = \mathbf{Z}$, *Dirichlet's theorem* handles infinite arithmetic progressions $V = a\mathbf{Z} + r$, and a somewhat complicated result of Green, Tao, and Ziegler [**GrTa2010, GrTa2012, GrTaZi2012**] handles "nondegenerate" affine lattices in $\mathbf{Z}^d$ of rank two or more (such as the lattice of length d arithmetic progressions), but there are a few other positive results known that are not based on the above cases (though we will note the remarkable theorem of Friedlander and Iwaniec [**FrIw1998**] that there are infinitely many primes of the form $a^2 + b^4$, and the related result of Heath-Brown [**He2001**] that there are infinitely many primes of the form $a^3 + 2b^3$, as being in a kindred spirit to the above results, though they are not explicitly associated to an orbit of a reasonable action as far as I know).

On the other hand, much more is known if one is willing to replace the primes by the larger set of *almost primes* — integers with a small number of prime factors (counting multiplicity). Specifically, for any $r \geq 1$, let us call an *r-almost prime* an integer which is the product of at most r primes, and possibly by the unit -1 as well. Many of the above sorts of questions which are open for primes, are known for r-almost primes for r sufficiently large. For instance, with regards to the twin prime conjecture, it is a result of Chen [**Ch1973**] that there are infinitely many pairs $p, p+2$ where p is a prime and $p+2$ is a 2-almost prime; in a similar vein, it is a result of Iwaniec [**Iw1978**] that there are infinitely many 2-almost primes of the form $n^2 + 1$. On the other hand, it is still open for any fixed r whether there are infinitely[1] many Mersenne numbers $2^n - 1$ which are r-almost primes.

The main tool that allows one to count almost primes in orbits is *sieve theory*. The reason for this lies in the simple observation that in order to ensure that an integer n of magnitude at most x is an r-almost prime, it suffices to guarantee that n is not divisible by any prime less than $x^{1/(r+1)}$. Thus, to create r-almost primes, one can start with the integers up to some large threshold x and remove (or "sieve out") all the integers that are multiples of any prime p less than $x^{1/(r+1)}$. The difficulty is then to ensure that a sufficiently nontrivial quantity of integers remain after this process, for the purposes of finding points in the given set V.

The most basic sieve of this form is the *sieve of Eratosthenes*, which when combined with the inclusion-exclusion principle gives the *Legendre sieve* (or *exact sieve*), which gives an exact formula for quantities such as the number $\pi(x, z)$ of natural numbers less than or equal to x that are not divisible by any prime less than or equal to a given threshold z. Unfortunately, when one tries to evaluate this formula, one encounters error terms which grow exponentially in z, rendering this sieve useful only for very small thresholds z (of logarithmic size in x). To improve the sieve level up to a small power of x such as $x^{1/(r+1)}$, one has to replace the exact sieve by *upper bound sieves* and *lower bound sieves* which only seek to obtain upper or lower bounds on quantities such as $\pi(x, z)$, but contain a polynomial number of terms rather than an exponential number. There are a variety of such sieves, with the two most common such sieves being *combinatorial sieves* (such as the *beta sieve*), based on various combinatorial truncations of the inclusion-exclusion formula, and the *Selberg upper bound sieve*, based on upper bounds that are the square of a divisor sum. (There is also the *large sieve*, which is somewhat different in nature and based on L^2 almost orthogonality considerations, rather than on any actual sieving, to obtain upper bounds.) We will primarily work with a specific sieve in this chapter, namely the beta sieve, and we will not attempt to optimise all the parameters of this sieve (which ultimately means that the almost primality parameter r in our results will be somewhat large). For a more detailed study of sieve theory, see [**HaRi1974**], [**IwKo2004**], [**FrIw2010**].

Very roughly speaking, the end result of sieve theory is that excepting some degenerate and "exponentially thin" settings (such as those associated with the Mersenne primes), all the orbits which are expected to have a large number of primes, can be proven to at least have a large number of r-almost

---

[1] For the superficially similar situation with the numbers $2^n + 1$, it is in fact believed (but again unproven) that there are only finitely many r-almost primes for any fixed r (the *Fermat prime conjecture*).

primes for some[2] finite r. One formulation of this principle was established by Bourgain, Gamburd, and Sarnak [**BoGaSa2010**]:

Theorem 7.0.1 (Bourgain-Gamburd-Sarnak). *Let Γ be a subgroup of $\mathrm{SL}_2(\mathbf{Z})$ which is not virtually solvable. Let $f\colon \mathbf{Z}^4 \to \mathbf{Z}$ be a polynomial with integer coefficients obeying the following* primitivity *condition: for any positive integer q, there exists $A \in \Gamma \subset \mathbf{Z}^4$ such that $f(A)$ is coprime to q. Then there exists an $r \geq 1$ such that there are infinitely many $A \in \Gamma$ with $f(A)$ nonzero and r-almost prime.*

This is not the strongest version of the Bourgain-Gamburd-Sarnak theorem, but it captures the general flavour of their results. Note that the theorem immediately implies an analogous result for orbits $\Gamma b \subset \mathbf{Z}^2$, in which f is now a polynomial from $\mathbf{Z}^2$ to $\mathbf{Z}$, and one uses $f(Ab)$ instead of $f(A)$. It is in fact conjectured that one can set $r = 1$ here, but this is well beyond current technology. For the purpose of reaching $r = 1$, it is very natural to impose the primitivity condition, but as long as one is content with larger values of r, it is possible to relax the primitivity condition somewhat; see [**BoGaSa2010**] for more discussion.

By specialising to the polynomial $f : \begin{pmatrix} a & b \\ c & d \end{pmatrix} \to abcd$, we conclude as a corollary that as long as Γ is primitive in the sense that it contains matrices with all coefficients coprime to q for any given q, then Γ contains infinitely many matrices whose elements are all r-almost primes for some r depending only on Γ. For further applications of these sorts of results, for instance to *Appolonian packings*; see [**BoGaSa2010**].

It turns out that to prove Theorem 7.0.1, the Cayley expansion results in $\mathrm{SL}_2(\mathbf{F}_p)$ from the previous chapter are not quite enough; one needs a more general Cayley expansion result in $\mathrm{SL}_2(\mathbf{Z}/q\mathbf{Z})$ where q is square-free but not necessarily prime. The proof of this expansion result uses the same basic methods as in the $\mathrm{SL}_2(\mathbf{F}_p)$ case, but is significantly more complicated technically, and we will only discuss it briefly here. As such, we do not give a complete proof of Theorem 7.0.1, but hopefully the portion of the argument presented here is still sufficient to give an impression of the ideas involved.

7.1. Combinatorial sieving

In this section we set up the combinatorial sieve needed to establish Theorem 7.0.1. To motivate this sieve, let us focus first on a much simpler model problem, namely the task of estimating the number $\pi(x,z)$ of natural numbers less than or equal to a given threshold x which are not divisible by any prime less than or equal to z. Note that for z between $\sqrt{x}$ and x, $\pi(x,z)$ is

---

[2]Unfortunately, there is a major obstruction, known as the *parity problem*, which prevents sieve theory from lowering r all the way to 1; see [**Ta2008**, §2.10] for more discussion.

7.1. Combinatorial sieving

simply the number of primes in the interval $(z, x]$; but for z less than $\sqrt{x}$, $\pi(x, z)$ also counts some almost primes in addition to genuine primes. This quantity can be studied quite precisely by a variety of tools, such as those coming from multiplicative number theory; see for instance [**GrSo2004**] for some of the most precise results currently known in this direction.

The quantity $\pi(x, z)$ is easiest to estimate when z is small. For instance, $\pi(x, 1)$ is simply the number of natural numbers less than x, and so

$$\pi(x, 1) = x + O(1).$$

Similarly, $\pi(x, 2)$ is the number of odd numbers less than x, and so

$$\pi(x, 2) = \frac{1}{2}x + O(1).$$

Carrying this further, $\pi(x, 3)$ is the number of numbers less than x that are coprime to 6, and so

$$\pi(x, 3) = \frac{1}{3}x + O(1)$$

(but note that the implied constant in the $O(1)$ error is getting increasingly large). Continuing this analysis, it is not hard to see that

$$\pi(x, z) = (\prod_{p \leq z}(1 - \frac{1}{p}))x + O_z(1)$$

for any fixed z; note from *Mertens' theorem* that

(7.1) $$\prod_{p \leq z}(1 - \frac{1}{p}) = \frac{e^\gamma + o(1)}{\log z}$$

leading to the heuristic approximation

$$\pi(x, z) \approx e^\gamma \frac{x}{\log z}$$

where $\gamma = 0.577\ldots$ is the *Euler-Mascheroni constant*. Note though that this heuristic should be treated with caution when z is large; for instance, from the prime number theorem we see that we have the conflicting asymptotic

$$\pi(x, z) = (1 + o(1))\frac{x}{\log x}$$

when $\sqrt{x} \leq z \leq o(x)$. This is already a strong indication that one needs to pay careful attention to the error terms in this analysis. (Indeed, many false "proofs" of conjectures in analytic number theory, such as the twin prime conjecture, have been based on a cavalier attitude to such error terms, and their asymptotic behaviour under various limiting regimes.)

Let us thus work more carefully to control the error term $O_z(1)$. Write $P(z) := \prod_{p \leq z} p$ for the product of all the primes less than or equal to z (this

quantity is also known as the *primorial* of z). Then we can write
$$\pi(x,z) = \sum_{n \leq x} 1_{(n,P(z))=1}$$
where the sum ranges over natural numbers n less than x, and $(n, P(z))$ is the greatest common divisor of n and $P(z)$. The function $1_{(n,P(z))=1}$ is periodic of period $P(z)$, and is equal to 1 on $(\prod_{p \leq z}(1-\frac{1}{p}))P(z)$ of the residue classes modulo $P(z)$, which leads to the crude bound

(7.2) $$\pi(x,z) = \left(\prod_{p \leq z}(1-\frac{1}{p})\right) x + O(P(z)).$$

However, this error term is too large for most applications: from the *prime number theorem*, we see that $P(z) = \exp((1+o(1))z)$, so the error term grows exponentially in z. In particular, this estimate is only nontrivial in the regime $z = O(\log x)$.

One can do a little better than this by using the *inclusion-exclusion principle*, which in this context is also known as the *Legendre sieve*. Consider, for instance, $\pi(x, 3)$, which counts the number of natural numbers $n \leq x$ coprime to $P(3) = 2 \times 3$. We can compute this quantity by first counting all numbers less than x, then subtracting those numbers divisible by 2 and by 3, and then adding back those numbers divisible by both 2 and 3. A convenient way to describe this procedure in general is to introduce the *Möbius function* $\mu(n)$, defined to equal $(-1)^k$ when n is the product of k distinct primes for some $k \geq 0$. The key point is that

(7.3) $$1_{n=1} = \sum_{d|n} \mu(d)$$

for any natural number n, where d ranges over the divisors of n; indeed, this identity can be viewed as an alternate way to define the Möbius function. In particular, $1_{(n,P(z))=1} = \sum_{d|P(z)} \mu(d) 1_{d|n}$, leading to the *Legendre identity*
$$\pi(x,z) = \sum_{d|P(z)} \mu(d) \sum_{n \leq x; d|n} 1.$$
The inner sum can be easily estimated as

(7.4) $$\sum_{n \leq x; d|n} 1 = \frac{x}{d} + O(1);$$

since $P(z)$ has $2^{\pi(z)}$ distinct factors, where $\pi(z)$ is the number of primes less than or equal to z, we conclude that
$$\pi(x,z) = \sum_{d|P(z)} \mu(d)\frac{x}{d} + O(2^{\pi(z)}).$$

7.1. Combinatorial sieving

The main term here can be factorised as

(7.5) $$\sum_{d|P(z)} \mu(d)\frac{x}{d} = x \prod_{p \leq z}(1 - \frac{1}{p})$$

leading to the following slight improvement

$$\pi(x, z) = \left(\prod_{p \leq z}(1 - \frac{1}{p})\right) x + O(2^{\pi(z)})$$

to (7.2). Note from the prime number theorem that

$$2^{\pi(z)} = O(\exp(O(z/\log z))),$$

so this error term is asymptotically better than the one in (7.2); the bound here is now nontrivial in the slightly larger regime $z = O(\log x \log \log x)$. But this is still not good enough for the purposes of counting almost primes, which would require z as large as a power of x.

To do better, we will replace the exact identity (7.3) by combinatorial truncations

(7.6) $$\sum_{d|n: d \in \mathcal{D}_-} \mu(d) \leq 1_{n=1} \leq \sum_{d|n: d \in \mathcal{D}_+} \mu(d)$$

of that identity, where n divides $P(z)$ and $\mathcal{D}_-, \mathcal{D}_+$ are sets to be specified later, leading to the upper bound sieve

(7.7) $$\pi(x, z) \leq \sum_{d|P(z); d \in \mathcal{D}_+} \mu(d)\frac{x}{d} + O(|\mathcal{D}_+|)$$

and the lower bound sieve

(7.8) $$\pi(x, z) \geq \sum_{d|P(z); d \in \mathcal{D}_-} \mu(d)\frac{x}{d} + O(|\mathcal{D}_-|).$$

The key point will be that $\mathcal{D}_+$ and $\mathcal{D}_-$ can be chosen to be only polynomially large in z, rather than exponentially large, without causing too much damage to the main terms $\sum_{d|P(z); d \in \mathcal{D}_\pm} \mu(d)\frac{x}{d}$, which lead to upper and lower bounds on $\pi(x, z)$ that remain nontrivial for moderately large values of z (e.g., $z = x^{1/(r+1)}$ for some fixed r).

We now turn to the task of locating reasonably small sets $\mathcal{D}_+, \mathcal{D}_-$ obeying (7.6). We begin with (7.3), which we rewrite as

(7.9) $$1_{n=1} = \sum_{d|P(z)} \mu(d) 1_{d|n}$$

for n a divisor of $P(z)$. One can view the divisors of $P(z)$ as a $\pi(z)$-dimensional combinatorial cube, with the right-hand side in (7.9) being a sum over that cube; the idea is then to hack off various subcubes of that cube in a way that only serves to increase (for the upper bound sieve) or decrease (for the lower bound sieve) that sum, until only a relatively small portion of the cube remains.

Now we turn to the details. Our starting point will be the identity

(7.10) $$\sum_{d|P(z):d=p_1\ldots p_k d', d'|P(p_k)} \mu(d) 1_{d|n} = (-1)^k 1_{n=p_1\ldots p_k}$$

whenever $z \geq p_1 > p_2 > \ldots > p_k$ are primes, which follows easily from applying (7.3) to $n/(p_1 \ldots p_k)$ when $p_1, \ldots, p_k$ divide n. One can view the left-hand side of (7.10) as a subsum of the sum in (7.9), and (7.10) implies that this subsum is nonnegative when k is even and nonpositive when k is odd. In particular, we see that (7.6) will hold when $\mathcal{D}_+$ is formed from the "cube" $\{d : d|P(z)\}$ by removing some disjoint "subcubes" of the form $\{d = p_1 \ldots p_k d' : d'|P(p_k)\}$ for $z \geq p_1 > \ldots > p_k$ and k odd, and similarly for $\mathcal{D}_-$ but with k now required to be even instead of odd.

Observe that the subcube $\{d = p_1 \ldots p_k d' : d'|P(p_k)\}$ consists precisely of those divisors d of $P(z)$ whose top k prime factors are $p_1, \ldots, p_k$. We now have the following general inequality:

Lemma 7.1.1 (Combinatorial sieve). *Let $z > 0$. For each natural number k, let $A_k(p_1, \ldots, p_k)$ be a predicate pertaining to k decreasing primes $z \geq p_1 > \ldots > p_k$ (thus $A_k(p_1, \ldots, p_k)$ is either true or false for each choice of $p_1, \ldots, p_k$). Let $\mathcal{D}_+$ be the set of all natural numbers $n|P(z)$ which, when factored as $n = p_1 \ldots p_r$ for $z \geq p_1 > \ldots > p_r$, is such that $A_k(p_1, \ldots, p_k)$ holds for all odd $1 \leq k \leq r$. Similarly, define $\mathcal{D}_-$ by requiring k to be even instead of odd. Then (7.6) holds for all $n|P(z)$.*

Proof. $\mathcal{D}_+$ is formed from $\{d : d|P(z)\}$ by removing those subcubes of the form $\{d = p_1 \ldots p_k d' : d'|P(p_k)\}$ for $z \geq p_1 > \ldots > p_k$, k odd, and such that $A_{k'}(p_1, \ldots, p_{k'})$ holds for all odd $1 \leq k' < k$ but fails for $k' = k$. These subcubes are all disjoint, and so the claim for $\mathcal{D}_+$ follows from the preceding discussion. Similarly for $\mathcal{D}_-$. □

This gives us the upper and lower bounds (7.7), (7.8) for $\pi(x, z)$. To make these bounds useful, we need to choose $\mathcal{D}_\pm$ so that the partial sums $\sum_{d|P(z); d \in \mathcal{D}_\pm} \mu(d) \frac{x}{d}$ are close to

$$\sum_{d|P(z)} \mu(d) \frac{x}{d} = x \prod_{p \leq z} (1 - \frac{1}{p}).$$

To do this, one must select the predicates $A_k(p_1, \ldots, p_k)$ carefully. The best choices for these predicates are not immediately obvious; but after much trial and error, it was discovered that one fairly efficient choice is to let

7.1. Combinatorial sieving

$A_k(p_1, \ldots, p_k)$ be the predicate

$$p_1 \ldots p_{k-1} p_k^{\beta+1} < y$$

for some moderately large parameter $\beta \geq 2$ (we will eventually take $\beta := 10$) and some parameter $y := z^s$ for some $s > \beta$ to be optimised in later (we will eventually take it to be almost as large as x). The use of this choice is referred to as the *beta sieve*.

Let us now estimate the errors

(7.11)
$$\left| \sum_{d|P(z)} \mu(d) \frac{x}{d} - \sum_{d|P(z): d \in \mathcal{D}_\pm} \mu(d) \frac{x}{d} \right|.$$

For the sake of argument let us work with $\mathcal{D}_-$, as the $\mathcal{D}_+$ case is almost identical. By the triangle inequality, we can bound this error by

$$\sum_{k \text{ even}} \sum_{*} \left| \sum_{d = p_1 \ldots p_k d' : d' | P(p_k)} \mu(d) \frac{x}{d} \right|$$

where k ranges over positive even integers, and $\sum_*$ denotes a sum over primes $z \geq p_1 > \ldots > p_k$ ranges over primes such that

(7.12)
$$p_1 \ldots p_{k'-1} p_{k'}^{\beta+1} < y$$

for all even $k' < k$, but

(7.13)
$$p_1 \ldots p_{k-1} p_k^{\beta+1} \geq y.$$

Since $p_1, \ldots, p_k \leq z$ and $y = z^s$, this in particular gives the bound

$$k \geq s - \beta.$$

From (7.12) we have

$$p_1 \ldots p_{k'-1} p_{k'}^\beta < y$$

for all $1 \leq k' < k$ (not necessarily even); note that the case $k' = 1$ follows from the hypothesis $y > z^\beta$. We can rewrite this inequality as

$$\frac{y}{p_1 \ldots p_{k'}} > \left(\frac{y}{p_1 \ldots p_{k'-1}} \right)^{\frac{\beta-1}{\beta}}$$

and hence by induction

$$\frac{y}{p_1 \ldots p_{k'}} > y^{(\frac{\beta-1}{\beta})^{k'-1}}$$

for all $1 \leq k' < k$. From (7.13) we then have

$$p_k > y^{\frac{1}{\beta+1}(\frac{\beta-1}{\beta})^{k-1}} > y^{\frac{1}{\beta}(\frac{\beta-1}{\beta})^k} > z^{(\frac{\beta-1}{\beta})^k}.$$

We conclude that the error (7.11) is bounded by

$$\sum_{k=1}^{\infty} \sum_{z \geq p_1 > \ldots > p_k > z^{(\frac{\beta-1}{\beta})^k}} \left| \sum_{d = p_1 \ldots p_k d' : d' | P(p_k)} \mu(d) \frac{x}{d} \right|$$

in the $\mathcal{D}_+$; a similar argument also gives this bound in the $\mathcal{D}_-$ case. The inner sum can be computed as

$$\left| \sum_{d = p_1 \ldots p_k d' : d' | P(p_k)} \mu(d) \frac{x}{d} \right| = \frac{x}{p_1 \ldots p_k} \prod_{p < p_k} (1 - \frac{1}{p})$$

and thus by Mertens' theorem (7.1) and the bound $p_k > z^{(\frac{\beta-1}{\beta})^k}$ we have

$$\left| \sum_{d = p_1 \ldots p_k d' : d' | P(p_k)} \mu(d) \frac{x}{d} \right| \ll (\frac{\beta}{\beta - 1})^k \frac{x}{p_1 \ldots p_k \log z}.$$

We have thus bounded (7.11) by

$$\ll \frac{x}{\log z} \sum_{k \geq s - \beta} \left(\frac{\beta}{\beta - 1} \right)^k \sum_{z \geq p_1 > \ldots > p_k > z^{(\frac{\beta-1}{\beta})^k}} \frac{1}{p_1 \ldots p_k}.$$

The inner sum can be bounded by

$$\frac{1}{k!} \left(\sum_{z \geq p > z^{(\frac{\beta-1}{\beta})^k}} \frac{1}{p} \right)^k.$$

By another of Mertens' theorems (or by taking logarithms of (7.1)) one has

$$\sum_{z \geq p > z^{(\frac{\beta-1}{\beta})^k}} \frac{1}{p} \leq k \log \frac{\beta}{\beta - 1} + O(1)$$

and so (7.11) is bounded by

$$\ll \frac{x}{\log z} \sum_{k \geq s - \beta} \frac{1}{k!} \left(k \frac{\beta}{\beta - 1} \log \frac{\beta}{\beta - 1} + O(1) \right)^k.$$

Using the crude bound $k! \geq \frac{k^k}{e^k}$ (as can be seen by considering the kth term in the Taylor expansion of e^k) we conclude the bound

$$\ll \frac{x}{\log z} \sum_{k \geq s - \beta} \left(e \frac{\beta}{\beta - 1} \log \frac{\beta}{\beta - 1} + O(\frac{1}{k}) \right)^k.$$

7.1. Combinatorial sieving

If β is large enough ($\beta = 10$ will suffice), then the expression $e\frac{\beta}{\beta-1} \log \frac{\beta}{\beta-1}$ is less than $1/e$; since $(1 + O(1/k))^k = O(1)$, this leads to the bound

$$\ll \frac{x}{\log z} \sum_{k \geq s-\beta} e^{-k}$$

which after summing the geometric series becomes

$$\ll e^{-s} \frac{x}{\log z}$$

(allowing implied constants to depend on β). From this bound on (7.11) and (7.5), (7.1) we have

$$\sum_{d \mid P(z): d \in \mathcal{D}_\pm} \mu(d) \frac{x}{d} = \frac{x}{\log z}(e^\gamma + O(e^{-s}) + o(1)).$$

Finally, if $d = p_1 \ldots p_k$ is an element of $\mathcal{D}_\pm$, then by (7.12) and the hypothesis $\beta \geq 2$ we have

$$d = p_1 \ldots p_k \leq y$$

and so we have the crude upper bounds $|\mathcal{D}_\pm| \leq y$. From (7.7), (7.8) and recalling that $y = z^s$, we thus have

$$\pi(x, z) = \frac{x}{\log z}(e^\gamma + O(e^{-s}) + o(1)) + O(z^s).$$

If $x > z^{11}$, we may optimise in s by setting $s := \frac{\log x}{\log z} - 1$ (in order to make the final error term much less than x), leading to the bound

$$\pi(x, z) = \frac{x}{\log z}(e^\gamma + O(e^{-\log x / \log z}) + o(1)).$$

In particular, we have

(7.14) $$\frac{x}{\log z} \ll \pi(x, z) \ll \frac{x}{\log z}$$

whenever $2 \leq z \leq x^\varepsilon$ for some sufficiently small absolute constant $\varepsilon > 0$.

Remark 7.1.2. The bound (7.14) implies, among other things, that there exists an absolute constant r such that the number of r-almost primes less than x is $\gg x/\log x$, which is a very weak version of the prime number theorem. Note though that the upper bound in (7.14) does *not* directly imply a corresponding upper bound on this count of r-almost primes, because r-almost primes are allowed to have prime factors that are less than x. Indeed, a routine computation using Mertens' theorem shows that for any fixed r, the number of r-almost primes less than x is comparable to $\frac{x}{\log x}(\log \log x)^{r-1}$.

We can generalise the above argument as follows:

Exercise 7.1.1 (Beta sieve). Let a_n be an absolutely convergent sequence of nonnegative reals for $n \geq 1$. Let $x > 1$, $\kappa \geq 1$, and $\varepsilon > 0$. Let $g\colon \mathbf{N} \to \mathbf{R}^+$ be a multiplicative function taking values between 0 and 1, with $g(p) < 1$ for all primes p. Assume the following axioms:

(i) (Control in arithmetic progressions) For any $d \leq x^\varepsilon$, one has
$$\sum_{n : d \mid n} a_n = g(d) x + O(x^{1-\varepsilon}).$$

(ii) (Mertens type theorem) For all $2 \leq z \leq x^\varepsilon$, one has

(7.15) $$\frac{1}{\log^\kappa z} \ll \prod_{p \leq z} (1 - g(p)) \ll \frac{1}{\log^\kappa z}.$$

Conclude that there is an $\varepsilon' > 0$ depending only on κ, ε, and the implied constants in the above axioms, such that
$$\frac{x}{\log^\kappa z} \ll \sum_{n : (n, P(z)) = 1} a_n \ll \frac{x}{\log^\kappa z}$$

whenever $2 \leq z \leq x^{\varepsilon'}$, and the implied constants may depend on κ, ε, and the implied constants in the above axioms. (Note that (7.14) corresponds to the case when $\kappa := 1$, $g(n) := 1/n$, and $a_n := 1_{n \leq x}$.)

Exercise 7.1.2. Suppose we have the notation and hypotheses of the preceding exercise, except that the estimate (7.15) is replaced by the weaker bound

(7.16) $$g(p) \leq \frac{\kappa}{p} + O\!\left(\frac{1}{p^2}\right)$$

for all sufficiently large p. (For small p, note that we still have the bound $g(p) < 1$.) Show that we still have the lower bound
$$\sum_{n : (n, P(z)) = 1} a_n \gg \frac{x}{\log^\kappa z}$$

whenever $2 \leq z \leq x^{\varepsilon'}$ for sufficiently small ε' (which may depend on g), where the implied constant is now allowed to depend on g. (*Hint:* The main trick here is to extract out a common factor of $\prod_{p \leq z}(1 - g(p))$ from the analysis first, and then use the bound (7.16) to upper bound quantities such as $\prod_{p_k \leq p \leq z}(1 - g(p))^{-1}$.)

One can weaken the axioms somewhat and still obtain nontrivial results from the beta sieve, but this somewhat crude version of the sieve will suffice for our purposes. Another, more abstract, formalisation of the above argument (involving a construction of sets $\mathcal{D}_\pm$ obeying (7.6) and a number of other desirable properties) is sometimes referred to as *the fundamental lemma of sieve theory*; see, e.g., [**FrIw2010**].

7.1. Combinatorial sieving

Exercise 7.1.3 (Twin almost primes). Let $\pi_2(x, z)$ be the number of integers n between 1 and x such that n and $n+2$ are both coprime to $P(z)$.

(i) Show that
$$\frac{x}{\log^2 z} \ll \pi_2(x, z) \ll \frac{x}{\log^2 z}$$
if $2 \leq z \leq x^\varepsilon$, and $\varepsilon > 0$ is a sufficiently small absolute constant.

(ii) Show that there exists an $r \geq 1$ such that there are infinitely many pairs $n, n+2$ which are both r-almost primes. (Indeed, the argument here allows one to take $r = 20$ without much effort, and by working considerably harder to optimise everything, one can lower r substantially, although the parity problem mentioned earlier prevents one from taking r below 2.)

(iii) Establish *Brun's theorem* that the sum of reciprocals of the twin primes is convergent.

Exercise 7.1.4 (Landau conjecture for almost primes). Let $\pi_*(x, z)$ be the number of integers n between 1 and x such that $n^2 + 1$ is coprime to $P(z)$.

(i) Show that
$$\frac{x}{\log z} \ll \pi_2(x, z) \ll \frac{x}{\log z}$$
if $2 \leq z \leq x^\varepsilon$, and $\varepsilon > 0$ is a sufficiently small absolute constant. (*Hint:* You will need the fact that -1 is a quadratic residue mod p if and only if $p \neq 3 \mod 4$, and Merten's theorem for arithmetic progressions, which among other things asserts that $\sum_{p \leq x : p = 1 \mod 4} \frac{1}{p} = \frac{1}{2} \log x + O(1)$.)

(ii) Show that there exists an $r \geq 1$ such that there are infinitely natural numbers n such that $n^2 + 1$ is an r-almost prime.

Exercise 7.1.5. Let $P \colon \mathbf{Z} \to \mathbf{Z}$ be a polynomial with integer coefficients and degree k. Assume that P is *primitive* in the sense that for each natural number q, there exists a natural number n such that $P(n)$ is coprime to q. Show that there exists an r depending only on P such that for all sufficiently large x, there are at least $\gg_P x/\log^k x$ natural numbers n less than x such that $P(n)$ is an r-almost prime.

In many cases (e.g., if P is irreducible) one can decrease the power of $\log x$ here (as in Exercise 7.1.4), by using tools such as *Landau's prime ideal theorem*; see [**Ta2013c**, §7.3] for some related discussion.

Remark 7.1.3. The combinatorial sieve is not the only type of sieve used in sieve theory. Another popular choice is the *Selberg upper bound sieve*, in which the starting point is not the combinatorial inequalities (7.6), but rather the variant

$$1_{n=1} \leq (\sum_{d|n} \lambda_d)^2$$

where the λ_d are arbitrary real parameters with $\lambda_1 := 1$, typically supported up to some level $d < y$. By optimising the choice of weights λ_d, the Selberg sieve can lead to upper bounds on quantities such as $\pi(x, z)$ which are competitive with the beta sieve (particularly when z is moderately large), although it is more difficult for this sieve to produce matching lower bounds. A somewhat different type of sieve is the *large sieve*, which does not upper bound or lower bound indicator functions such as $1_{n=1}$ directly, but rather controls the size of a function that avoids many residue classes by exploiting the L^2 properties of these residue classes, such as almost orthogonality phenomena or Fourier uncertainty principles. See [**FrIw2010**] for a much more thorough discussion and comparison of these sieves.

7.2. The strong approximation property

For any natural number q, let $\pi_q \colon \mathrm{SL}_2(\mathbf{Z}) \to \mathrm{SL}_2(\mathbf{Z}/q\mathbf{Z})$ be the obvious projection homomorphism. An easy application of *Bezout's theorem* (or the *Euclidean algorithm*) shows that this map is surjective. From the *Chinese remainder theorem*, we also have $\mathrm{SL}_2(\mathbf{Z}/q\mathbf{Z}) \equiv \mathrm{SL}_2(\mathbf{Z}/q_1\mathbf{Z}) \times \mathrm{SL}_2(\mathbf{Z}/q_2\mathbf{Z})$ whenever $q = q_1 q_2$ and q_1, q_2 are coprime.

To set up the sieve needed to establish Theorem 7.0.1, we need to understand the images $\pi_q(\Gamma)$ of a nonvirtually-solvable subgroup Γ of $\mathrm{SL}_2(\mathbf{Z})$. Clearly this is a subgroup of $\mathrm{SL}_2(\mathbf{Z}/q\mathbf{Z})$. Given that Γ is fairly "large" (in particular, such groups can be easily seen to be Zariski-dense in SL_2), we expect that in most cases $\pi_q(\Gamma)$ is in fact all of $\mathrm{SL}_2(\mathbf{Z}/q\mathbf{Z})$. This type of belief is formalised in general as the *strong approximation property*. We will not prove the most general instance of this property, but instead focus on the model case of $\mathrm{SL}_2(\mathbf{Z}/q\mathbf{Z})$ for q square-free, in which one can proceed by *ad hoc* elementary arguments. The general treatment of the strong approximation property was first achieved in [**MaVeWe1984**] using the classification of finite simple groups; a subsequent paper of Nori [**No1987**] gave an alternate treatment that avoided the use of this classification.

In Remark 6.0.5 it was already observed that $\pi_p(\Gamma) = \mathrm{SL}_2(\mathbf{Z}/p\mathbf{Z})$ for all sufficiently large primes p. (Indeed, Γ did not need to be free for this to hold; it was enough that Γ not be virtually solvable.) To extend from the prime case to the (square-free) composite case, we will need some basic group theory, and in particular, the theory of composition factors.

7.2. The strong approximation property

Define a *composition series* for a group G to be a finite sequence

$$\{1\} = H_0 \triangleleft H_1 \triangleleft \ldots \triangleleft H_n = G$$

of subgroups, where each H_i is a normal subgroup of H_{i+1}, and the quotients H_{i+1}/H_i are all simple[3]. The quotients H_{i+1}/H_i for $i = 1, \ldots, n-1$ are referred to as the *composition factors* of this series.

Exercise 7.2.1. Show that every finite group has at least one composition series.

A key fact about composition factors, known as the *Jordan-Holder theorem*, asserts that, up to permutation and isomorphism, they are independent of the choice of series:

Theorem 7.2.1 (Jordan-Holder theorem). *Let*

$$\{1\} = H_0 \triangleleft H_1 \triangleleft \ldots \triangleleft H_n = G$$

and

$$\{1\} = K_0 \triangleleft K_1 \triangleleft \ldots \triangleleft K_m = G$$

be two composition series of the same group G. Then there is a bijection $\sigma : \{0, \ldots, n-1\} \to \{0, \ldots, m-1\}$ such that for each $i = 0, \ldots, n-1$, H_{i+1}/H_i is isomorphic to $K_{\phi(i+1)}/K_{\phi(i)}$. (In particular, n and m must be equal.)

Proof. By symmetry we may assume that $n \leq m$. Fix $0 \leq i < n$. Let $\pi_i \colon H_{i+1} \to H_{i+1}/H_i$ be the quotient map, and consider the groups $A_j^{(i)} := \pi_i(H_{i+1} \cap K_j) \equiv (H_{i+1} \cap K_j)/(H_i \cap K_j)$ for $j = 0, \ldots, m$. These are an increasing family of subgroups of H_{i+1}/H_i, with $A_0^{(i)} = \{1\}$ and $A_m^{(i)} = H_{i+1}/H_i$. Since each K_j is a normal subgroup of K_{j+1}, we see that $A_j^{(i)}$ is a normal subgroup of $A_{j+1}^{(i)}$. As $A_m^{(i)}$ is simple, this implies that there is a unique element $\sigma(i)$ of $\{0, \ldots, m-1\}$ such that $A_j^{(i)}$ is trivial for $j \leq \sigma(i)$ and $A_j^{(i)}$ is equal to H_{i+1}/H_i for $j > \sigma(i)$.

Now we claim that σ is a bijection. Suppose this is not the case. Since $n \leq m$, there thus exists $j \in \{0, \ldots, m-1\}$ which is not in the range of σ. This implies that $A_j^{(i)} = A_{j+1}^{(i)}$ for all i. An induction on i then shows that $H_i \cap K_j = H_i \cap K_{j+1}$ for all i, and thus $K_j = K_{j+1}$, contradicting the assumption that K_{j+1}/K_j is simple.

Finally, fix $i_0 \in \{0, \ldots, n-1\}$, and let $j_0 := \sigma(i_0)$. Then we have $A_{j_0}^{(i)} = A_{j_0+1}^{(i)}$ for all $i \neq i_0$, while $A_{j_0}^{(i_0)} = \{1\}$ and $A_{j_0+1}^{(i_0)} \equiv H_{i_0+1}/H_{i_0}$. From this and induction we see[4] that $(H_i \cap K_{j_0+1})/(H_i \cap K_{j_0})$ is trivial for

---

[3] By convention, we do not consider the trivial group to be simple.
[4] Here we are basically relying on a special case of the *Zassenhaus lemma*.

$i \leq i_0$ but isomorphic to H_{i_0+1}/H_{i_0} for $i > i_0$. In particular, K_{j_0+1}/K_{j_0} is isomorphic to H_{i_0+1}/H_{i_0}, and the claim follows. □

In view of this theorem, we can assign to each finite group a set (or more precisely, *multiset*) of composition factors of simple groups, which are unique up to permutation and isomorphism. This is somewhat analogous to how the fundamental theorem of arithmetic assigns to each positive integer a multiset of prime numbers, which are unique up to permutation. (Indeed, the latter can be viewed as the special case of the former in the case of cyclic groups.)

Exercise 7.2.2. Show that for $p \geq 5$ a prime, the composition factors of $\mathrm{SL}_2(\mathbf{F}_p)$ are (up to isomorphism and permutation) the cyclic group $\mathbf{Z}/2\mathbf{Z}$ and the projective special linear group $P\mathrm{SL}_2(\mathbf{F}_p)$. What happens instead when $p = 2$ or $p = 3$?

Also, show that the only normal subgroup of $\mathrm{SL}_2(\mathbf{F}_p)$ (other than the trivial group and all of $\mathrm{SL}_2(\mathbf{F}_p)$) is the center $Z(\mathrm{SL}_2(\mathbf{F}_p)) \equiv \mathbf{Z}/2\mathbf{Z}$ of the group. Thus, we see (in contrast with the fundamental theorem of arithmetic) that one cannot permute the composition factors arbitrarily.

Exercise 7.2.3. Let N be a normal subgroup of a finite group G. Show that the set of composition factors of G is equal to (up to isomorphism, and counting multiplicity) the union of the set of composition factors of N, and the set of composition factors of G/N. In particular, the set of composition factors of N and of G/N are subsets of the set of composition factors of G (again up to isomorphism, and counting multiplicity). As another corollary, we see that the composition factors of a direct product $G \times H$ or semidirect product $G \ltimes H$ of two finite groups G, H is the union of the set of composition factors of G and H separately (again up to isomorphism, and counting multiplicity).

Knowing the composition factors of a group can assist in classifying its subgroups; in particular, groups which are "coprime" in the sense of having no composition factors in common are difficult to "join" together[5]. Here is an example of this which will be of importance in our application:

Lemma 7.2.2. *Let $p \geq 5$ be a prime, and G a finite group which does not have a copy of $P\mathrm{SL}_2(\mathbf{F}_p)$ amongst its composition factors. Let H be a subgroup of $G \times \mathrm{SL}_2(\mathbf{F}_p)$ whose projections to G and $\mathrm{SL}_2(\mathbf{F}_p)$ are surjective. Then H is all of $G \times \mathrm{SL}_2(\mathbf{F}_p)$.*

Proof. We apply *Goursat's lemma* (see Exercise 7.2.4 below). Thus, if $N_1 := \{g \in G : (g,1) \in H\}$ and $N_2 := \{h \in \mathrm{SL}_2(\mathbf{F}_p) : (1,h) \in H\}$, then

---
[5]Interestingly, the phenomenon of "coprimality" implying "disjointness" also shows up in ergodic theory, in the theory of *joinings*, but we will not discuss this further here.

N_1, N_2 are normal subgroups of $G, \mathrm{SL}_2(\mathbf{F}_p)$ respectively such that G/N_1 is isomorphic to $\mathrm{SL}_2(\mathbf{F}_p)/N_2$. From Exercise 7.2.2 we see that N_2 is either trivial, all of $\mathrm{SL}_2(\mathbf{F}_p)$, or is the center $Z(\mathrm{SL}_2(\mathbf{F}_p))$.

If N_2 is trivial, then $\mathrm{SL}_2(\mathbf{F}_p)$ is isomorphic to a quotient of G, and thus by Exercise 7.2.3 the composition factors of $\mathrm{SL}_2(\mathbf{F}_p)$ are a subset of those of G. But this is a contradiction, since $P\mathrm{SL}_2(\mathbf{F}_p)$ is a composition factor of $\mathrm{SL}_2(\mathbf{F}_p)$ but not of G. Similarly, if N_2 is the center, since $\mathrm{SL}_2(\mathbf{F}_p)/N_2$ is then isomorphic to $P\mathrm{SL}_2(\mathbf{F}_p)$. So the only remaining case is when N_2 is all of $\mathrm{SL}_2(\mathbf{F}_p)$. But then as H surjects onto G, we see that H is all of $G \times \mathrm{SL}_2(\mathbf{F}_p)$ and we are done. □

Exercise 7.2.4 (Goursat's lemma). Let G_1, G_2 be groups, and let H be a subgroup of $G_1 \times G_2$ whose projections to G_1, G_2 are surjective. Let $N_1 := \{g_1 \in G_1 : (g_1, 1) \in H\}$ and $N_2 := \{g_2 \in G_2 : (1, g_2) \in H\}$. Show that N_1, N_2 are normal subgroups of G_1, G_2, and that G_1/N_1 and G_2/N_2 are isomorphic. (Indeed, after quotienting out by $N_1 \times N_2$, H becomes a graph of such an isomorphism.) Conclude that the set of composition factors of H are a subset of the union of the set of composition factors of G_1 and the set of composition factors of G_2 (up to isomorphism and counting multiplicity, as usual).

As such, we have the following satisfactory description of the images $\pi_q(\Gamma)$ of a free group Γ:

Corollary 7.2.3 (Strong approximation). *Let Γ be a subgroup of $\mathrm{SL}_2(\mathbf{Z})$ which is not virtually solvable. Let $M \geq 1$ be an integer. Then there exists a multiple q_1 of M with the following property: whenever q is of the form $q = dp_1 \ldots p_k$ with $d|q_1$ and $p_1, \ldots, p_k$ distinct and coprime to q_1, one has*

$$\pi_q(\Gamma) = \pi_d(\Gamma) \times \mathrm{SL}_2(F_{p_1}) \times \ldots \times \mathrm{SL}_2(F_{p_k})$$

(after using the Chinese remainder theorem to identify $\mathrm{SL}_2(F_q)$ with $\mathrm{SL}_2(F_d) \times \mathrm{SL}_2(F_{p_1}) \times \ldots \times \mathrm{SL}_2(F_{p_k})$). In particular, one has

$$\pi_q(\Gamma) = \pi_d(\Gamma) \times \mathrm{SL}_2(\mathbf{Z}/p_1 \ldots p_k \mathbf{Z}).$$

The parameter M will not actually be needed in our application, but is useful in the more general setting in which f has rational coefficients instead of integer coefficients.

Proof. We already know that $\pi_p(\Gamma) = \mathrm{SL}_2(\mathbf{F}_p)$ for all but finitely many primes p. Let q_0 be the product of M with all the exceptional primes, as well as 2 and 3, thus $p \geq 5$ and $\pi_p(\Gamma) = \mathrm{SL}_2(\mathbf{F}_p)$ for all p coprime to q_0. By repeated application of Lemma 7.2.2 this implies that $\pi_{p_1 \ldots p_k}(\Gamma) = \mathrm{SL}_2(F_{p_1}) \times \ldots \times \mathrm{SL}_2(F_{p_k})$ for any distinct primes $p_1, \ldots, p_k$ coprime to q_0

(the key point being that the groups $PSL_2(\mathbf{F}_p)$ for primes $p \geq 5$ are all nonisomorphic to each other and to $\mathbf{Z}/2\mathbf{Z}$ by cardinality considerations).

The finite group $\pi_{q_0}(\Gamma)$ may contain copies of $PSL_2(\mathbf{F}_p)$ amongst their composition factors for a finite number of primes p coprime to q_0; let q_1 be the product of q_0 with all these primes. By many applications of Exercise 7.2.4, we see that the set of composition factors of $\pi_{q_1}(\Gamma)$ are contained in the union of the set of composition factors of $\pi_{q_0}(\Gamma)$, and the set of composition factors of $\pi_p(\Gamma) = SL_2(\mathbf{F}_p)$ for all p dividing q_1 but not q_0. As a consequence, we see that $PSL_2(\mathbf{F}_p)$ is *not* a composition factor of $\pi_{q_1}(\Gamma)$ for any p coprime to q_1; by Exercise 7.2.3, $PSL_2(\mathbf{F}_p)$ is also not a composition factor of $\pi_d(\Gamma)$ for any d dividing q_1 and p coprime to q_1. By many applications of Lemma 7.2.2, we then obtain the claim. □

As a simple application of the above corollary, we observe that we may reduce Theorem 7.0.1 to the case when Λ is a free group on two generators. Indeed, if Λ is not virtually solvable, then by the *Tits alternative* (Theorem 6.1.1)), Λ contains a subgroup Λ' which is a free group on two generators (and in particular, continues to not be virtually solvable). Now the polynomial f need not be primitive on Λ', so we cannot deduce Theorem 7.0.1 for Λ, f from its counterpart for Λ', f. However, by Corollary 7.2.3 we have an integer $q_1 \geq 1$ such that

$$(7.17) \qquad \pi_q(\Gamma') = \pi_d(\Gamma') \times SL_2(\mathbf{Z}/p_1 \ldots p_k \mathbf{Z})$$

whenever $q = dp_1 \ldots p_k$ with $d|q_1$ and $p_1, \ldots, p_k$ are distinct primes coprime to q_1.

As f is primitive with respect to Λ, we may find $a \in \Gamma$ such that $f(a)$ is coprime to q_1. By translating f by a, we obtain a new polynomial f' for which $f'(1)$ is coprime to q_1. In particular, for any $d|q_1$, we have $f'(1)$ coprime to d. By (7.17), this implies that for any square-free q (and hence for arbitrary q), we can find $a \in \Gamma'$ with $f'(a)$ coprime to q. Thus f' is primitive with respect to Λ', and so we may deduce Theorem 7.0.1 for Λ, f from its counterpart for Λ', f'.

7.3. Sieving in thin groups

We can now deduce Theorem 7.0.1 from the following expander result:

Theorem 7.3.1 (Uniform expansion). *Let a, b generate a free group Λ in $SL_2(\mathbf{Z})$. Then, as q runs through the square-free integers, $\mathrm{Cay}(\pi_q(\Lambda), \pi_q(\{a, b, a^{-1}, b^{-1}\}))$ form a two-sided expander family.*

When q is restricted to be prime, this result follows from Theorem 6.0.4. The extension of this theorem to nonprime q is more difficult, and will be

7.3. Sieving in thin groups

discussed later. For now, let us assume Theorem 7.3.1 and see how we can use it, together with the beta sieve, to imply Theorem 7.0.1.

As discussed in the preceding section, to show Theorem 7.0.1 we may assume without loss of generality that Λ is a free group on two generators a, b. Let $\mu := \frac{1}{4}(\delta_a + \delta_b + \delta_{a^{-1}} + \delta_{b^{-1}})$ be the generator of the associated random walk, and let T be a large integer. Then $\mu^{(T)}$ will be supported on elements of Λ whose coefficients have size $O(\exp(O(T)))$, where we allow implied constants to depend on a, b. In particular, for x in this support, $f(x)$ will be an integer of size $O(\exp(O(T)))$, where we allow implied constants to depend on f also. On the other hand, $\mu^{(T)}$ has an ℓ^∞ norm that decreases exponentially in T (by Exercise 6.1.4). If we then set $z := \exp(\varepsilon' T)$ for a sufficiently small absolute constant $\varepsilon' > 0$, it will then suffice to show that, with probability $\gg T^{-O(1)}$, an element x drawn from Λ with distribution $\mu^{(T)}$ is such that $f(x)$ is nonzero and coprime to $P(z)$.

It will be convenient to knock out a few exceptional primes. From Corollary 7.2.3, we may find an integer q_1 with the property that

$$\pi_q(\Gamma) = \pi_d(\Gamma) \times \mathrm{SL}_2(F_{p_1}) \times \ldots \times \mathrm{SL}_2(F_{p_k})$$

whenever $q = dp_1 \ldots p_k$ with $d | q_1$ and $p_1, \ldots, p_k$ distinct and coprime to q_1. As f is primitive, we may find a residue class $x_1 \in \pi_{q_1}(\Gamma)$ such that $f(x_1)$ is coprime to q_1. For each integer n, let a_n denote the quantity

$$a_n := \sum_{x \in \Lambda : f(x) = n; x = x_1 \bmod q_1} \mu^{(T)}(x).$$

It will suffice to show that

$$\sum_{n:(n,P(z))=1} a_n \gg T^{-O(1)}.$$

To do this, we will use the beta sieve. Indeed, by Exercise 7.1.2 it suffices to establish a bound of the form

$$(7.18) \qquad \sum_{n:d|n} a_n = \frac{1}{|\pi_{q_1}(\Lambda)|} g(d) + O(\exp(-\varepsilon T))$$

for all square-free $1 \leq d \leq \exp(\varepsilon T)$, some constant $\varepsilon > 0$, and some multiplicative function g obeying the bounds

$$g(p) < 1$$

and

$$g(p) \ll 1/p$$

for all primes p.

By choice of x_1, the quantity a_n vanishes whenever n is not coprime to q_1. So we will set $g(p) = 0$ for the primes p dividing q_1, and it will suffice to

establish (7.18) for d coprime to q_1. The left-hand side of (7.18) can then be expressed as
$$\sum_{x \in \pi_d(\Lambda): f(x)=0} ((\pi_{q_1 d})_* \mu)^{(T)}(x_1, x),$$
where we descend the polynomial $f \colon \mathrm{SL}_2(\mathbf{Z}) \to \mathbf{Z}$ to a polynomial
$$f \colon \mathrm{SL}_2(\mathbf{Z}/d\mathbf{Z}) \to \mathbf{Z}/d\mathbf{Z}$$
in the obvious fashion. However, in view of Theorem 7.3.1 (and the random walk interpretation of expansion), we have
$$((\pi_{q_1 d})_* \mu)^{(T)}(x) = |\pi_{q_1 d}(\Lambda)|^{-1} + O(\exp(-cT))$$
for some $c > 0$ independent of ε. Note that
$$|\pi_d(\Lambda)| \leq |\mathrm{SL}_2(\mathbf{Z}/d\mathbf{Z})| \ll d^{O(1)} \ll \exp(O(\varepsilon T))$$
while from Corollary 7.2.3 we have
$$|\pi_{q_1 d}(\Lambda)| = |\pi_{q_1}(\Lambda)| |\pi_d(\Lambda)|$$
and thus
$$\sum_{n : d|n} a_n = \frac{1}{|\pi_{q_1}(\Lambda)|} g(d) + O(\exp(-\varepsilon T))$$
for $\varepsilon > 0$ small enough, where $g(d)$ is defined for d coprime to q_1 as
$$g(d) := \frac{1}{|\pi_d(\Gamma)|} |\{x \in \mathrm{SL}_2(\mathbf{Z}/d\mathbf{Z}) : f(x) = 0\}|.$$
As f is primitive, we have $g(d) < 1$ for all such d; from Corollary 7.2.3 we see that g is multiplicative for such d. Finally, from the Schwarz-Zippel lemma (see Exercise 5.4.3) we have
$$g(p) \ll 1/p$$
and Theorem 7.0.1 follows.

Remark 7.3.2. One can obtain more precise bounds on $g(p)$ using the *Lang-Weil theorem*; see Section 9. Such results would, however, be needed if one wanted more quantitative information than Theorem 7.0.1; see [**BoGaSa2010**] for details.

It remains to establish Theorem 7.0.1. In the case when q is prime, this was achieved in previous chapters using the ingredients of quasirandomness, product theorems, and nonconcentration. In [**BoGaSa2010**], these ingredients were extended to the square-free case by hand, which led to a fairly lengthy argument. In the subsequent paper of Varju [**Va2012**], it was shown that each of these ingredients can in fact be more or less automatically bootstrapped from the prime case to the square-free case by using tools such as the Chinese remainder theorem (or strong approximation) to "factor" the

7.3. Sieving in thin groups

latter case into copies of the former, thus simplifying the extension to the square-free case significantly. We will not give the full argument here, but just to convey a taste of these sorts of product arguments, we will discuss the product structure of just one of the three ingredients, namely quasirandomness. (The extension of this ingredient to the square-free setting was already observed in [**BoGaSa2010**].)

As a consequence of Example 2.1.4, the following claim was shown:

Proposition 7.3.3. *Let G be a $|G|^\alpha$-quasirandom finite group, S is a symmetric set of generators of G not containing the identity of cardinality k, and let $\mu = \frac{1}{|S|}\sum_{s\in S} \delta_s$ be such that*

$$\|\mu^{*n}\|_{\ell^2(G)} \leq |G|^{-1/2+\alpha/4}$$

(say) for some $n = O(\log|G|)$ with $\mu := \frac{1}{|S|}\sum_{s\in S}\delta_s$, then $\mathrm{Cay}(G,S)$ is a two-sided ε-expander for some ε depending only on α, k and the implied constants in the $O()$ notation.

It turns out that this fact can be extended to product groups:

Proposition 7.3.4. *Proposition 7.3.3 continues to hold if the hypothesis that G is $|G|^\alpha$-quasirandom is replaced with the hypothesis that $G = G_1 \times \ldots \times G_n$ for some $n \geq 0$ and some finite groups $G_1, \ldots, G_n$, with each G_i being $|G_i|^\alpha$-quasirandom.*

The key point here is that the expansion constant ε does not depend on the number n of groups in this factorisation.

Proof (Sketch). For technical reasons it is convenient to allow S to have multiplicity and to possibly contain the identity; this will require generalising the notion of Cayley graph, and of expansion in such generalised graphs.

Let f be a nonconstant eigenfunction of the adjacency operator, thus $f * \mu = \lambda f$ for some real λ. The objective is to prove that $|\lambda| \leq 1 - \varepsilon$ for some sufficiently small $\varepsilon > 0$ independent of n.

The claim $n = 0$ is trivial, so assume inductively that $n \geq 1$ and that the claim is proven for all smaller values of n (with a fixed choice of ε). For each G_i, we can partition the eigenspaces of the adjacency operator into those functions which are invariant in the G_i direction, and those functions which have mean zero in each coset of G_i. These partitions are compatible with each other as i varies (basically because the operations of averaging in G_i and averaging in G_j commute). Thus, without loss of generality, we may assume that the eigenfunction f is such that for each i, either f is G_i-invariant, or has mean zero on each G_i coset.

Suppose that f was G_n-invariant, then the eigenvalue λ would also persist after projecting S down from G to $G_1 \times G_{n-1}$ (possibly picking up the identity or some multiplicity in the process). The claim then follows from the induction hyopthesis. Similarly, if f was G_i-invariant for any other value of i. Thus we may assume that f has mean zero in each G_i coset.

One can show that every irreducible unitary representation of G splits as a tensor product of irreducible unitary representations of the G_i. If one lets V be the subspace of $\ell^2(G)$ spanned by f and its left translates, we thus see that V contains at least one such tensor product; but as every element of V will have mean zero in each G_i coset, the factors in this tensor product will all be nontrivial. Using quasirandomness, the ith factor will have dimension at least $|G_i|^\alpha$, and so V must have dimension at least $|G|^\alpha$. At this point, one can use a trace formula to relate V to $\|\mu^{*n}\|_{\ell^2(G)}^2$ to conclude the argument. □

Exercise 7.3.1. Develop the above sketch into a complete proof of the proposition.

Part 2

Related Articles

Chapter 8

Cayley graphs and the algebra of groups

This is a sequel to the article "Cayley graphs and the geometry of groups" from [**Ta2013b**, §2.3]. In that article, the concept of a *Cayley graph* of a group G was used to place some geometry on that group G. In this chapter, we explore a variant of that theme, in which (fragments of) a Cayley graph on G is used to describe the basic algebraic structure of G, and in particular on elementary word identities in G. Readers who are familiar with either *category theory* or *group cohomology* will recognise these concepts lurking not far beneath the surface; we will remark briefly on these connections later in this article. However, no knowledge of categories or cohomology is needed for the main discussion, which is primarily focused on elementary group theory.

Throughout this chapter, we fix a single group $G = (G, \cdot)$, which is allowed to be nonabelian and/or infinite. All our graphs will be directed, with loops and multiple edges permitted.

In [**Ta2013b**, §2.3], we drew the entire Cayley graph of a group G. Here, we will be working much more locally, and will only draw the portions of the Cayley graph that are relevant to the discussion. In this graph, the vertices are elements x of the group G, and one draws a directed edge from x to xg labeled (or "coloured") by the group element g for any $x, g \in G$; the graph consisting of all such vertices and edges will be denoted $\mathrm{Cay}(G, G)$. Thus, a typical edge in $\mathrm{Cay}(G, G)$ is as shown in Figure 1.

One usually does not work with the complete Cayley graph $\mathrm{Cay}(G, G)$. It is customary to instead work with smaller Cayley graphs $\mathrm{Cay}(G, S)$, in

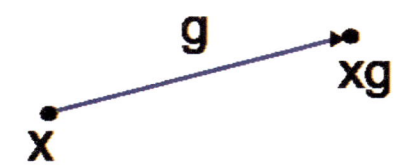

Figure 1. An edge of a Cayley graph.

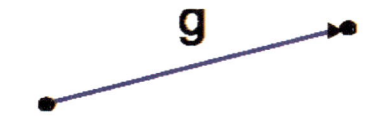

Figure 2. An edge of a Cayley graph with unlabeled vertices.

which the edge colours g are restricted to a smaller subset of G, such as a set of generators for G. As we will be working locally, we will in fact work with even smaller fragments of $\text{Cay}(G, G)$ at a time; in particular, we only use a handful of colours (no more than nine, in fact, for any given diagram), and we will not require these colours to generate the entire group (we do not care if the Cayley graph is connected or not, as this is a global property rather than a local one).

Cayley graphs are left-invariant: for any $a \in G$, the left translation map $x \mapsto ax$ is a graph isomorphism. To emphasise this left invariance, we will usually omit the vertex labels, and leave only the coloured directed edge, as in Figure 2.

This is analogous to how, in undergraduate mathematics and physics, vectors in Euclidean space are often depicted as arrows of a given magnitude and direction, with the initial and final points of this arrow being of secondary importance only. (Indeed, this depiction of vectors in a vector space can be viewed as an abelian special case of the more general depiction of group elements used in this article.)

Let us define a *diagram* to be a finite directed graph $H = (V, E)$, with edges coloured by elements of G, which has at least one graph homomorphism into the complete Cayley graph $\text{Cay}(G, G)$ of G; thus there exists a map $\phi: V \to G$ (not necessarily injective) with the property that $\phi(w) = \phi(v)g$ whenever (v, w) is a directed edge in H coloured by a group element $g \in G$. Informally, a diagram is a finite subgraph of a Cayley graph with the vertex labels omitted, and with distinct vertices permitted to represent the same group element. Thus, for instance, the single directed edge displayed in Figure 2 is a very simple example of a diagram. An even simpler example of a diagram would be a depiction of the identity element; see

8. Cayley graphs and the algebra of groups

Figure 3. The identity diagram.

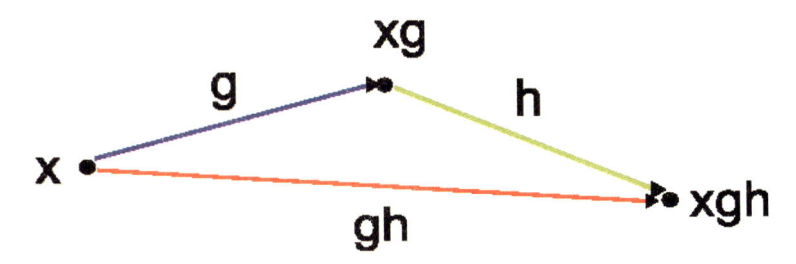

Figure 4. Group multiplication.

Figure 3. We will, however, omit the identity loops in our diagrams in order to reduce clutter.

We make the obvious remark that any directed edge in a diagram can be coloured by at most one group element g, since $y = xg, y = xh$ implies $g = h$. This simple observation provides a way to prove group theoretic identities using diagrams: to show that two group elements g, h are equal, it suffices to show that they connect together (with the same orientation) the same pair of vertices in a diagram.

Remark 8.0.1. One can also interpret these diagrams as *commutative diagrams* in a category in which all the objects are copies of G, and the morphisms are right-translation maps. However, we will deviate somewhat from the *category theoretic* way of thinking here by focusing on the geometric arrangement and shape of these diagrams, rather than on their abstract combinatorial description. In particular, we view the arrows more as distorted analogues of vector arrows, than as the abstract arrows appearing in category theory.

Just as vector addition can be expressed via concatenation of arrows, group multiplication can be described by concatenation of directed edges. Indeed, for any $x, g, h \in G$, the vertices x, xg, xgh can be connected by the triangular diagram in Figure 4. In a similar spirit, inversion is described by the diagram in Figure 5.

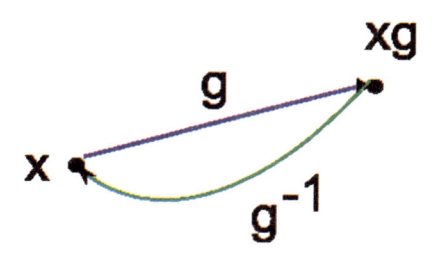

Figure 5. Group inversion.

We make the pedantic remark though that we do not consider a g^{-1} edge to be the reversal of the g edge, but rather as a distinct edge that just happens to have the same initial and final endpoints as the reversal of the g edge. (This will be of minor importance later, when we start integrating "1-forms" on such edges.)

A fundamental operation for us will be that of *gluing* two diagrams together.

Lemma 8.0.2 ((Labeled) gluing). *Let $D_1 = (V_1, E_1), D_2 = (V_2, E_2)$ be two diagrams of a given group G. Suppose that the intersection $D_1 \cap D_2 := (V_1 \cap V_2, E_1 \cap E_2)$ of the two diagrams connects all of $V_1 \cap V_2$ (i.e., any two elements of $V_1 \cap V_2$ are joined by a path in $D_1 \cap D_2$). Then the union $D_1 \cup D_2 := (V_1 \cup V_2, E_1 \cup E_2)$ is also a diagram of G.*

Proof. By hypothesis, we have graph homomorphisms $\phi_1 \colon D_1 \to \mathrm{Cay}(G, G)$, $\phi_2 \colon D_2 \to \mathrm{Cay}(G, G)$. If they agree on $D_1 \cap D_2$, then one simply glues together the two homomorphisms to create a new graph homomorphism $\phi \colon D_1 \cup D_2 \to \mathrm{Cay}(G, G)$. If they do not agree, one can apply a left translation to either ϕ_1 or ϕ_2 to make the two diagrams agree on at least one vertex of $D_1 \cap D_2$; then by the connected nature of $D_1 \cap D_2$ we see that they now must agree on all vertices of $D_1 \cap D_2$, and then we can form the glued graph homomorphism as before. □

The above lemma requires one to specify the label the vertices of D_1, D_2 (in order to form the intersection $D_1 \cap D_2$ and union $D_1 \cup D_2$). However, if one is presented with two diagrams D_1, D_2 with unlabeled vertices, one can identify some partial set of vertices of D_1 with a partial set of vertices of D_2 of matching cardinality. Provided that the subdiagram common to D_1 and D_2 after this identification connects all of the common vertices together, we may use the above lemma to create a glued diagram D.

For instance, if a diagram D contains two of the three edges in the triangular diagram in Figure 4, one can "fill in" the triangle by gluing in the third edge; see Figure 6.

8. Cayley graphs and the algebra of groups

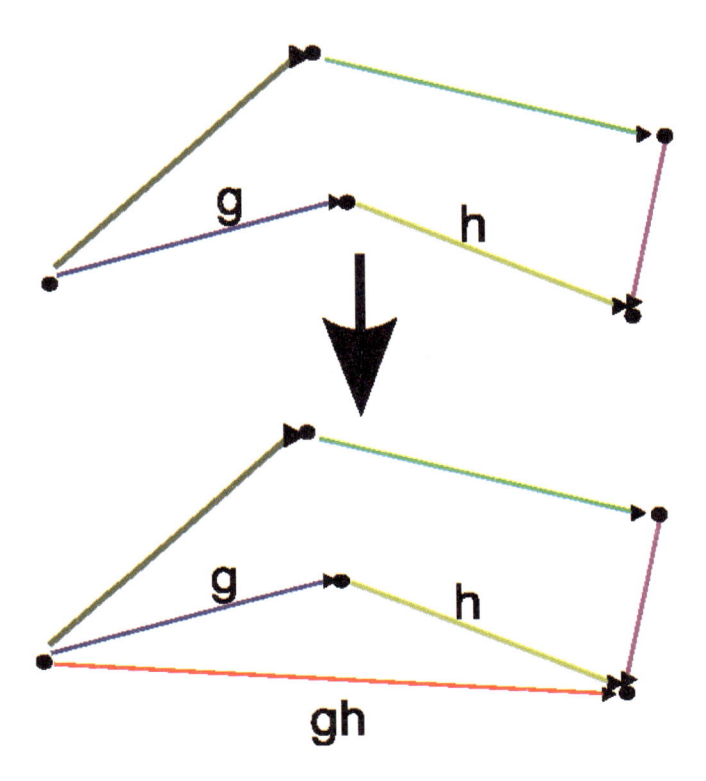

Figure 6. Filling in a triangle.

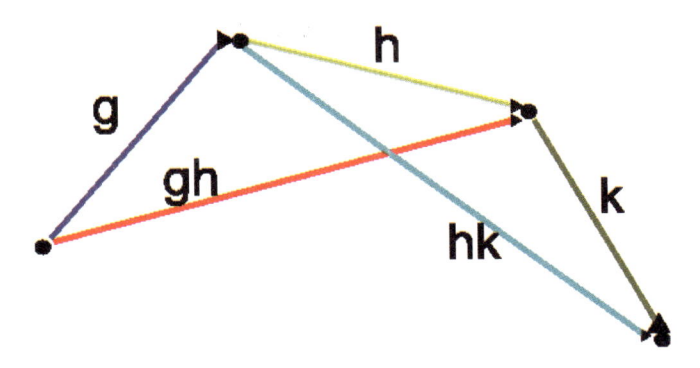

Figure 7. A partial demonstration of the associative law.

One can use glued diagrams to demonstrate various basic group-theoretic identities. For instance, by gluing together two copies of the triangular diagram in Figure 4 to create the glued diagram in Figure 7. Filling in two more triangles, we obtain a tetrahedral diagram that demonstrates the associative law $(gh)k = g(hk)$; see Figure 8.

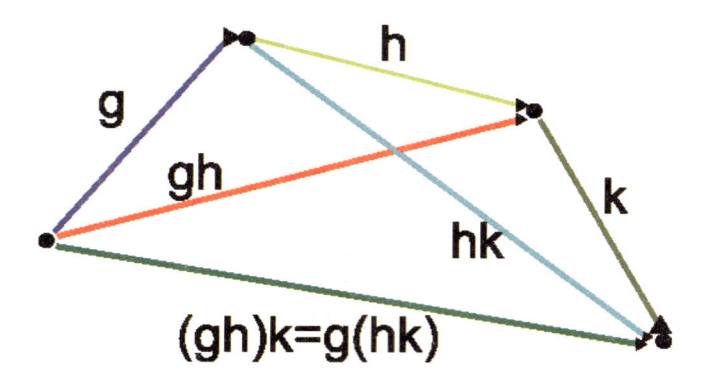

Figure 8. Completion of the proof of the associative law.

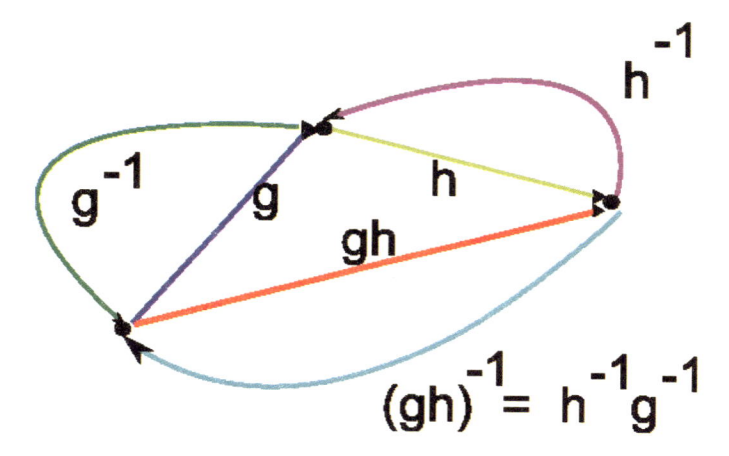

Figure 9. The Abel identity.

Similarly, by gluing together two copies of Figure 4 with three copies of Figure 5 in an appropriate order, we can demonstrate the Abel identity $(gh)^{-1} = h^{-1}g^{-1}$; see Figure 9.

In addition to gluing, we will also use the trivial operation of *erasing*: if D is a diagram for a group G, then any subgraph of D (formed by removing vertices and/or edges) is also a diagram of G. This operation is not strictly necessary for our applications, but serves to reduce clutter in the pictures.

If two group elements g, h commute, then we obtain a parallelogram as a diagram, exactly as in the vector space case; see Figure 10.

In general, of course, two arbitrary group elements g, h will fail to commute, and so this parallelogram is no longer available. However, various substitutes for this diagram exist. For instance, if we introduce the *conjugate* $g^h := h^{-1}gh$ of one group element g by another, then we have the slightly distorted parallelogram in Figure 11.

8. Cayley graphs and the algebra of groups

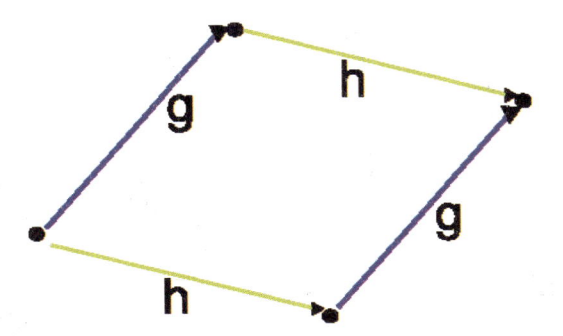

Figure 10. Two commuting group elements.

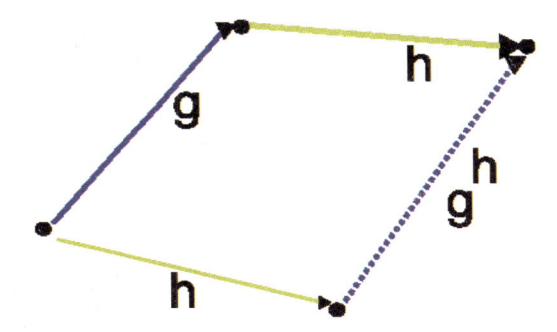

Figure 11. Two nearly commuting group elements.

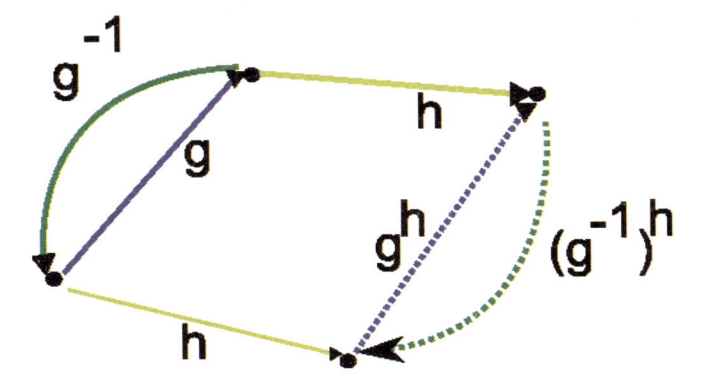

Figure 12. A partial demonstration of the fact that conjugation is a homomorphism.

By appropriate gluing and filling, this can be used to demonstrate the homomorphism properties of a conjugation map $g \mapsto g^h$; see Figures 12 and 13.

174   8. Cayley graphs and the algebra of groups

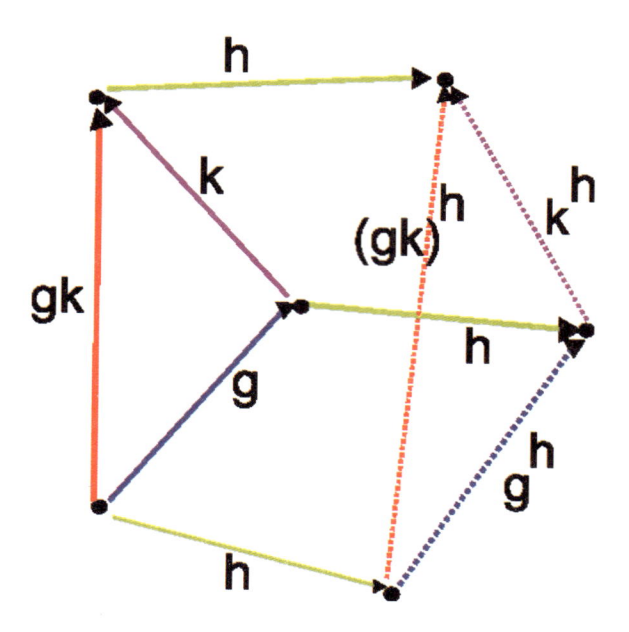

Figure 13. Completion of the demonstration that conjugation is a homomorphism.

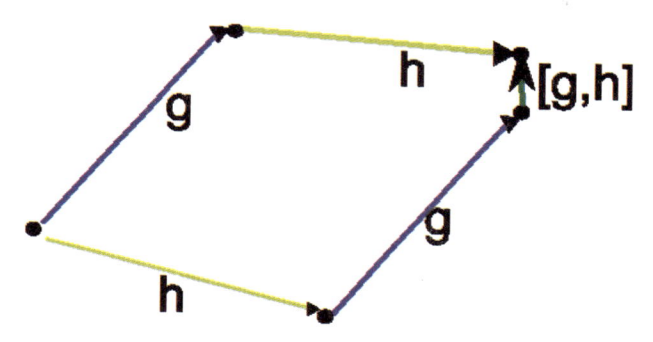

Figure 14. The commutator of two elements.

Another way to replace the parallelogram in Figure 10 is to introduce the *commutator* $[g, h] := g^{-1}h^{-1}gh$ of two elements, in which case we can perturb the parallelogram into a pentagon; see Figure 14.

We will tend to depict commutator edges as being somewhat shorter than the edges generating that commutator, reflecting[1] a "perturbative" or "nilpotent" philosophy. We will also be adopting a "Lie" perspective of interpreting groups as behaving like perturbations of vector spaces, in particular, by trying to draw all edges of the same colour as being approximately

---

[1]Of course, to fully reflect a nilpotent perspective, one should orient commutator edges in a different dimension from their generating edges, but of course the diagrams drawn here do not have enough dimensions to display this perspective easily.

8. Cayley graphs and the algebra of groups

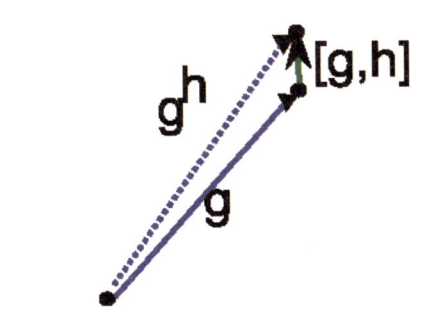

Figure 15. Relating the conjugate with the commtuator.

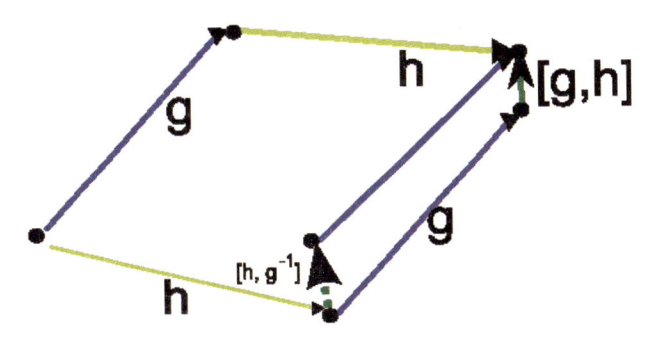

Figure 16. Relating two commutators together.

(though not perfectly) parallel to each other (and with approximately the same length).

Gluing the above pentagon with the conjugation parallelogram and erasing some edges, we discover a "commutator-conjugate" triangle, describing the basic identity $g^h = g[g,h]$; see Figure 15.

Other gluings can also give the basic relations between commutators and conjugates. For instance, by gluing the pentagon in Figure 14 with its reflection, we see that $[g,h] = [h,g]^{-1}$. The diagram in Figure 16, obtained by gluing together copies of Figures 11 and 15, demonstrates that $[h,g^{-1}] = [g,h]^{g^{-1}}$, while Figure 17 demonstrates that $[g,hk] = [g,k][g,h]^k$.

Now we turn to a more sophisticated identity, the *Hall-Witt identity*

$$[[g,h],k^g][[k,g],h^k][[h,k],g^h] = 1,$$

which is the fully noncommutative version of the more well-known *Jacobi identity* for Lie algebras.

The full diagram for the Hall-Witt identity resembles a slightly truncated parallelopiped. Drawing this truncated paralleopiped in full would result in a rather complicated looking diagram, so I will instead display three components of this diagram separately, and leave it to the reader to mentally

176   8. *Cayley graphs and the algebra of groups*

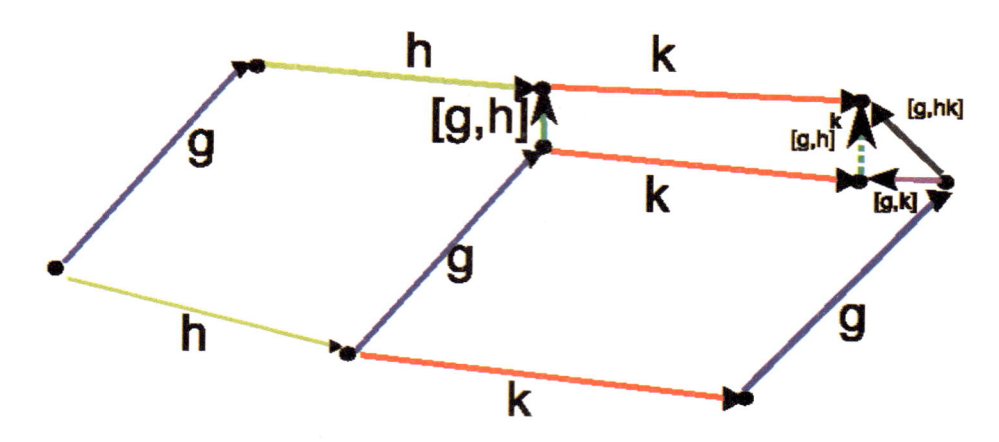

Figure 17. Relating three commutators together.

glue these three components back to form the full parallelopiped. The first component of the diagram is formed by gluing together three pentagons from Figure 14, and is depicted in Figure 18. This should be thought of as the "back" of the truncated parallelopiped needed to establish the Hall-Witt identity.

While it is not needed for proving the Hall-Witt identity, we also observe for future reference that we may also glue in some distorted parallelograms and obtain a slightly more complicated diagram; see Figure 19.

To form the second component, let us now erase all interior components of Figure 18 or Figure 19 to obtain Figure 20. Then we fill in three distorted parallelograms; see Figure 21. This is the second component, and is the "front" of the truncated parallelopiped, minus the portions exposed by the truncation.

Finally, we turn to the third component. We begin by erasing the outer edges from the second component in Figure 21 to obtain Figure 22, then glue in three copies of the commutator-conjugate triangle from Figure 15 to obtain Figure 23.

But now we observe that we can fill in three pentagons, and obtain a small triangle with edges $[[g,h], k^g][[k,g], h^k][[h,k], g^h]$:

Erasing everything except this triangle gives the Hall-Witt identity. Alternatively, one can glue together Figures 18, 21, and 24 to obtain a truncated parallelopiped which one can view as a geometric representation of the *proof* of the Hall-Witt identity.

Among other things, I found these diagrams to be useful to visualise group cohomology; I give a simple example of this below, developing an analogue of the Hall-Witt identity for 2-cocycles.

8.1. A Hall-Witt identity for 2-cocycles 177

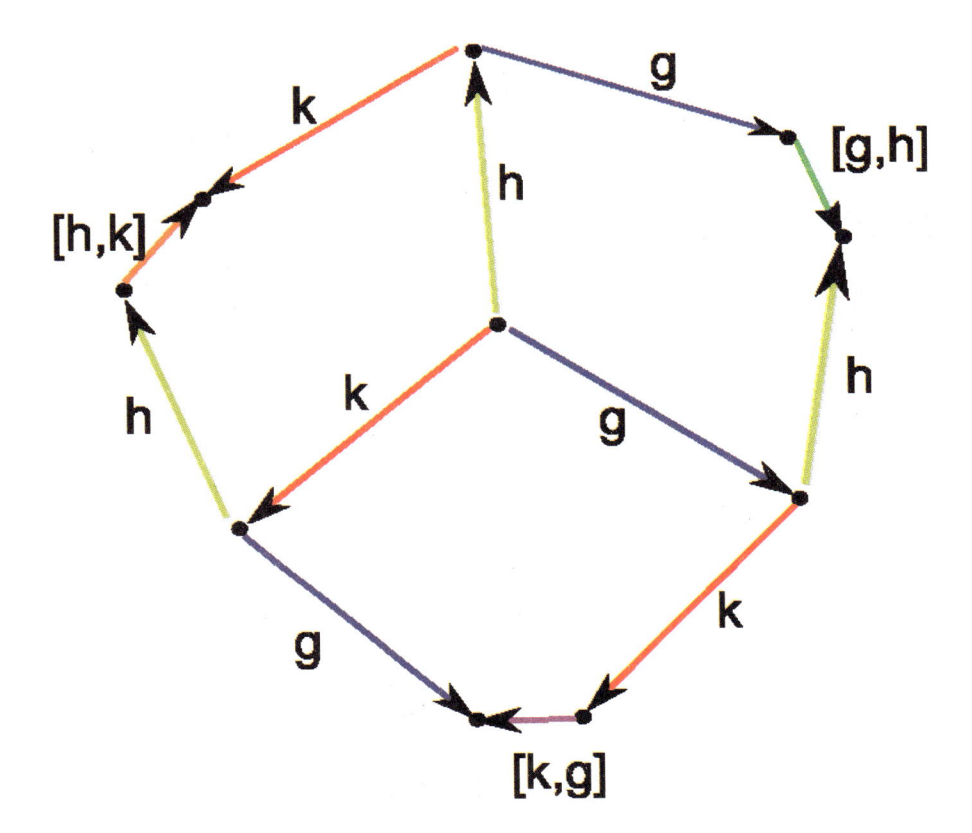

Figure 18. Back portion of the Hall-Witt identity.

8.1. A Hall-Witt identity for 2-cocycles

It is instructive to start interpreting the basic building blocks of *group homology* and *group cohomology* in terms of these diagrams; among other things, this interpretation highlights the close relationship between group cohomology and other types of cohomology, such as *simplicial cohomology* and *de Rham cohomology*. We will not do so systematically here, but present just a small fragment of group cohomology in this setting, to give the flavour of things.

To warm up, let's begin with the easy theory of 1-cohomology. Fix some coefficient ring U (e.g., the integers $\mathbf{Z}$, the reals $\mathbf{R}$, or a cyclic group such as $\mathbf{Z}/2\mathbf{Z}$; for the elementary cohomology topics we will be presenting here, the exact choice of the coefficient ring U will not be important). A 1-*cochain* is just a map $f \colon G \to U$. Using our diagram perspective, we can interpret a 1-cochain in a "de Rham" way as a "1-form" that assigns the element $\int_e f := f(g)$ of U to any oriented edge e in a diagram that has colour g. By definition, we also define the integral $\int_{-e} f$ along the reversal of an oriented

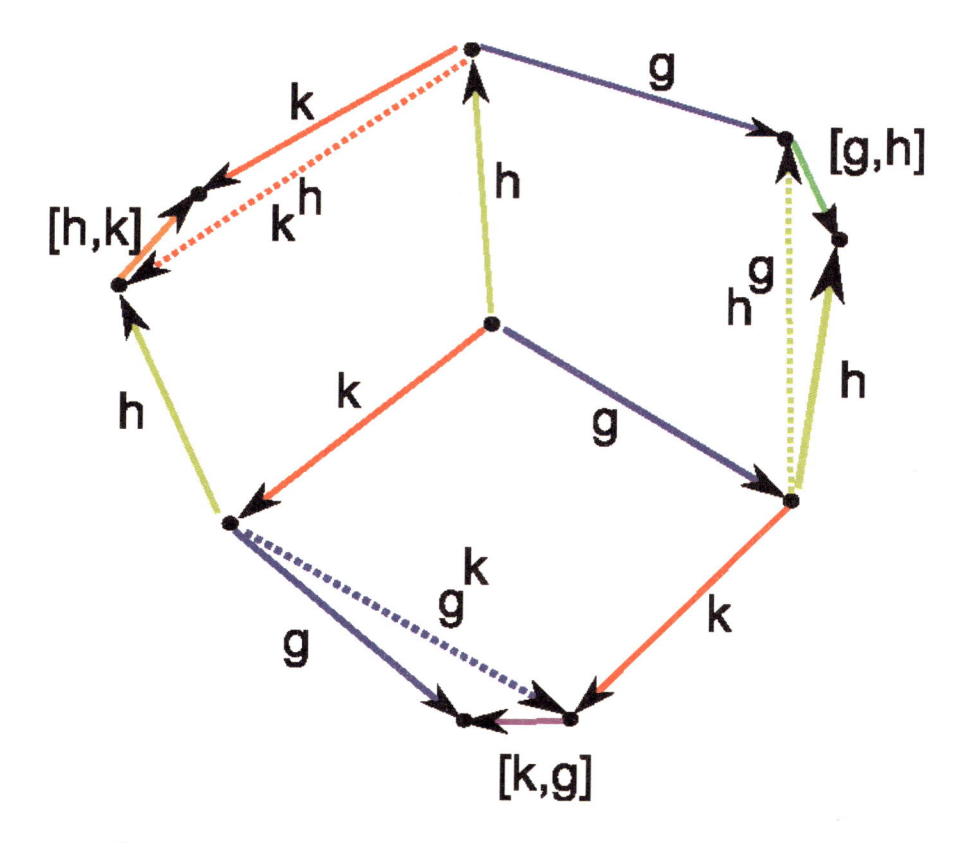

Figure 19. More detail of the back portion of the Hall-Witt identity.

edge e in a diagram by $\int_{-e} f := -\int_e f$. Note though that the integral along a g^{-1} edge is not necessarily the negation of the integral along a g edge, since we may have $f(g^{-1}) \neq -f(g)$. This explains the previous remark that we do not view a g^{-1} edge as the reversal of a g edge. Similarly, since $f(1)$ need not equal 0, the integral of f on a loop need not be nonzero. Thus one has to take a little bit of care with the analogy between group cohomology and de Rham cohomology. However, if the 1-chain is normalised in the sense that $f(1) = 0$ and $f(g^{-1}) = -f(g)$ (which is for instance the case with the 1-cocycles discussed below), then the analogy becomes more accurate.

Given any oriented path γ in a diagram consisting of a sequence $e_1, \ldots, e_k$ of edges (which either are aligned with, or have the opposite orientation from, the edge in the diagram), we can then define the "line integral" $\int_\gamma f$ of a 1-cochain f as the sum of the individual edge integrals $\int_{e_1} f, \ldots, \int_{e_k} f$. A key point here is that of *translation invariance*: if two paths γ, γ' are translates of each other, in the sense that they have the same length and the colours of their edges match, then they have the same line integral with respect to f.

8.1. A Hall-Witt identity for 2-cocycles

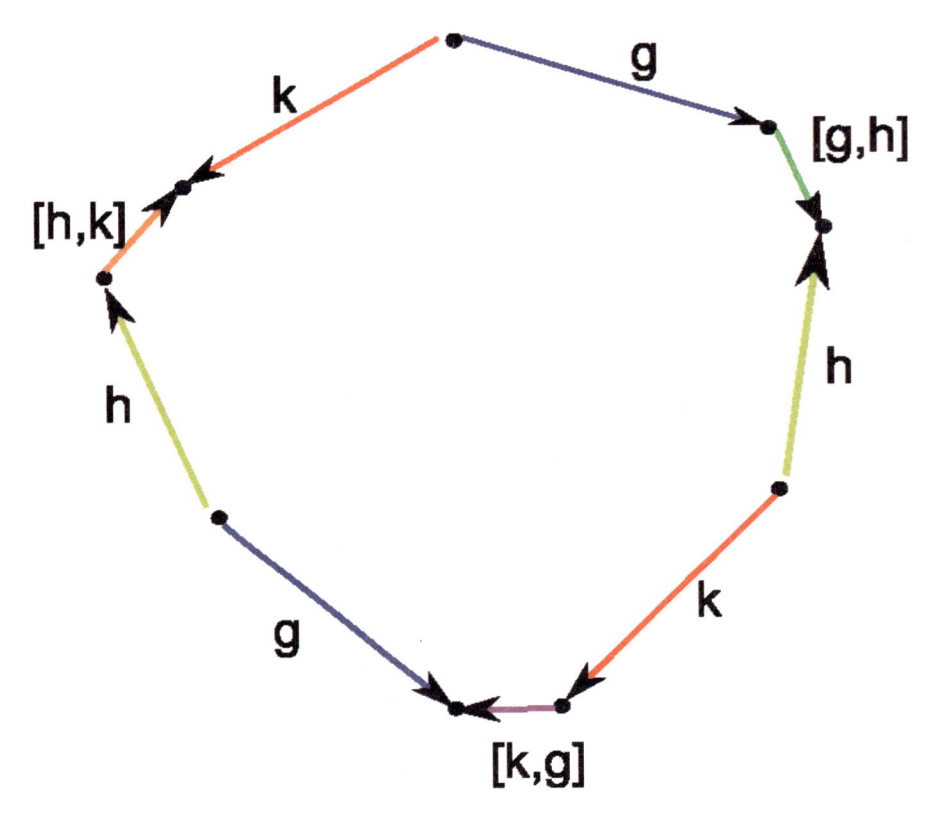

Figure 20. The border between the front and back portions of the Hall-Witt identity.

Remark 8.1.1. One can of course more generally integrate 1-cocycles against 1-*chains*, namely formal linear combinations of oriented edges with coefficients in U, which is the starting point for the diagrammatic interpretation of group homology, but we will not use this formalism here.

A 1-*cocycle* is a 1-cochain $f \colon G \to U$ which obeys the identity

$$f(g) - f(gh) + f(h) = 0$$

for all $g, h \in G$. Of course, this equation is nothing more than the assertion that f is a group homomorphism from G to U; but let us pretend that we are unaware of this interpretation of 1-cocycles, and instead interpret the 1-cocycle condition diagrammatically, as the fact that the line integral around any triangle (Figure 4) vanishes. Because any closed loop in a diagram can be triangulated (possibly after first filling in some more edges), we see more generally that a 1-cocycle is nothing more than a 1-cochain which is *closed* in the sense that its integral on any closed loop is zero. On the other hand, we also have translation invariance of the 1-cocycle, which leads to some

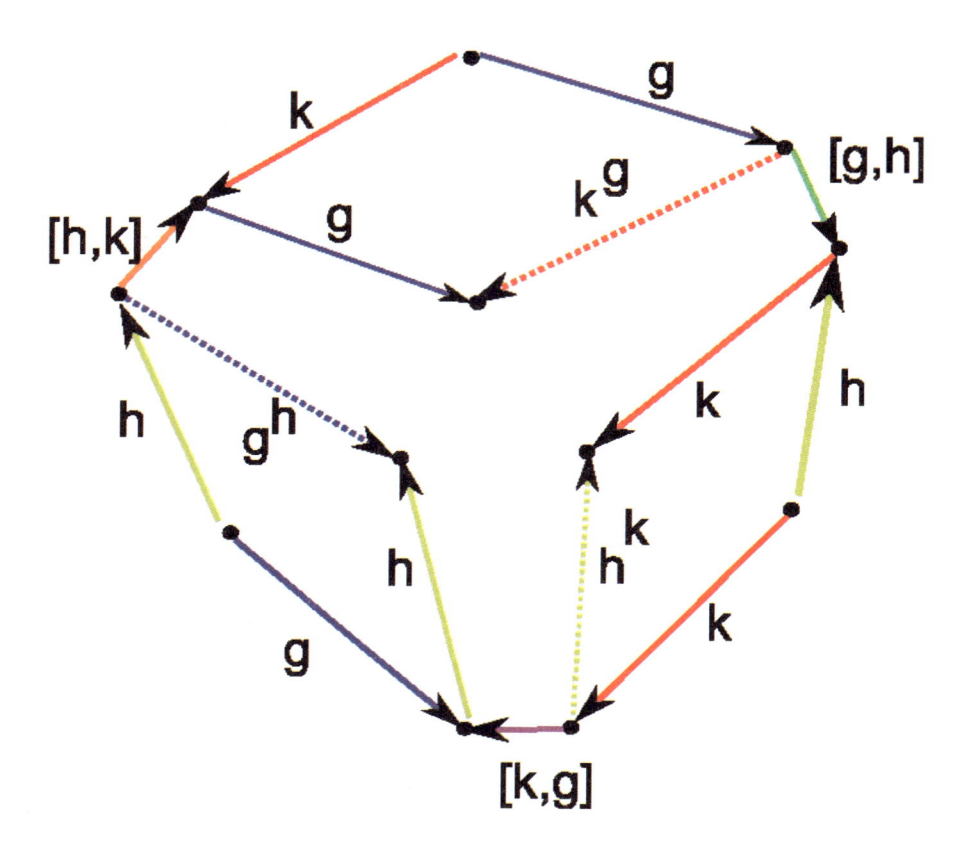

Figure 21. The front portions of the Hall-Witt identity, minus the centre.

additional cancellations. For instance, by integrating a 1-cocycle against the pentagon in Figure 14, the contribution of the edges of matching colour cancel each other out, leaving one with the conclusion that

(8.1) $$f([g,h]) = 0$$

for all g, h and any 1-cocycle f. Of course, this fact was already obvious from the group homomorphism interpretation, but the point is that it can also be observed "geometrically" by inspection of a relevant diagram. Similarly, from Figure 11 we have $f(g) = f(g^h)$ for any $g, h \in G$ and any 1-cocycle f.

Next, we define a 2-*cochain* to be a function $\rho \colon G \times G \to U$. Just as 1-cochains can be viewed as "1-forms" that can be integrated on oriented edges and thus oriented paths, we can view 2-cochains as "2-forms" that can be integrated on oriented triangles and thus triangulated surfaces, though we make the technical restriction that our triangles must be of the form in Figure 4, i.e., not all arrows are oriented in the same direction. We can then interpret $\rho(g, h)$ as the integral $\int_\Delta \rho$ of ρ on the triangle in Figure 4, endowed with the clockwise orientation; reversing the orientation of this triangle leads

8.1. A Hall-Witt identity for 2-cocycles

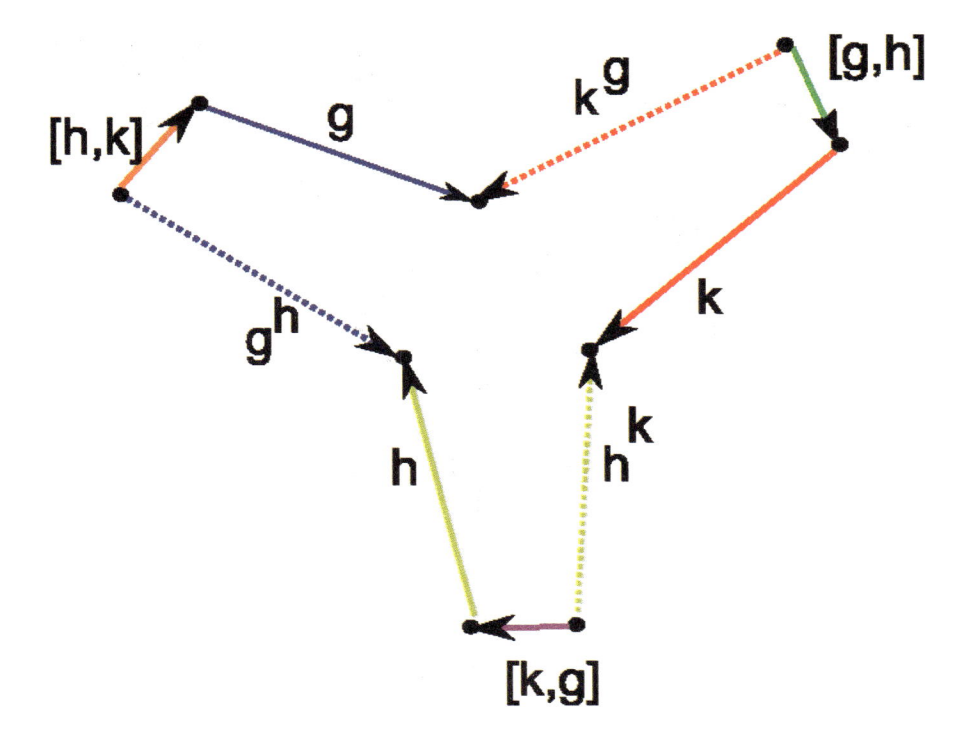

Figure 22. The inner front portions of the Hall-Witt identity.

to a negation of the integral. One can then integrate ρ on any oriented triangulated surface (or, more generally, 2-chains, as mentioned previously) in a diagram in the obvious fashion, provided that in each triangle, the arrows are not all oriented in the same direction.

A *2-cocycle* is a 2-cochain which obeys the identity

(8.2) $$\rho(g,h) - \rho(g,hk) + \rho(gh,k) - \rho(h,k) = 0$$

for all $g, h, k \in G$. As we will recall later, 2-cocycles can be viewed as coordinatisations of central U-extensions of G, but we will again pretend that we are unaware of this interpretation of 2-cocycles, and instead take a diagrammatic interpretation (which has the advantage over the central extension interpretation that it extends much more readily to higher orders of cohomology than 2-cohomology). The cocycle identity (8.2) is then asserting that the integral of ρ on the tetrahedron in Figure 8 (or, in fact, any other tetrahedron) vanishes. Because any closed oriented triangulated 2-surface on a diagram can be broken up into tetrahedra (again after filling in some edges if necessary), we conclude that a 2-cocycle is nothing more than a 2-cochain which is *closed* in the sense that its integral on any closed triangulated 2-surface vanishes. Among other things, this now allows one

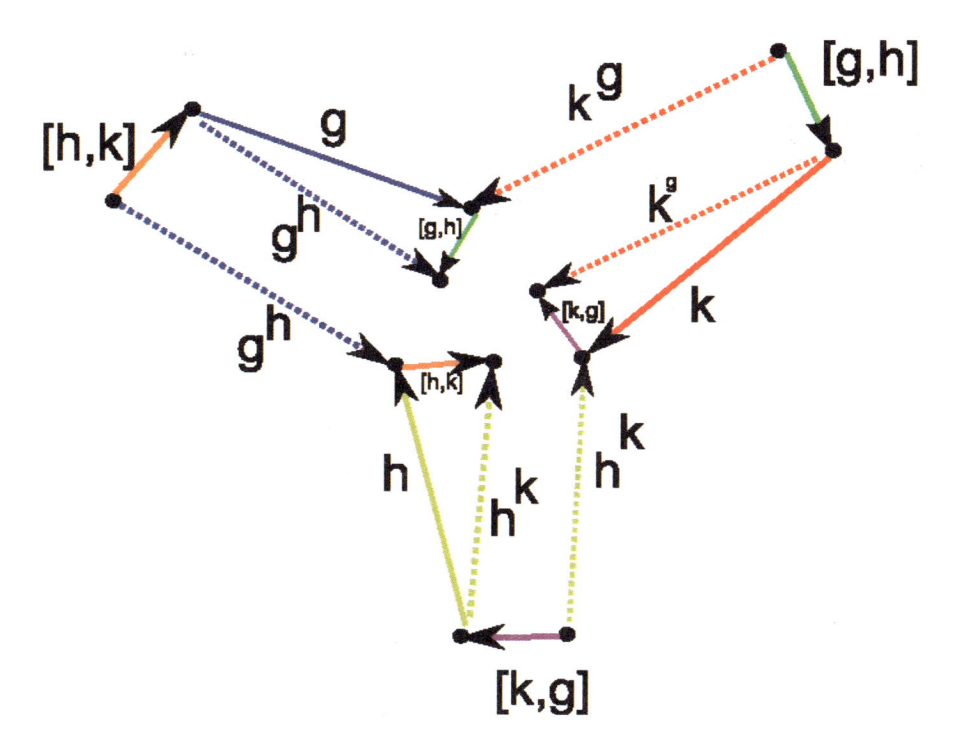

Figure 23. More detail on the inner front portions of the Hall-Witt identity.

to define the integral $\int_S \rho$ of a 2-cocycle on any oriented surface S (not necessarily closed or triangulated) which is bordered by some closed loop γ in a diagram, by replacing S by some triangulated oriented surface with the same oriented boundary γ as S. For instance, we can integrate a 2-cocycle ρ on the pentagon P in Figure 14 with the counterclockwise orientation to obtain an element $\int_P \rho$ of ρ; by selecting a suitable triangulation of this pentagon, this integral can be expressed explicitly as

(8.3) $$\int_P \rho = \rho(h,g) + \rho(hg, [g,h]) - \rho(g,h)$$

but we can choose other triangulations to obtain other representations of the same integral, e.g.,

$$\int_P \rho = -\rho(g, g^{-1}h) + \rho(g^{-1}h, g) + \rho(h^g, [g,h]).$$

Again, a key point is translation invariance: if two surfaces S, S' are translates of each other in the sense that their boundaries are translates of each other, then their integrals against any 2-cocycle ρ will agree.

8.1. A Hall-Witt identity for 2-cocycles

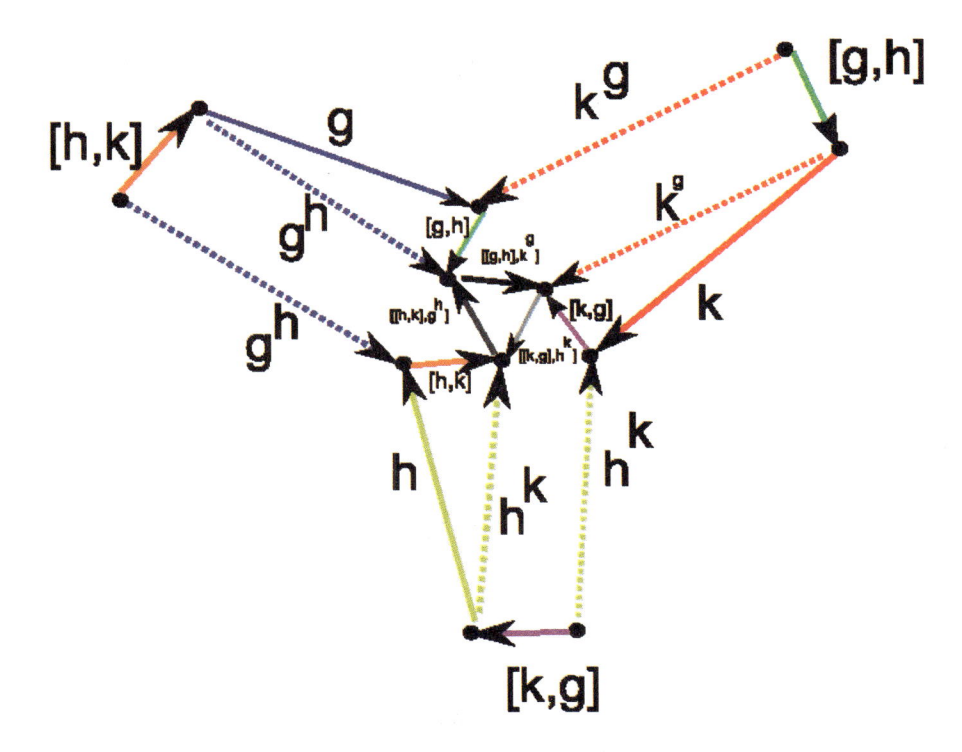

Figure 24. The central portion of the Hall-Witt.

A special case of a 2-cocycle is a *2-coboundary*, defined as a 2-cochain df of the form

$$df(g,h) := f(g) - f(gh) + f(h)$$

for some 1-cochain $f\colon G \to U$. In other words, the integral of df on any (oriented) triangle such as that in Figure 4 is equal to the integral of f on the boundary of that triangle, and more generally we have the "Stokes theorem"

$$\int_S df = \int_{\partial S} f$$

for any oriented surface S in a diagram with boundary ∂S. One can quotient the space $Z^2(G,U)$ of all 2-cocycles (which is a U-module) by the space of all 2-coboundaries $B^2(G,U)$, leading to the 2-*cohomology group* $H^2(G,U)$, which is then seen to be closely analogous to the 2-cohomology group in either *simplicial cohomology* or *de Rham cohomology*. (One can of course perform the same construction for any order. For instance, in the case of 1-cohomology, the space of 1-coboundaries turns out to be trivial in group cohomology, and so the first cohomology group $H^1(G,U)$ is isomorphic to the space $Z^1(G,U)$ of 1-cocycles, or in other words the space $\mathrm{Hom}(G,U)$ of homomorphisms from G to U.)

184   8. Cayley graphs and the algebra of groups

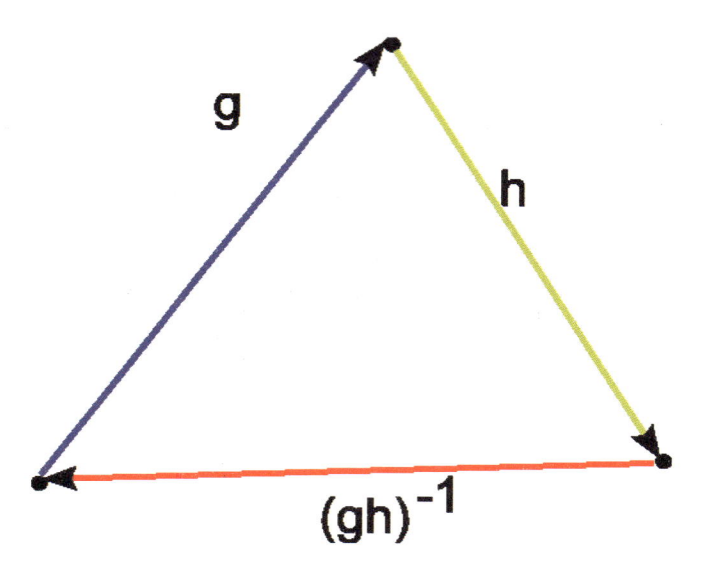

Figure 25. A triangle with a consistent arrow orientation.

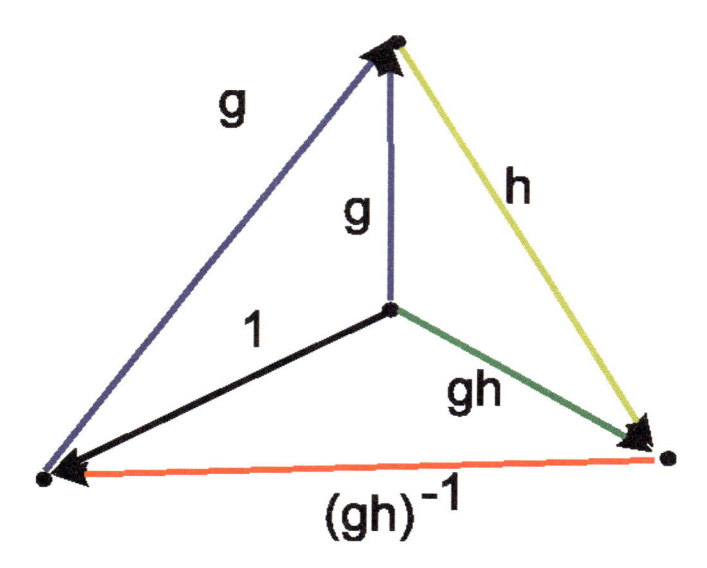

Figure 26. A partition of the previous triangle.

We will need to integrate a 2-cocycle ρ on triangles Δ in which all arrows point in the same direction, as depicted in Figure 25.

This can be done by adding another point to decompose the triangle into one which we can already integrate; see Figure 26.

8.1. A Hall-Witt identity for 2-cocycles

Thus we see (if we give Δ the clockwise orientation) that

(8.4) $$\int_\Delta \rho = \rho(g,h) + \rho(1,g) + \rho(gh, (gh)^{-1});$$

of course, other expressions for $\int_\Delta \rho$ are possible by performing other oriented triangulations of Figure 25. In practice we can simplify these sorts of expressions by normalising the 2-cocycle to the conditions

(8.5) $$\rho(1,g) = \rho(g,1) = \rho(g, g^{-1}) = 0$$

by subtracting the coboundary df, where $f(g) := (\rho(g, g^{-1}) + \rho(1,1))/2$; a brief calculation using the cocycle equation (8.2) reveals that any 2-cocycle will obey (8.5) after subtracting off df. When one achieves this normalisation, then the integral of ρ on the triangle Δ is simply $\rho(g,h)$; also, the integral on triangles in which one of the edges is the identity is automatically zero, and the integral on the loop in Figure 5 is also zero. Thus, 2-cocycles which are normalised by (8.5) can be viewed as being quite analogous to closed 2-forms in de Rham cohomology.

Now we apply the above formalism to the truncated parallelopiped used to prove the Hall-Witt identity. We glue together Figures 18, 21, and 24 to obtain a closed 2-surface. If ρ is a 2-cocycle normalised by (8.5), the integral on this surface vanishes. On the other hand, we see that there is a lot of cancellation in this integral; in particular, all of the distorted parallelograms and triangles that appear in Figure 18, also appear (with the opposite orientation) in either Figure 21 or Figure 24. Cancelling out these faces, we are left with the three distorted parallelograms in Figure 24, together with the central triangle in Figure 24. Evaluating these integrals, we conclude the Hall-Witt identity for (normalised) 2-cocycles:

$$\begin{aligned}
&(\rho(g^h, [h,k]) + \rho(g^h[h,k], [[h,k], g^h]) - \rho([h,k], g^h)) \\
&+ (\rho(h^k, [k,g]) + \rho(h^k[k,g], [[k,g], h^k]) - \rho([k,g], h^k)) \\
&+ (\rho(k^g, [g,h]) + \rho(k^g[g,h], [[g,h], k^g]) - \rho([g,h], k^g)) \\
&- \rho([[h,k], g^h], [[k,g], h^k]) \\
&= 0.
\end{aligned}$$

Thus, for instance, if g,h and g,k commute (so that $g, [h,k]$ also commute), the Hall-Witt identity tells us that

$$\rho(g, [h,k]) = \rho([h,k], g)$$

or, in other words, that the integral of ρ on any parallelogram with edges $g, [h,k]$ vanishes. This can also be seen by noting how the Hall-Witt truncated parallelopiped degenerates in the presence of so much commutativity.

The Hall-Witt identity for cocycles can also be derived from the group extension interpretation of a 2-cocycle. Observe that if $\rho \colon G \times G \to U$ is a

2-cocycle, then we can form a new group G' whose elements are pairs (g, u) with $g \in G$ and $u \in U$, and whose group multiplication law is given by

$$(g, u)(h, v) := (gh, u + v + \rho(g, h));$$

one can easily verify that this law is associative when ρ is a 2-cocycle, and defines a group structure on G' with identity $(1, -\rho(1, 1))$ and inverse map

$$(g, u)^{-1} := (g^{-1}, -u - \rho(g, g^{-1}) - \rho(1, 1)).$$

(If we normalise ρ to obey (8.5), then the identity simplifies to $(1, 0)$, and the inverse operation simplifies to $(g, u)^{-1} = (g^{-1}, -u)$.) The group G' is then a *central group extension* of G by U, and indeed it is not difficult to see that all central group extensions of G by U arise in this manner (with the extensions being isomorphic relative to the base group G if the underlying 2-cocycles differ by a 2-coboundary). One can then deduce the Hall-Witt identity for cocycles $\rho \in Z^2(G, U)$ by applying the ordinary Hall-Witt identity to the elements $(g, 0), (h, 0), (k, 0)$ in the extended group G'; we omit the details. However, it appears that the group extension interpretation of 2-cohomology does not easily extend to higher cohomology (unless perhaps one works with more general notions than groups, such as *n-groups*), whereas the simplicial approach given in this article has a more obvious extension to higher cohomology.

Chapter 9

The Lang-Weil bound

Let F be a finite field, with *algebraic closure* $\overline{F}$, and let V be an (affine) *algebraic variety* defined over $\overline{F}$, by which I mean a set of the form

$$V = \{x \in \overline{F}^d : P_1(x) = \ldots = P_m(x) = 0\}$$

for some ambient dimension $d \geq 0$, and some finite number of polynomials $P_1, \ldots, P_m \colon \overline{F}^d \to \overline{F}$. In order to reduce the number of subscripts later on, let us say that V has *complexity at most* M if d, m, and the degrees of the $P_1, \ldots, P_m$ are all less than or equal to M. Note that we do not require at this stage that V be irreducible (i.e., not the union of two strictly smaller varieties), or defined over F, though we will often specialise to these cases later in this chapter. (Also, everything said here can also be applied with almost no changes to projective varieties, but we will stick with affine varieties for the sake of concreteness.)

One can consider two crude measures of how "big" the variety V is. The first measure, which is algebraically geometric in nature, is the *dimension* $\dim(V)$ of the variety V, which is an integer between 0 and d (or, depending on convention, $-\infty$, -1, or undefined, if V is empty) that can be defined in a large number of ways (e.g., it is the largest r for which the generic linear projection from V to $\overline{F}^r$ is dominant, or the smallest r for which the intersection with a generic codimension r subspace is nonempty). The second measure, which is number-theoretic in nature, is the number $|V(F)| = |V \cap F^d|$ of F-points of V, i.e., points $x = (x_1, \ldots, x_d)$ in V all of whose coefficients lie in the finite field, or equivalently the number of solutions to the system of equations $P_i(x_1, \ldots, x_d) = 0$ for $i = 1, \ldots, m$ with variables $x_1, \ldots, x_d$ in F.

These two measures are linked together in a number of ways. For instance, we have the basic *Schwarz-Zippel type bound* (which, in this qualitative form, goes back at least to [**LaWe1954**, Lemma 1]).

Lemma 9.0.1 (Schwarz-Zippel type bound). *Let V be a variety of complexity at most M. Then we have $|V(F)| \ll_M |F|^{\dim(V)}$.*

This lemma may be compared with Exercise 5.4.3, which gives a more precise bound in the case that V is a hypersurface.

Proof (Sketch). For the purposes of exposition, we will not carefully track the dependencies of implied constants on the complexity M, instead we simply assume that all of these quantities remain controlled throughout the argument[1].

We argue by induction on the ambient dimension d of the variety V. The $d = 0$ case is trivial, so suppose $d \geq 1$ and that the claim has already been proven for $d - 1$. By breaking up V into irreducible components we may assume that V is irreducible (this requires some control on the number and complexity of these components, but this is available, as discussed in [**Ta2012**, §2.1]). For each $x_1, \ldots, x_{d-1} \in \overline{F}$, the fibre $\{x_d \in \overline{F} : (x_1, \ldots, x_{d-1}, x_d) \in V\}$ is either one-dimensional (and thus all of $\overline{F}$) or zero-dimensional. In the latter case, one has $O_M(1)$ points in the fibre from the fundamental theorem of algebra (indeed one has a bound of D in this case), and $(x_1, \ldots, x_{d-1})$ lives in the projection of V to $\overline{F}^{d-1}$, which is a variety of dimension at most $\dim(V)$ and controlled complexity, so the contribution of this case is acceptable from the induction hypothesis. In the former case, the fibre contributes $|F|$ F-points, but $(x_1, \ldots, x_{d-1})$ lies in a variety in $\overline{F}^{d-1}$ of dimension at most $\dim(V) - 1$ (since otherwise V would contain a subvariety of dimension at least $\dim(V) + 1$, which is absurd) and controlled complexity, and so the contribution of this case is also acceptable from the induction hypothesis. □

One can improve the bound on the implied constant to be linear in the degree of V (see, e.g., [**DvKoLo2012**, Claim 7.2], [**ElObTa2010**, Lemma A.3], or the classical Schwarz-Zippel lemma [**Sc1980**], [**Zi1979**]), but we will not be concerned with these improvements here.

---

[1]If one wished, one could obtain ineffective bounds on these quantities by an ultralimit argument, as discussed in [**Ta2012**, §2.1], or equivalently by moving everything over to a nonstandard analysis framework; one could also obtain such uniformity using the machinery of schemes.

9. The Lang-Weil bound

Without further hypotheses on V, the above upper bound is sharp (except for improvements in the implied constants). For instance, the variety

$$V := \{(x_1, \ldots, x_d) \in \overline{F}^d : \prod_{j=1}^{D}(x_d - a_j) = 0\},$$

where $a_1, \ldots, a_D \in F$ are distinct, is the union of D distinct hyperplanes of dimension $d-1$, with $|V(F)| = D|F|^{d-1}$ and complexity $\max(D,d)$; similar examples can easily be concocted for other choices of $\dim(V)$. In the other direction, there is also no nontrivial lower bound for $|V(F)|$ without further hypotheses on V. For a trivial example, if a is an element of $\overline{F}$ that does not lie in F, then the hyperplane

$$V := \{(x_1, \ldots, x_d) \in \overline{F}^d : x_d - a = 0\}$$

clearly has no F-points whatsoever, despite being a $d-1$-dimensional variety in $\overline{F}^d$ of complexity d. For a slightly less nontrivial example, if a is an element of F that is not a quadratic residue, then the variety

$$V := \{(x_1, \ldots, x_d) \in \overline{F}^d : x_d^2 - a = 0\},$$

which is the union of two hyperplanes, still has no F-points, even though this time the variety is defined over F instead of $\overline{F}$ (by which we mean that the defining polynomial(s) have all of their coefficients in F). There is, however, the important *Lang-Weil bound* [**LaWe1954**] that allows for a much better estimate as long as V is both defined over F and irreducible:

Theorem 9.0.2 (Lang-Weil bound). *Let V be a variety of complexity at most M. Assume that V is defined over F, and that V is irreducible as a variety over $\overline{F}$ (i.e., V is geometrically irreducible or absolutely irreducible). Then*

$$|V(F)| = (1 + O_M(|F|^{-1/2}))|F|^{\dim(V)}.$$

Again, more explicit bounds on the implied constant here are known, but will not be the focus of this chapter. As the previous examples show, the hypotheses of definability over F and geometric irreducibility are both necessary.

The Lang-Weil bound is already nontrivial in the model case $d = 2$, $\dim(V) = 1$ of plane curves:

Theorem 9.0.3 (Hasse-Weil bound). *Let $P : \overline{F}^2 \to \overline{F}$ be an irreducible polynomial of degree D with coefficients in F. Then*

$$|\{(x,y) \in F^2 : P(x,y) = 0\}| = |F| + O_D(|F|^{1/2}).$$

Thus, for instance, if $a, b \in F$, then the elliptic curve $\{(x,y) \in F^2 : y^2 = x^3 + ax + b\}$ has $|F| + O(|F|^{1/2})$ F-points, a result first established by Hasse [**Ha1936**]. The Hasse-Weil bound is already quite nontrivial, being the analogue of the Riemann hypothesis for plane curves. For hyper-elliptic curves, an elementary proof (due to Stepanov) is discussed in [**Ta2010b**, §2.5]. For general plane curves, the first proof was by Weil [**We1949**] (leading to his famous *Weil conjectures*); there is also a nice version of Stepanov's argument due to Bombieri [**Bo1974**] covering this case which is a little less elementary (relying crucially on the *Riemann-Roch theorem* for the upper bound, and a lifting trick to then get the lower bound), which I briefly summarise later in this chapter. The full Lang-Weil bound is deduced from the Hasse-Weil bound by an induction argument using generic hyperplane slicing, as I will also summarise later in this chapter.

The hypotheses of definability over F and geometric irreducibility in the Lang-Weil can be removed after inserting a geometric factor:

Corollary 9.0.4 (Lang-Weil bound, alternate form). *Let V be a variety of complexity at most M. Then one has*

$$|V(F)| = (c(V) + O_M(|F|^{-1/2}))|F|^{\dim(V)}$$

where $c(V)$ is the number of top-dimensional components of V (i.e., geometrically irreducible components of V of dimension $\dim(V)$) that are definable over F, or equivalently are invariant with respect to the Frobenius endomorphism $x \mapsto x^{|F|}$ *that defines F.*

Proof. By breaking up a general variety V into components (and using Lemma 9.0.1 to dispose of any lower-dimensional components), it suffices to establish this claim when V is itself geometrically irreducible. If V is definable over F, the claim follows from Theorem 9.0.2. If V is not definable over F, then it is not fixed by the Frobenius endomorphism Frob (since otherwise one could produce a set of defining polynomials that were fixed by Frobenius and thus defined over F by using some canonical basis (such as a *reduced Grobner basis*) for the associated ideal), and so $V \cap \mathrm{Frob}(V)$ has strictly smaller dimension than V. But $V \cap \mathrm{Frob}(V)$ captures all the F-points of V, so in this case the claim follows from Lemma 9.0.1. □

Note that if V is reducible but is itself defined over F, then the Frobenius endomorphism preserves V itself, but may permute the components of V around. In this case, $c(V)$ is the number of fixed points of this permutation action of Frobenius on the components. In particular, $c(V)$ is always a natural number between 0 and $O_M(1)$; thus we see that regardless of the geometry of V, the normalised count $|V(F)|/|F|^{\dim(V)}$ is asymptotically

9. The Lang-Weil bound

restricted to a bounded range of natural numbers (in the regime where the complexity stays bounded and $|F|$ goes to infinity).

Example 9.0.5. Consider the variety
$$V := \{(x,y) \in \overline{F}^2 : x^2 - ay^2 = 0\}$$
for some nonzero parameter $a \in F$. Geometrically (by which we basically mean "when viewed over the algebraically closed field $\overline{F}$"), this is the union of two lines, with slopes corresponding to the two square roots of a. If a is a quadratic residue, then both of these lines are defined over F, and are fixed by Frobenius, and $c(V) = 2$ in this case. If a is not a quadratic residue, then the lines are not defined over F, and the Frobenius automorphism permutes the two lines while preserving V as a whole, giving $c(V) = 0$ in this case.

Corollary 9.0.4 effectively computes (at least to leading order) the number-theoretic size $|V(F)|$ of a variety in terms of geometric information about V, namely its dimension $\dim(V)$ and the number $c(V)$ of top-dimensional components fixed by Frobenius. It turns out that with a little bit more effort, one can extend this connection to cover not just a single variety V, but a family of varieties indexed by points in some base space W. More precisely, suppose we now have two affine varieties V, W of bounded complexity, together with a *regular map* $\phi \colon V \to W$ of bounded complexity[2]. It will be convenient to assume that the base space W is irreducible. If the map ϕ is a *dominant map* (i.e., the image $\phi(V)$ is Zariski dense in W), then standard algebraic geometry results tell us that the fibres $\phi^{-1}(\{w\})$ are an unramified family of $\dim(V) - \dim(W)$-dimensional varieties outside of an exceptional subset W' of W of dimension strictly smaller than $\dim(W)$ (and with $\phi^{-1}(W')$ having dimension strictly smaller than $\dim(V)$); see, e.g., [**Sh1997**, §I.6.3].

Now suppose that V, W, and ϕ are defined over F. Then, by Lang-Weil, $W(F)$ has $(1+O(|F|^{-1/2}))|F|^{\dim(W)}$ F-points, and by Schwarz-Zippel, for all but $O(|F|^{\dim(W)-1})$ of these F-points w (the ones that lie in the subvariety W'), the fibre $\phi^{-1}(\{w\})$ is an algebraic variety defined over F of dimension $\dim(V) - \dim(W)$. By using ultraproduct arguments (see, e.g., [**BrGrTa2011**, Lemma 3.7]), this variety can be shown to have bounded complexity, and thus by Corollary 9.0.4, has
$$(c(\phi^{-1}(\{w\})) + O(|F|^{-1/2})|F|^{\dim(V)-\dim(W)}$$
F-points. One can then ask how the quantity $c(\phi^{-1}(\{w\}))$ is distributed. A simple but illustrative example occurs when $V = W = F$ and $\phi \colon F \to F$ is

---

[2] The definition of complexity of a regular map is a bit technical; see, e.g., [**BrGrTa2011**] for details. But one can think, for instance, of a polynomial or rational map of bounded degree as a good example of a map of bounded complexity.

the polynomial $\phi(x) := x^2$. Then $c(\phi^{-1}(\{w\})$ equals 2 when w is a nonzero quadratic residue and 0 when w is a nonzero quadratic nonresidue (and 1 when w is zero, but this is a negligible fraction of all w). In particular, in the asymptotic limit $|F| \to \infty$, $c(\phi^{-1}(\{w\}))$ is equal to 2 half of the time and 0 half of the time.

Now we describe the asymptotic distribution of the $c(\phi^{-1}(\{w\}))$. We need some additional notation. Let w_0 be an F-point in $W \backslash W'$, and let $\pi_0(\phi^{-1}(\{w_0\}))$ be the connected components of the fibre $\phi^{-1}(\{w_0\})$. As $\phi^{-1}(\{w_0\})$ is defined over F, this set of components is permuted by the Frobenius endomorphism $Frob$. But there is also an action by monodromy of the fundamental group $\pi_1(W \backslash W')$ (this requires a certain amount of *étale machinery* to properly set up, as we are working over a positive characteristic field rather than over the complex numbers, but I am going to ignore this rather important detail here). This fundamental group may be infinite, but (by the étale construction) is always *profinite*, and in particular has a *Haar probability measure*, in which every finite index subgroup (and their cosets) are measurable. Thus we may meaningfully talk about elements drawn uniformly at random from this group, so long as we work only with the profinite σ-algebra on $\pi_1(W \backslash W')$ that is generated by the cosets of the finite index subgroups of this group (which will be the only relevant sets we need to measure when considering the action of this group on finite sets, such as the components of a generic fibre).

Theorem 9.0.6 (Lang-Weil with parameters). *Let V, W be varieties of complexity at most M with W irreducible, and let $\phi \colon V \to W$ be a dominant map of complexity at most M. Let w_0 be an F-point of $W \backslash W'$. Then, for any natural number a, one has $c(\phi^{-1}(\{w\})) = a$ for $(\mathbf{P}(X = a) + O_M(|F|^{-1/2}))|F|^{\dim(W)}$ values of $w \in W(F)$, where X is the random variable that counts the number of components of a generic fibre $\phi^{-1}(w_0)$ that are invariant under $g \circ \mathrm{Frob}$, where g is an element chosen uniformly at random from the étale fundamental group $\pi_1(W \backslash W')$. In particular, in the asymptotic limit $|F| \to \infty$, and with w chosen uniformly at random from $W(F)$, $c(\phi^{-1}(\{w\}))$ (or, equivalently, $|\phi^{-1}(\{w\})(F)|/|F|^{\dim(V) - \dim(W)}$) and X have the same asymptotic distribution.*

This theorem generalises Corollary 9.0.4 (which is the case when W is just a point, so that $\phi^{-1}(\{w\})$ is just V and g is trivial). Informally, the effect of a nontrivial parameter space W on the Lang-Weil bound is to push around the Frobenius map by monodromy for the purposes of counting invariant components, and a randomly chosen set of parameters corresponds to a randomly chosen loop on which to perform monodromy.

Example 9.0.7. Let $V = W = F$ and $\phi(x) = x^m$ for some fixed $m \geq 1$; to avoid some technical issues let us suppose that m is coprime to $|F|$.

Then W' can be taken to be $\{0\}$, and for a base point $w_0 \in W\backslash W'$ we can take $w_0 = 1$. The fibre $\phi^{-1}(\{1\})$—the mth roots of unity—can be identified with the cyclic group $\mathbf{Z}/m\mathbf{Z}$ by using a primitive root of unity. The étale fundamental group $\pi(W\backslash W') = \pi(\overline{F}\backslash 0)$ is (I think) isomorphic to the profinite closure $\hat{\mathbf{Z}}$ of the integers $\mathbf{Z}$ (excluding the part of that closure coming from the characteristic of F). Not coincidentally, the integers $\mathbf{Z}$ are the fundamental group of the complex analogue $\mathbf{C}\backslash\{0\}$ of $W\backslash W'$. (Brian Conrad points out to me though that for more complicated varieties, such as covers of $\overline{F}\backslash\{0\}$ by a power of the characteristic, the etale fundamental group is more complicated than just a profinite closure of the ordinary fundamental group, due to the presence of *Artin-Schreier covers* that are only ramified at infinity.) The action of this fundamental group on the fibres $\mathbf{Z}/m\mathbf{Z}$ can be given by translation. Meanwhile, the Frobenius map Frob on $\mathbf{Z}/m\mathbf{Z}$ is given by multiplication by m. A random element $g \circ$ Frob then becomes a random affine map $x \mapsto |F|x + b$ on $\mathbf{Z}/m\mathbf{Z}$, where b chosen uniformly at random from $\mathbf{Z}/m\mathbf{Z}$. The number of fixed points of this map is equal to the greatest common divisor $(|F|-1, m)$ of $|F|-1$ and m when b is divisible by $(|F|-1, m)$, and equal to 0 otherwise. This matches up with the elementary number fact that a randomly chosen nonzero element of F will be an mth power with probability $1/(|F|-1, m)$, and when this occurs, the number of mth roots in F will be $(|F|-1, m)$.

Example 9.0.8. (Thanks to Jordan Ellenberg for this example.) Consider a random elliptic curve $E = \{y^2 = x^3 + ax + b\}$, where a, b are chosen uniformly at random, and let $m \geq 1$. Let $E[m]$ be the m-torsion points of E (i.e., those elements $g \in E$ with $mg = 0$ using the elliptic curve addition law); as a group, this is isomorphic to $\mathbf{Z}/m\mathbf{Z} \times \mathbf{Z}/m\mathbf{Z}$ (assuming that F has sufficiently large characteristic, for simplicity), and consider the number of F points of $E[m]$, which is a random variable taking values in the natural numbers between 0 and m^2. In this case, the base variety W is the *modular curve* $X(1)$, and the covering variety V is the modular curve $X_1(m)$. The generic fibre here can be identified with $\mathbf{Z}/m\mathbf{Z} \times \mathbf{Z}/m\mathbf{Z}$, the monodromy action projects down to the action of $\text{SL}_2(\mathbf{Z}/m\mathbf{Z})$, and the action of Frobenius on this fibre can be shown to be given by a 2×2 matrix with determinant $|F|$ (with the exact choice of matrix depending on the choice of fibre and of the identification), so the distribution of the number of F-points of $E[m]$ is asymptotic to the distribution of the number of fixed points X of a random linear map of determinant $|F|$ on $\mathbf{Z}/m\mathbf{Z} \times \mathbf{Z}/m\mathbf{Z}$.

Theorem 9.0.6 seems to be well-known "folklore" among arithmetic geometers, though I do not know of an explicit reference for it. It can be derived from the ordinary Lang-Weil theorem and the moment method; see below.

Many thanks to Brian Conrad and Jordan Ellenberg for helpful discussions on these topics.

9.1. The Stepanov-Bombieri proof of the Hasse-Weil bound

We now give (most of) the Stepanov-Bombieri proof of Theorem 9.0.3, following the exposition of Bombieri [**Bo1974**] (see also [**Ko2010**]), focusing on the model case of curves of the form $\{(x, y) : y^d = P(x)\}$. In this section all implied constants are allowed to depend on the degree D of the polynomial $P(x, y)$.

Let $C(F)$ be the F-points of the curve $C := \{(x, y) \in \overline{F}^2 : P(x, y) = 0\}$; the hypothesis that P is irreducible means that the curve C is also irreducible. Our task is to establish the upper bound

(9.1) $$|C(F)| \leq |F| + O(|F|^{1/2})$$

and the lower bound

(9.2) $$|C(F)| \geq |F| - O(|F|^{1/2}).$$

For technical reasons, we only prove these bounds directly when $|F|$ is a perfect square; the general case then follows by using the explicit formula for $C[F_{p^r}]$ as a function of r and the tensor power trick, as explained in [**Ta2008**, §1.9].

Now we prove the upper bound. One uses Stepanov's *polynomial method*. The basic idea here is to bound the size of a finite set $C(F)$ by constructing a nontrivial polynomial of controlled degree that vanishes to high order at each point of $C(F)$. Stepanov's original argument projected the curve C onto an affine line $\overline{F}$ so that one could use ordinary one-dimensional polynomials, but Bombieri's approach works directly on the curve C. For this it becomes convenient to work in the language of *divisors* of rational functions, rather than zeroes of polynomials (and also it becomes slightly more convenient to work projectively rather than affinely, though I will gloss over this minor detail). Given a rational function f on the curve C (with coefficients in $\overline{F}$) that is not identically zero or identically infinite on C, one can define the divisor (f) to be the formal sum $\sum_Q m_Q Q$ of the zeroes and poles Q of f on C, weighted by the multiplicity m_Q of the zero or pole (with poles being viewed as having negative multiplicity; one has to be a little careful defining multiplicity at singular points of C, but this can be done by using a sufficiently algebraic formalism). The irreducibility of C (which, of course, must be used at some juncture to prove the Hasse-Weil bound) ensures that this formal sum only has finitely many nonzero terms. One can show that the total signed multiplicities of zeroes and poles of (f) always add up to

9.1. The Stepanov-Bombieri proof of the Hasse-Weil bound

zero: $\sum_Q m_Q = 0$. (This generalises the fundamental theorem of algebra, which is the case when C is a projective line.)

Fix a point P_∞ of C (which one can think of as being at infinity), and for each degree M, let $L(MP_\infty)$ be the space of those rational functions f which are either zero, or have $(f) \geq -MP_\infty$, i.e., f has a pole of order M at P_∞ but no other poles; this generalises the space of polynomials of degree at most M, which corresponds to the case when C is a projective line and P_∞ is the point at infinity. It is easy to see that $L(MP_\infty)$ is a vector space (over $\overline{F}$), which is nondecreasing in M. In the case when C is a line, this space clearly has dimension $M+1$. The *Riemann-Roch theorem* is the generalisation of this to the case of a general curve. In its most precise form, it asserts the identity

$$\dim L(MP_\infty) + \dim L(K - MP_\infty) = M - g + 1$$

where g is the *genus* of C, and K is the *canonical divisor* of C (i.e., the divisor of the canonical line bundle of C). For our application, we will need two specific corollaries of the Riemann-Roch theorem, coming from the nonnegativity and monotonicity properties of L. The first is the *Riemann inequality*

$$M - g + 1 \leq \dim L(MP_\infty) \leq M + 1$$

(so in particular $\dim L(MP_\infty) = M + O(1)$), and the second is the inequality

$$\dim L(MP_\infty) \leq \dim L((M+1)P_\infty) \leq \dim L(MP_\infty) + 1$$

for all M. In particular, if we set $\mathcal{N}$ to be the set of all degrees M such that $\dim L(MP_\infty) = \dim L((M-1)P_\infty) + 1$, then $\mathcal{N}$ consists of all but at most g elements of the natural numbers, and we can find a sequence $\{e_M : M \in \mathcal{N}\}$ of rational functions on C with $e_M \in \dim(MP_\infty) \setminus \dim((M-1)P_\infty)$ (i.e., e_M has a pole of order exactly M at P_∞ and no other poles), with each $L(MP_\infty)$ being spanned by the $f_{M'}$ with $M' \leq M$. The e_M can be viewed as a generalisation of the standard monomial basis $1, x, x^2, \ldots$ of the polynomials of one variable. A key point here is that the degrees of this basis are all distinct; this will be crucial later on for ensuring that the polynomial we will be using for Stepanov's method does not vanish identically.

We will not prove the Riemann-Roch theorem here, and simply assume it (and its consequences) as a black box. (A proof may be found for instance in [**Gr1994**, p. 245].) Next, we consider the Frobenius map $\text{Frob}\colon \overline{F}^2 \to \overline{F}^2$ defined by $\text{Frob}(x,y) := (x^{|F|}, y^{|F|})$. As C is defined over F, this map preserves C, and the fixed points of this map are precisely the F-points of C.

The plan is now to find a nontrivial element $f \in L(M_*P_\infty)$ for some controlled M_* that vanishes to order at least m at each fixed point of Frob

on C for some m. As the total number of zeroes and poles of f must agree, this forces
$$m|C(F)| \leq M_*$$
and by optimising the parameters m, M_* this should lead to (9.1).

The trick is to pick rational functions f of a specific form, namely
$$(9.3) \qquad f(x,y) = \sum_{M \in \mathcal{N}: M \leq M_0} f_M(x,y)^m e_M(\mathrm{Frob}(x,y))$$
where M_0 is a parameter to be optimised later, and for each M, f_M lies in $L(M_1 P_\infty)$ for another parameter M_1 to be optimised in later. We also pick m to divide $|F|$, thus m is a power of the characteristic p of F that is less than or equal to $|F|$. There are several reasons to working with polynomials of this form. The first is that because of the Frobenius endomorphism identity $e_M(\mathrm{Frob}(x,y)) = e_M(x,y)^{|F|}$, and m divides $|F|$ by hypothesis, each term in $f(x,y)$ is an mth power, and hence (since $a^m + b^m = (a+b)^m$ in characteristic p) f is itself the mth power of some rational function. As such, whenever f vanishes at a point, it automatically vanishes to order at least m.

Secondly, in order for f to vanish at every fixed point of the Frobenius map on C (and thus vanish to order m at each such point, by the above discussion), it is of course sufficient that
$$(9.4) \qquad \sum_{M \in \mathcal{N}: M \leq M_0} f_M(x,y)^m e_M(x,y) = 0$$
where we view the left-hand side as a rational function in two indeterminates x, y. This is a constraint on the f_M which is linear over F_p (because the map $x \mapsto x^m$ is linear over F_p, though not over F), and so we can try to enforce this constraint by linear algebra. Note that if each f_M lies in $L(M_1 P_\infty)$, then the rational function on the left-hand side of of (9.4) lies in $L((mM_1 + M_0)P_\infty)$, so the vanishing (9.4) imposes $r \dim(L((mM_1 + M_0)P_\infty)) = r(mM_1 + M_0 + O(1))$ homogeneous linear constraints (possibly dependent) over F_p on the f_M, where $|F| =: p^r$. On the other hand, the dimension (over F_p) of the space of all possible f_M is
$$r|\{M \in \mathcal{N} : M \leq M_0\}| \dim(M_1 P_\infty) = r(M_0 + O(1))(M_1 + O(1)).$$
Thus, by linear algebra, we can find a collection of $f_M \in L(M_1 P_\infty)$, not all zero, obeying (9.4) as long as we have
$$(9.5) \qquad (M_0 - C)_+ (M_1 - C)_+ > mM_1 + M_0 + C.$$

We are not done yet, because even if the f_M are are not all zero, it is conceivable that the combined function (9.3) could still vanish. But observe that if M is the largest index for which f_M is nonzero, then the rational function $f_M(x,y)^m e_M(\mathrm{Frob}(x,y))$ has a pole of order at least $M|F|$ at P_∞, and all the other terms have a pole of order at most $(M-1)|F| + mM_1$

(here it is crucial that the e_M have poles of different orders at P_∞, thanks to Riemann-Roch). So as long as we have

$$(9.6) \qquad mM_1 < |F|,$$

we have that f does not vanish identically. As it vanishes to degree at least m at every point of $C(F)$, and lies in $L((mM_1 + |F|M_0)P_\infty)$, and so we obtain the upper bound

$$|C(F)| \leq \frac{mM_1 + |F|M_0}{m}.$$

Now we optimise in the parameters m, M_0, M_1 subject to the constraints (9.5), (9.6). As $|F|$ is assumed to be a perfect square and is sufficiently large, then a little work shows that one can satisfy the constraints with $m := |F|^{1/2}$, $M_1 := |F|^{1/2} - 1$, and $M_0 := |F|^{1/2} + C'$ for some sufficiently large C', leading to the desired bound (9.1).

Remark 9.1.1. When $|F|$ is not a perfect square, but is not a prime, one can still choose other values of m and obtain a weaker version of (9.1) that is still nontrivial. But it is curious that Bombieri's version of Stepanov's argument breaks down completely when F has prime order. As mentioned previously, one can still recover this case *a posteriori* by the tensor product trick, but this requires the explicit formula for $|C[F_{p^r}]|$ which is not entirely trivial (it relies again on the Riemann-Roch theorem, see, e.g., [**IwKo2004**, Ch. 11]). On the other hand, other versions of Stepanov's argument do work in the prime order case, so this obstruction may be purely artificial in nature.

Now we need to pass from the upper bound (9.1) to the lower bound (9.2). In model cases, such as curves of the form $C = \{(x,y) : y^d = P(x)\}$ with $d-1$ not divisible by the characteristic of F, one can proceed by observing that the multiplicative group $F^\times$ foliates into d cosets $g_1H, \ldots, g_dH$ of the subgroup $H := \{x^d : x \in F^\times\}$ of dth powers. Because of this, we see that the union $C' := C_1 \cup \ldots \cup C_d$ of the curves $C_i := \{(x,y) : g_iy^d = P(x)\}$ contains exactly d F-points on all but $O(1)$ vertical lines $\{x = a\}$, $a \in F$, and hence the union has cardinality $d|F| + O(1)$ F-points. (In other words, the projection $(x,y) \mapsto x$ from C' to $\overline{F}$ is generically d-to-one. Also, as the C_i are all dilations of C, they will be irreducible if C is, and thus by (9.1) they each have at most $|F| + O(|F|^{1/2})$ F-points; from subtracting all but one of the curves from $C_1 \cup \ldots \cup C_d$ we then see that each of the C_i has at least $|F| - O(|F|^{1/2})$ F-points also. As we can take one of the g_i to be the identity, the claim (9.2) follows in this case. The general case follows in a similar fashion, using Galois theory to lifting a general curve C to another curve C' whose projection to some coordinate line $\overline{F}$ has fixed multiplicity (basically by ensuring that the function field of the former is a Galois extension of the function field of the latter); see [**Bo1974**] for details.

9.2. The proof of the Lang-Weil bound

Now we prove the Lang-Weil bound. We allow all implied constants to depend on the complexity bound M, and will ignore the issues of how to control the complexity of the various algebraic objects that arise in our analysis (as before, one can obtain such bounds more or less "for free" from an ultraproduct argument, if desired).

The proof proceeds by induction on the dimension $\dim(V)$ of the variety V. The first nontrivial case occurs at dimension one. If V is a plane curve (so $\dim(V) = 1$, and the ambient dimension is 2), then the claim follows directly from the Hasse-Weil bound. If V is a higher-dimensional curve, we can use *resultants* to eliminate all but two of the variables (or, alternatively, one can apply the *primitive element theorem* to the *function field* of the curve), and convert the curve to a *birationally equivalent* (over F) plane curve, to reduce to the plane curve case (noting that the singular points of the birational transformation only influence the number of F-points by $O(1)$ at most).

Now suppose inductively that $\dim(V) \geq 2$, and that the claim has already been proven for irreducible varieties of dimension $\dim(V) - 1$. The induction step is then based on the following two parallel facts about hyperplane slicing, the first in the algebraic geometric category, and the second in the combinatorial category:

Lemma 9.2.1 (Bertini's theorem). *Let V be an irreducible variety in $\overline{F}^d$ of dimension at least two. Then, for a generic affine hyperplane H (over $\overline{F}$), the slice $V \cap H$ is an irreducible variety of dimension $\dim(V) - 1$.*

Lemma 9.2.2 (Random sampling). *Let E be a subset of F^d for some $d = O(1)$. Then, for a affine hyperplane H (defined over F) chosen uniformly at random, the random variable $|E \cap H|$ has mean $|E|/|F|$ and variance $O(|E|/|F|)$. In particular, by Chebyshev's inequality, the median value of $|E \cap H|$ is $|E|/|F| + O((|E|/|F|)^{1/2})$.*

Let us see how the two lemmas conclude the induction. There are $|F|\frac{|F|^d-1}{|F|-1} = (1 + O(|F|^{-1}))|F|^d$ different affine hyperplanes H over F. The set of hyperplanes H over $\overline{F}$ for which $V \cap H$ fails to be an irreducible variety of dimension $\dim(V)$ is, by Bertini's theorem, a proper algebraic variety in the Grassmannian space, which one can show to have bounded complexity; and so by Lemma 9.0.1, only $O(1/|F|)$ of all such hyperplanes

9.2. The proof of the Lang-Weil bound

over F fall into this variety. For the remaining $1 - O(1/|F|)$ such hyperplanes, we may apply the induction hypothesis and conclude that $|V(F) \cap H| = (1 + O(|F|^{-1/2}))|F|^{\dim(V)-1}$. In particular, if H is an affine hyperplane chosen uniformly at random, then the median value of $|V(F) \cap H|$ is $(1 + O(|F|^{-1/2}))|F|^{\dim(V)-1}$. But this is consistent with Lemma 9.2.2 only when $|V(F)| = (1 + O(|F|^{-1/2}))|F|^{\dim(V)}$, closing the induction.

Remark 9.2.3. One can also proceed without the variance bound in Lemma 9.2.2 by using Lemma 9.0.1 as a crude estimate for $|V(F) \cap H|$ for the exceptional H, after first making an easy reduction to the case where $V(F)$ is not already contained inside a single hyperplane. However, I think the variance bound is illuminating, as it illustrates the concentration of measure that occurs when $|E|$ becomes much larger than $|F|$, which only occurs in dimensions two and higher. In contrast, if one attempts to slice a curve by a hyperplane, one typically gets zero or multiple points, so there is no concentration of measure and irreducibility usually fails. So the slicing method can reduce the dimension of V down to one, but not to zero, so the need to invoke Hasse-Weil cannot be avoided by this method.

We first show Lemma 9.2.2, which is a routine first and second moment calculation. Any given point x in F^n lies in precisely $1/|F|$ of the hyperplanes H, so summing in x we obtain $\mathbf{E}|E \cap H| = |E|/|F|$. Next, any two distinct points x, y in F^n lie in $\frac{1}{|F|}\frac{|F|^{d-1}-1}{|F|^d-1} = \frac{1}{|F|^2} + O(|F|^{-d-1})$ of the hyperplanes H, and so

$$\mathbf{E}|E \cap H|^2 = |E|/|F| + |E|(|E|-1)(\frac{1}{|F|^2} + O(|F|^{-d-1}))$$
$$= (\mathbf{E}|E \cap H|)^2 + O(|E|/|F|)$$

giving the variance bound.

Now we show Bertini's theorem. A dimension count shows that the space of pairs (p, H) where H is a hyperplane and p is a singular point of V in H has dimension at most $\dim(V) - 1 + d - 1$, and so for a generic hyperplane, the space of singular points of V in H is contained in a variety of dimension at most $\dim(V) - 2$. Similarly, the space of pairs (p, H) where p is a smooth point of V and H is a hyperplane containing the tangent space to V at p is at most $\dim(V) + d - \dim(V) - 1$, and so the space of smooth points of V in H whose tangent space is not transverse to a generic H also has dimension at most $\dim(V) - 2$. Thus, for generic H, the slice $V \cap H$ is a $\dim(V) - 1$-dimensional variety, which is generically comes from smooth points of V whose tangent space is transverse to H. The remaining task is to establish irreducibility for generic H. As irreducibility is an algebraic condition, $V \cap H$ is either generically irreducible or generically reducible. Suppose for contradiction that $V \cap H$ is generically reducible. Consider the set S of triples (p, q, H) where p, q are distinct smooth points of V, and H is a hyperplane through p, q that is transverse to the tangent spaces of both p and q, with $H \cap V$ reducible. This is a quasiprojective variety of dimension $d + (\dim(V) - 1) + (\dim(V) - 1)$. It can be decomposed into two top-dimensional components: one where p, q lie in the same component of $V \cap H$, and one where they lie in distinct components. (The hypothesis $\dim(V) \geq 2$ is crucial to ensure that the first component is nonempty.) As such, S is disconnected in the Zariski topology. On the other hand, for fixed p, q, the set of all H contributing to S is connected, while the set of all distinct smooth pairs p, q is also connected, and so S (and hence is closure) is also connected, a contradiction.

Remark 9.2.4. See also [**Gr1994**, p. 174] for a slight variant of this argument (thanks to Jordan Ellenberg for pointing this out). In that reference, it is also noted that the claim can also be derived from the *Lefschetz hyperplane theorem* in the case that V is smooth.

9.3. Lang-Weil with parameters

Now we prove Theorem 9.0.6. Again we allow all implied constants to depend on the complexity parameter M, and gloss over the task of making sure that all algebraic objects have complexity $O_M(1)$.

We will proceed by the moment method. Observe that $c(w)$ is a bounded natural number random variable, so to show that $c(w)$ has the asymptotic distribution of X up to errors of $1 + O(|F|^{-1/2})$, it suffices to show that

$$\mathbf{E}_{w \in W(F) \setminus W'(F)} c(w)^k = \mathbf{E} X^k + O(|F|^{-1/2})$$

for any fixed natural number $k = O(1)$, or equivalently (by the Lang-Weil bound for $W(F)$ and for the fibres $\phi^{-1}(\{w\})$)

(9.7) $$\sum_{w \in W(F) \setminus W'(F)} |\phi^{-1}(\{w\})(F)|^k = (\mathbf{E} X^k + O(|F|^{-1/2}))$$
$$\times |F|^{\dim(W) + k(\dim(V) - \dim(W))}.$$

Fix k. We form the k-fold fibre product V_k of V over $W \setminus W'$, consisting of all k-tuples $(v_1, \ldots, v_k) \in V^k$ such that $\phi(v_1) = \ldots = \phi(v_k)$ lies in $W \setminus W'$. This fibre product is a quasiprojective variety of dimension $\dim(W) + k(\dim(V) - \dim(W))$, is defined over F, and the left-hand side of (9.7) is precisely equal to the number of F-points of V_k. Thus, by the Lang-Weil bound for V_k, it suffices to show that

$$c(V_k) = \mathbf{E} X^k,$$

i.e., that V_k has precisely $\mathbf{E} X^k$ top-dimensional components which are Frobenius-invariant.

We will in fact prove a more general statement: if $\phi\colon U \to W$ is any dominant map which has smooth unramified fibres on $W\backslash W'$, then the number $c(U)$ of top-dimensional components of U that are Frobenius-invariant is equal to the expected number of top-dimensional components of $g \circ \text{Frob}$ on a generic fibre $\phi^{-1}(\{w_0\})$, where g is selected uniformly at random from $\pi_1(W\backslash W')$. Specialising this to the fibre product V_k (noting that the generic fibre of V_k is just the k-fold Cartesian power of the generic fibre of k, and similarly for the action of Frobenius and the fundamental group), we obtain the claim.

Now we prove this more general statement. By decomposing into irreducible components, we may certainly assume that U is irreducible. If U is not Frobenius-invariant, then $\text{Frob}(U)$ is generically disjoint from U and so there certainly can be no components of $\phi^{-1}(\{w_0\})$ that are fixed by $g \circ \text{Frob}$. Thus we may assume that U is Frobenius-invariant (i.e., defined over F). Then the situation becomes very close to that of *Burnside's lemma*. If there are K components $S_1, \ldots, S_K$ of $\phi^{-1}(\{w_0\})$, the connectedness of U implies that the fundamental group $\pi_1(W\backslash W')$ acts transitively on these components, and so for any single component S_i and uniformly chosen $g \in \pi_1(W\backslash W')$, gS_i will be uniformly distributed amongst the $S_1, \ldots, S_K$, as will $g(\text{Frob}(S_i))$. In particular, each S_i has a $1/K$ chance of being a fixed point of $g \circ \text{Frob}$, so the expected number of fixed points of $g \circ \text{Frob}$ is 1 as required.

Exercise 9.3.1. Obtain the following extension of Corollary 9.0.6: if one has $k = O(1)$ different dominant maps $\phi_i\colon V_i \to W$ of varieties V_i, W of complexity $O(1)$, with unramified, nonsingular fibres on $W\backslash W'$, then the joint distribution of $c(\phi_1^{-1}(\{w\})), \ldots, c(\phi_k^{-1}(\{w\}))$ converges in distribution to the joint distribution of $X_1, \ldots, X_k$, where each X_i is the number of fixed points of $g \circ \text{Frob}$ on the components of a generic fibre, where g is independent of i and drawn uniformly from $\pi_1(W\backslash W')$.

Remark 9.3.1. While the moment computation is simple and cute, I would be interested in seeing either a heuristic or rigorous explanation of Theorem 9.0.6 (or Exercise 9.3.1) that did not proceed through moments. The theorem seems to be asserting a statement roughly of the following form (ignoring for now the fact that paths do not actually make much sense in positive characteristic): if one takes a random path γ in W connecting two random F-points, then the loop $\text{Frob}(\gamma) - \gamma$ behaves as if it is distributed uniformly in the fundamental group of $\pi_1(W)$ (or $\pi_1(W\backslash W')$, for any lower-dimensional set W'). But I do not know how to make this heuristic precise.

Chapter 10

The spectral theorem and its converses for unbounded self-adjoint operators

Let $L\colon H \to H$ be a self-adjoint operator on a finite-dimensional Hilbert space H. The behaviour of this operator can be completely described by the *spectral theorem* for finite-dimensional self-adjoint operators (i.e., Hermitian matrices, when viewed in coordinates), which provides a sequence $\lambda_1, \ldots, \lambda_n \in \mathbf{R}$ of eigenvalues and an orthonormal basis $e_1, \ldots, e_n$ of eigenfunctions such that $Le_i = \lambda_i e_i$ for all $i = 1, \ldots, n$. In particular, given any function $m\colon \sigma(L) \to \mathbf{C}$ on the spectrum $\sigma(L) := \{\lambda_1, \ldots, \lambda_n\}$ of L, one can then define the linear operator $m(L)\colon H \to H$ by the formula

$$m(L)e_i := m(\lambda_i)e_i,$$

which then gives a *functional calculus*, in the sense that the map $m \mapsto m(L)$ is a C^*-*algebra* isometric homomorphism from the algebra $BC(\sigma(L) \to \mathbf{C})$ of bounded continuous functions from $\sigma(L)$ to $\mathbf{C}$, to the algebra $B(H \to H)$ of bounded linear operators on H; in other words, the map $m \mapsto m(L)$ is a ring homomorphism such that $\overline{m}(L) = m(L)^*$ and $\|m(L)\|_{\mathrm{op}} = \sup_{z \in \sigma(L)} |m(z)|$. Thus, for instance, one can define heat operators e^{-tL} for $t > 0$, Schrödinger operators e^{itL} for $t \in \mathbf{R}$, resolvents $\frac{1}{L-z}$ for $z \notin \sigma(L)$, and (if L is positive) wave operators $e^{it\sqrt{L}}$ for $t \in \mathbf{R}$. These will be bounded operators (and, in the case of the Schrödinger and wave operators, unitary operators, and in the case of the heat operators with L positive, they will be contractions). Among other things, this functional calculus can then be used to solve differential equations such as the heat equation

(10.1) $$u_t + Lu = 0; \quad u(0) = f,$$

the Schrödinger equation

(10.2) $$u_t + iLu = 0; \quad u(0) = f,$$

the wave equation

(10.3) $$u_{tt} + Lu = 0; \quad u(0) = f; \quad u_t(0) = g,$$

or the Helmholtz equation

(10.4) $$(L - z)u = f.$$

The functional calculus can also be associated to a spectral measure. Indeed, for any vectors $f, g \in H$, there is a complex measure $\mu_{f,g}$ on $\sigma(L)$ with the property that

$$\langle m(L)f, g\rangle_H = \int_{\sigma(L)} m(x) d\mu_{f,g}(x);$$

indeed, one can set $\mu_{f,g}$ to be the discrete measure on $\sigma(L)$ defined by the formula

$$\mu_{f,g}(E) := \sum_{i:\lambda_i \in E} \langle f, e_i\rangle_H \langle e_i, g\rangle_H.$$

One can also view this complex measure as a coefficient

$$\mu_{f,g} = \langle \mu f, g\rangle_H$$

of a *projection-valued measure* μ on $\sigma(L)$, defined by setting

$$\mu(E)f := \sum_{i:\lambda_i \in E} \langle f, e_i\rangle_H e_i.$$

Finally, one can view L as unitarily equivalent to a multiplication operator $M: f \mapsto gf$ on $\ell^2(\{1, \ldots, n\})$, where g is the real-valued function $g(i) := \lambda_i$, and the intertwining map $U: \ell^2(\{1, \ldots, n\}) \to H$ is given by

$$U((c_i)_{i=1}^n) := \sum_{i=1}^n c_i e_i,$$

so that $L = UMU^{-1}$.

It is an important fact in analysis that many of these above assertions extend to operators on an infinite-dimensional Hilbert space H, so long as one is careful about what "self-adjoint operator" means; these facts are collectively referred to as the *spectral theorem*. For instance, it turns out that most of the above claims have analogues for *bounded* self-adjoint operators $L: H \to H$. However, in the theory of partial differential equations, one often needs to apply the spectral theorem to *unbounded, densely defined* linear operators $L\colon D \to H$, which (initially, at least), are only defined on a dense subspace D of the Hilbert space H. A very typical situation arises when $H = L^2(\Omega)$ is the square-integrable functions on some domain or manifold Ω (which may have a boundary or be otherwise "incomplete"), and $D = C_c^\infty(\Omega)$ are the smooth compactly supported functions on Ω, and L is some linear differential operator. It is then of interest to obtain the spectral theorem for such operators, so that one can build operators such as $e^{-tL}, e^{itL}, \frac{1}{L-z}, e^{it\sqrt{L}}$ or solve equations such as (10.1), (10.2), (10.3), (10.4).

In order to do this, some necessary conditions on the densely defined operator $L\colon D \to H$ must be imposed. The most obvious is that of *symmetry*, which asserts that

(10.5) $$\langle Lf, g\rangle_H = \langle f, Lg\rangle_H$$

for all $f, g \in D$. In some applications, one also wants to impose *positive definiteness*, which asserts that

(10.6) $$\langle Lf, f\rangle_H \geq 0$$

for all $f \in D$. These hypotheses are sufficient in the case when L is bounded and, in particular, when H is finite-dimensional. However, as it turns out, for unbounded operators these conditions are not, by themselves, enough to obtain a good spectral theory. For instance, one consequence of the spectral theorem should be that the resolvents $(L-z)^{-1}$ are well defined for any strictly complex z, which by duality implies that the image of $L-z$ should be dense in H. However, this can fail if one just assumes symmetry, or symmetry and positive definiteness. A well-known example occurs when H is the Hilbert space $H := L^2((0,1))$, $D := C_c^\infty((0,1))$ is the space of test functions, and L is the one-dimensional Laplacian $L := -\frac{d^2}{dx^2}$. Then L is symmetric and positive, but the operator $L - k^2$ does not have dense image for any complex k, since

$$\langle (L - \overline{k}^2)f, e^{\overline{k}x}\rangle_H = 0$$

for all test functions $f \in C_c^\infty((0,1))$, as can be seen from a routine integration by parts. As such, the resolvent map is not everywhere uniquely defined. There is also a lack of uniqueness for the wave, heat, and Schrödinger equations for this operator (note that there are no spatial boundary conditions specified in these equations).

Another example occurs when $H := L^2((0,+\infty))$, $D := C_c^\infty((0,+\infty))$, L is the momentum operator $L := i\frac{d}{dx}$. Then the resolvent $(L-z)^{-1}$ can be uniquely defined for z in the upper half-plane, but not in the lower half-plane, due to the obstruction

$$\langle (L - z)f, e^{i\overline{z}x}\rangle_H = 0$$

for all test functions f (note that the function $e^{i\bar{z}x}$ lies in $L^2((0, +\infty))$ when z is in the lower half-plane). For related reasons, the translation operators e^{itL} have a problem with either uniqueness or existence (depending on whether t is positive or negative), due to the unspecified boundary behaviour at the origin.

The key property that lets one avoid this bad behaviour is that of *essential self-adjointness*. Once L is essentially self-adjoint, then spectral theorem becomes applicable again, leading to all the expected behaviour (e.g., existence and uniqueness for the various PDEs given above).

Unfortunately, the concept of essential self-adjointness is defined rather abstractly, and is difficult to verify directly; unlike the symmetry condition (10.5) or the positive definite condition (10.6), it is not a "local" condition that can be easily verified just by testing L on various inputs, but is instead a more "global" condition. In practice, to verify this property, one needs to invoke one of a number of a partial converses to the spectral theorem, which roughly speaking, asserts that if at least one of the expected consequences of the spectral theorem is true for some symmetric densely defined operator L, then L is self-adjoint. Examples of "expected consequences" include:

(i) Existence of resolvents $(L-z)^{-1}$ (or equivalently, dense image for $L-z$);

(ii) Existence of a contractive heat propagator semigroup e^{tL} (in the positive case);

(iii) Existence of a unitary Schrödinger propagator group e^{itL};

(iv) Existence of a unitary wave propagator group $e^{it\sqrt{L}}$ (in the positive case);

(v) Existence of a "reasonable" functional calculus.

(vi) Unitary equivalence with a multiplication operator.

Thus, to actually verify essential self-adjointness of a differential operator, one typically has to first solve a PDE (such as the wave, Schrödinger, heat, or Helmholtz equation) by some nonspectral method (e.g., by a contraction mapping argument, or a perturbation argument based on an operator already known to be essentially self-adjoint). Once one can solve one of the PDEs, then one can apply one of the known converse spectral theorems to obtain essential self-adjointness, and then by the forward spectral theorem one can then solve all the other PDEs as well. But there is no getting out of that first step, which requires some input (typically of an ODE, PDE, or geometric nature) that is external to what abstract spectral theory can provide. For instance, if one wants to establish essential self-adjointness of the Laplace-Beltrami operator $L = -\Delta_g$ on a smooth Riemannian manifold (M, g) (using $C_c^\infty(M)$ as the domain space), it turns out (under reasonable regularity hypotheses) that essential self-adjointness is equivalent to *geodesic completeness* of the manifold, which is a global ODE condition rather than a local one: one needs geodesics to continue indefinitely in order to be able to (unitarily) solve PDEs such as the wave equation, which in turn leads to essential self-adjointness. (Note that the domains $(0,1)$ and $(0,+\infty)$ in the previous examples were not geodesically complete.) For this reason, essential self-adjointness of a differential operator is sometimes referred to as *quantum completeness* (with the completeness of the associated Hamilton-Jacobi flow then being the analogous *classical completeness*).

In this chapter, I will record the forward and converse spectral theorems, and to verify essential self-adjointness of the Laplace-Beltrami operator on geodesically complete manifolds. This material is quite standard, and can be found for instance in [**ReSi1975**].

10.1. Self-adjointness and resolvents

To begin, we study what we can abstractly say about a densely defined symmetric linear operator $L\colon D \to H$ on a Hilbert space H. To avoid some technical issues we shall assume that the Hilbert space is *separable* (i.e., it has a countable dense subset), which is the case typically encountered in applications (particularly in PDE). We will occasionally assume also that L is positive, but will make this hypothesis explicit whenever we are doing so.

All convergence in Hilbert spaces will be in the strong (i.e., norm) topology unless otherwise stated. Similarly, all inner products and norms will be over H unless otherwise stated.

For technical reasons, it is convenient to reduce to the case when L is *closed*, which means that the graph $\{(f, Lf) : f \in D\}$ is a closed subspace of $H \times H$. Equivalently, L is closed if whenever $f_n \in D$ is a sequence converging strongly to a limit $f \in H$, and Lf_n converges to a limit g in H, then $f \in D$ and $g = Lf$. Note from the *closed graph theorem* that an everywhere-defined linear operator is closed if and only if it is bounded.

Not every densely defined symmetric linear operator is closed (indeed, one could take a closed operator and restrict the domain of definition to a proper dense subspace). However, all such operators are *closable*, in that they have a closure:

Lemma 10.1.1 (Closure). *Let $L\colon D \to H$ be a densely defined symmetric linear operator. Then there exists a unique extension $\overline{L}\colon \overline{D} \to H$ of L as a closed, densely defined symmetric linear operator, such that the graph $\{(f, \overline{L}f) : f \in \overline{D}\}$ of $\overline{L}$ is the closure of the graph $\{(f, Lf) : f \in D\}$ of L.*

Proof. The key step is to show that the closure $\overline{\{(f, Lf) : f \in D\}}$ of the graph of L remains a graph, i.e., it obeys the vertical line test. If this failed, then by linearity one could find a sequence $f_n \in D$ converging to zero such that Lf_n converged to a nonzero limit g. Since D is dense, we can find $g' \in D$ such that $\langle g, g' \rangle \neq 0$. But then by symmetry

$$\langle f_n, Lg' \rangle = \langle Lf_n, g' \rangle = \langle g_n, g' \rangle \to \langle g, g' \rangle \neq 0.$$

On the other hand, as $f_n \to 0$, $\langle f_n, Lg' \rangle \to 0$. Thus $\overline{\{(f, Lf) : f \in D\}}$ is the graph of some function $\overline{L} : \overline{D} \to H$. It is easy to see that this is a densely defined symmetric linear operator extending L, and is the unique such operator. $\square$

Exercise 10.1.1. Show that $\overline{L}$ is positive if and only if L is positive.

Remark 10.1.2. We caution that $\overline{D}$ is not the closure or completion of D with respect to the usual norm $f \mapsto \|f\|$ on D. (Indeed, as D is a dense subspace of H, that completion of D is simply H.) However, it is the completion of D with respect to the modified norm $f \mapsto \|f\| + \|Lf\|$.

In PDE applications, the closure $\overline{L}$ tends to be defined on a Sobolev space of functions that behave well at the boundary, and is given by a distributional derivative. Here is a simple example of this:

Exercise 10.1.2. Let L be the Laplacian $L = -\frac{d^2}{dx^2}$, defined on the dense subspace $D := C_c^\infty((0,1))$ of $H := L^2((0,1))$. Show that the closure $\overline{L}$ is defined on the Sobolev space $H_0^2((0,1))$, defined as the closure of $C_c^\infty((0,1))$ under the Sobolev norm

$$\|f\|_{H_0^2((0,1))} := \|f\|_{L^2((0,1))} + \|Lf\|_{L^2((0,1))},$$

and that the action of $\overline{L}$ is given by the weak (distributional) derivative, $\overline{L} = -\frac{d^2}{dx^2}$.

Next, we define the *adjoint* $L^* : D^* \to H$ of $L : D \to H$, which informally speaking is the maximally defined operator for which one has the relationship

(10.7) $$\langle Lf, g \rangle = \langle f, L^* g \rangle$$

for $f \in D$ and $g \in D^*$. More formally, define D^* to be the set of all vectors $g \in H$ for which the map $f \mapsto \langle Lf, g \rangle$ is a bounded linear functional on D, which thus extends to the closure H of D. For such $g \in H$, we may apply the *Riesz representation theorem for Hilbert spaces* and locate a unique vector $L^* g \in H$ for which (10.7) holds. This is easily seen to define a linear operator $L^* : D^* \to H$. Furthermore, the claim that L is symmetric can be reformulated as the claim that L^* extends L.

If L is not symmetric, then L^* does not extend L, and need not be densely defined at all. However, it is still closed:

Exercise 10.1.3. Let $L\colon D \to H$ be a densely defined linear operator, and let $L^*\colon D^* \to H$ be its adjoint. Show that $\{(-L^*g, g) : g \in D^*\}$ is the orthogonal complement in $H \times H$ of (the closure of) $\{(f, Lf) : f \in D\}$. Conclude that L^* is always closed.

If L is symmetric, show that $L^{**} = \overline{L}$ and $L^* = (\overline{L})^*$.

Exercise 10.1.4. Construct an example of a densely defined linear operator $L\colon D \to H$ in a separable Hilbert space H for which L^* is only defined at $\{0\}$. (*Hint:* Build a dense linearly independent basis of H, let D be the algebraic span of that basis, and design L so that the graph of L is dense in $H \times H$.)

We caution that the adjoint $L^*\colon D^* \to H$ of a symmetric densely defined operator $L\colon D \to H$ need not be itself symmetric, despite extending the symmetric operator L:

Exercise 10.1.5. We continue Example 10.1.2. Show that for any complex number k, the functions $x \mapsto e^{kx}$ lie in D^* with $L^* e^{kx} = -k^2 e^{kx}$. Deduce that L^* is not symmetric, and not positive.

Intuitively, the problem here is that the domain of L is too "small" (it stays too far away from the boundary), which makes the domain of L^* too "large" (it contains too much stuff coming from the boundary), which ruins the integration by parts argument that gives symmetry.

Now we can define (essential) self-adjointness.

Definition 10.1.3. Let $L\colon D \to H$ be a densely defined linear operator.

(i) L is *self-adjoint* if $L = L^*$. (Note that this implies in particular that L is symmetric and closed.)

(ii) L is *essentially self-adjoint* if it is symmetric, and its closure $\overline{L}$ is self-adjoint.

Note that this extends the usual definition of self-adjointness for bounded operators. Conversely, from the closed graph theorem we also observe the *Hellinger-Toeplitz theorem*: an operator that is self-adjoint and everywhere defined, is necessarily bounded.

Exercise 10.1.6. Let $L\colon D \to H$ be a densely defined symmetric closed linear operator. Show that L is self-adjoint if and only if L^* is symmetric.

It is not immediately obvious what advantage self-adjointness gives. To see this, we consider the problem of inverting the operator $L - z\colon D \to H$ for some complex number z, where $L\colon D \to H$ is densely defined, symmetric,

and closed. Observe that if $f \in D$, then $\langle Lf, f \rangle = \langle f, Lf \rangle$ is necessarily real. In particular,
$$\operatorname{Im}\langle (L-z)f, f \rangle = -\operatorname{Im} z \|f\|^2$$
and hence by the Cauchy-Schwarz inequality
(10.8) $$\|(L-z)f\| \geq |\operatorname{Im} z| \|f\|.$$

In particular, if z is strictly complex (i.e., not real), then $L - z$ is injective. Furthermore, we see that if $f_n \in D$ is such that $(L-z)f_n$ is convergent, then by (10.8) f_n is convergent also, and hence Lf_n is convergent. As L was assumed to be closed, we conclude that f_n converges to a limit f in D, and $(L-z)f_n$ converges to $(L-z)f$. As a consequence, we see that the space $\operatorname{Image}(L-z) := \{(L-z)f : f \in D\}$ is a closed subspace of H. From (10.8) we then see that we can define an inverse $R(z)\colon \operatorname{Image}(L-z) \to D$ of $L - z$, which we call the *resolvent* of L with spectral parameter z; this is a bounded linear operator with norm $\|R(z)\|_{op} \leq \frac{1}{|\operatorname{Im}(z)|}$.

Exercise 10.1.7. If L is densely defined, symmetric, closed, and positive, and z is a complex number with $\operatorname{Re}(z) < 0$, show $R(z)$ is well defined on $\operatorname{Image}(L-z)$ with $\|R(z)\|_{op} \leq \frac{1}{|\operatorname{Re}(z)|}$.

Now we observe a connection between self-adjointness and the domain of the resolvent. We first need some basic properties of the resolvent:

Exercise 10.1.8. Let $L\colon D \to H$ be densely defined, symmetric, and closed.

(i) (Resolvent identity) If z, w are distinct strictly complex numbers with $R(z), R(w)$ everywhere defined, show that $(z-w)R(z)R(w) = R(z) - R(w)$. (*Hint:* Compute $R(z)(H-w)R(w) - R(z)(H-z)R(w)$ two different ways.)

(ii) If z is a strictly complex number with $R(z)$ and $R(z^*)$ everywhere defined, show that $R(z)^* = R(z^*)$.

Proposition 10.1.4. *Let $L\colon D \to H$ be densely defined, symmetric, and closed, let $L^*\colon D^* \to H$ be the adjoint, and let z be strictly complex.*

(i) *(Surjectivity is dual to injectivity) $R(z)$ is everywhere defined if and only if the operator $L^* - z^*\colon D^* \to H$ has trivial kernel.*

(ii) *(Self-adjointness implies surjectivity) If L is self-adjoint, then $R(z)$ is everywhere defined.*

(iii) *(Surjectivity implies self-adjointness) Conversely, if $R(z)$ and $R(z^*)$ are both everywhere defined, then L is self-adjoint.*

In particular, we see that we have a criterion for self-adjointness: a densely defined symmetric closed operator is self-adjoint if and only if $R(i)$

10.1. Self-adjointness and resolvents

and $R(-i)$ are both everywhere defined, or in other words, if $L+i$ and $L-i$ are both surjective.

Proof. We first prove (i). If $R(z)$ is not everywhere defined, then the closed subspace $\text{Image}(L-z) \subset H$ is not all of H. Thus there must be a nonzero vector $v \in H$ in the orthogonal complement of this subspace, thus $\langle (L-z)f, g\rangle = 0$ for all $f \in D$. In particular,

$$|\langle Lf, g\rangle| = |z||\langle f, g\rangle| \leq |z|\|f\|\|g\|$$

and hence, by definition of D^*, g lies in D^*. Now from (10.7) we have

$$\langle f, (L^* - z^*)g\rangle = \langle (L-z)f, g\rangle = 0$$

for all $f \in D$; since D is dense, we have $(L^* - z^*)g = 0$ as required. The converse implication follows by reversing these steps.

Now we prove (ii). Suppose that $R(z)$ was not everywhere defined. By (i) and self-adjointness, we conclude that $(L - z^*)f = 0$ for some nonzero $f \in D$. But this contradicts (10.8).

Now we prove (iii). Let $g \in D^*$; our task is to show that $g \in D$. From Exercise 10.1.8(ii) and (10.7) one has

$$\langle f, R(z^*)(L^* - z^*)g\rangle = \langle R(z)f, (L^* - z^*)g\rangle = \langle (L-z)R(z)f, g\rangle = \langle f, g\rangle$$

for all $f \in H$. We conclude that $g = R(z^*)(L^* - z^*)g$. Since $R(z^*)$ takes values in D, the claim follows. □

Exercise 10.1.9. Let $L: D \to H$ be densely defined, symmetric, and closed, and let z be strictly complex.

(i) If $R(z)$ is everywhere defined, show that $R(w)$ is everywhere defined whenever $|w - z| < \text{Im}(z)$. (*Hint:* Use *Neumann series*.)

(ii) If $R(z)$ is everywhere defined, show that $R(w)$ is everywhere defined whenever $\text{Im}(w)$ has the same sign as $\text{Im}(z)$.

(iii) If L is positive, and w is not a nonnegative real, show that $R(w)$ is everywhere defined if and only if L is self-adjoint.

As a particular corollary of the above exercise, we see that a densely defined, symmetric closed positive operator L is self-adjoint if and only if $R(-1)$ is everywhere defined, or in other words, if $1 + L$ is surjective.

Exercise 10.1.10. Let (X, μ) be a measure space with a countably generated σ-algebra (so that $L^2(X, \mu)$ is separable), let $m: X \to \mathbf{R}$ be a measurable function, and let D be the space of all $f \in L^2(X, \mu)$ such that $mf \in L^2(X, \mu)$. Show that the multiplier operator $L: D \to L^2(X, \mu)$ defined by $Lf := mf$ is a densely defined self-adjoint operator.

Exercise 10.1.11. Let $H := L^2((0,+\infty))$, $D := C_c^\infty((0,+\infty))$, and L is the momentum operator $L := i\frac{d}{dx}$. Show that L is densely defined and symmetric, and $R(-i)$ is everywhere defined, but $R(i)$ is only defined on the orthogonal complement of e^{-x}. (Here, we define the resolvent of a closable operator to be the resolvent of its closure.) In particular, L is not self-adjoint.

Exercise 10.1.12. Let L be densely defined and symmetric.

(i) Show that L is essentially self-adjoint if and only if $\text{Image}(L+i)$ and $\text{Image}(L-i)$ are dense in H.

(ii) If L is positive, show that L is essentially self-adjoint if and only if $\text{Image}(1+L)$ is dense in H.

Exercise 10.1.13. Let $a_1, a_2, \ldots$ and $b_1, b_2, \ldots$ be sequences of real numbers, with the b_n all nonzero. Define the *Jacobi operator* $T \colon \ell_c^2(\mathbf{N}) \to \ell^2(\mathbf{N})$ from the space $\ell_c^2(\mathbf{N})$ of compactly supported sequences $(x_n)_{n=1}^\infty$ to the space $\ell^2(\mathbf{N})$ of square-summable sequences $(y_n)_{n=1}^\infty$ by the formula

$$T(x_n)_{n=1}^\infty = (b_{n-1}x_{n-1} + a_n x_n + b_n x_{n+1})_{n=1}^\infty$$

with the convention that $b_{n-1}x_{n-1}$ vanishes for $n=1$.

(i) Show that T is densely defined and symmetric.

(ii) Show that T is essentially self-adjoint if and only if the (unique) solution ϕ_n to the recurrence

$$b_{n-1}\phi_{n-1} + (a_n - i)\phi_n + b_n\phi_{n+1} = 0$$

with $\phi_1 = 1$ (and the convention $b_0\phi_0 = 0$), is not square-summable.

This exercise shows that the self-adjointness of an operator, even one as explicit as a Jacobi operator, can depend in a rather subtle and "global" fashion on the behaviour of the coefficients of that oeprator.

10.2. Self-adjointness and spectral measure

We have seen that self-adjoint operators have everywhere-defined resolvents $R(z)$ for all strictly complex z. Now we use this fact to build spectral measures. We will need a useful tool from complex analysis, which places a one-to-one correspondence between finite nonnegative measures on $\mathbf{R}$ and certain analytic functions on the upper half-plane:

Theorem 10.2.1 (Herglotz representation theorem). *Let $F \colon \mathbf{H} \to \overline{\mathbf{H}}$ be an analytic function from the upper half-plane $\mathbf{H} := \{z \in \mathbf{C} : \text{Im}(z) > 0\}$ to the closed upper half-plane $\overline{\mathbf{H}} := \{z \in \mathbf{C} : \text{Im}(z) \geq 0\}$, obeying a bound of the* *form* $|F(z)| \leq C/\operatorname{Im}(z)$ *for all* $z \in \mathbf{H}$ *and some* $C > 0$. *Then there exists a finite nonnegative Radon measure* μ *on* $\mathbf{R}$ *such that*

$$F(z) = \int_{\mathbf{R}} \frac{1}{x-z} \, d\mu(x) \tag{10.9}$$

for all $z \in \mathbf{H}$. *Furthermore, one has* $yF(iy) \to \mu(\mathbf{R})$ *as* $y \to +\infty$.

We set the proof of the above theorem as an exercise below. Note in the converse direction that if μ is a finite nonnegative Radon measure, then the function F defined by (10.9) obeys all the hypotheses of the theorem. The Herglotz representation theorem, like the more well-known *Riesz representation theorem for measures*, is a useful tool to construct nonnegative Radon measures; later on we will also use *Bochner's theorem* for a similar purpose.

Exercise 10.2.1. Let F be as in the hypothesis of the Herglotz representation theorem. For each $\varepsilon > 0$, let $F_\varepsilon \colon \mathbf{R} \to \overline{\mathbf{H}}$ be the function $F_\varepsilon(x) := F(x + i\varepsilon)$.

(i) Show that one has $F_{\varepsilon+t} = F_\varepsilon * P_t$ for all $\varepsilon, t > 0$, where $P_t(x) := \frac{1}{\pi} \frac{t}{x^2+t^2}$ is the *Poisson kernel* and $*$ is the usual convolution operation. (*Hint:* Apply *Liouville's theorem* for harmonic functions to the difference between $F_{\varepsilon+t}$ and $F_\varepsilon * P_t$.)

(ii) Show that the nonnegative measures $\operatorname{Im} F_\varepsilon(x) \, dx$ have a finite mass independent of ε, and converge in the *vague topology* as $\varepsilon \to 0$ to a nonnegative finite measure μ.

(iii) Prove the Herglotz representation theorem.

Now we return to spectral theory. Let $L \colon D \to H$ be a densely-defined self-adjoint operator, and let $f \in H$. We consider the function

$$F_{f,f}(z) := \langle R(z)f, f \rangle$$

on the upper half-plane. We can use this function and the Herglotz representation theorem to construct spectral measures $\mu_{f,f}$:

Exercise 10.2.2. Let L, f, and $F_{f,f}$ be as above.

(i) Show that $F_{f,f}$ is analytic. (*Hint:* Use Neumann series and *Morera's theorem*.)

(ii) Show that $|F_{f,f}(z)| \leq \|f\|^2/\operatorname{Im}(z)$ for all z in the upper half-plane.

(iii) Show that $\operatorname{Im} F_{f,f}(z) \geq 0$ for all z in the upper half-plane. (*Hint:* You will find Exercise 10.1.8 to be useful.)

(iv) Show that $iyF_{f,f}(iy) \to \|f\|^2$ for all $f \in H$. (*Hint:* First show this for $f \in D$, writing $f = R(i)g$ for some $g \in H$.)

(v) Show that there is a nonnegative Radon measure $\mu_{f,f}$ of total mass $\|f\|^2$ such that
$$\langle R(z)f, f \rangle = \int_{\mathbf{R}} \frac{1}{x-z} \, d\mu_{f,f}(x)$$
for all z in either the upper or lower half-plane.

(vi) If L is positive, show that $\mu_{f,f}$ is supported in the right half-line $[0, +\infty)$.

We can *depolarise* these measures by defining
$$\mu_{f,g} := \frac{1}{4}(\mu_{f+g,f+g} - \mu_{f-g,f-g} + i\mu_{f+ig,f+ig} - i\mu_{f-ig,f-ig})$$
to obtain complex measures $\mu_{f,g}$ for any $f, g \in H$ such that
$$\langle R(z)f, g \rangle = \int_{\mathbf{R}} \frac{1}{x-z} \, d\mu_{f,g}(x)$$
for all z in either the upper or lower half-plane. From duality we see that this uniquely defines $\mu_{f,g}$. In particular, $\mu_{f,g}$ depends sesquilinearly on f, g, and $\mu_{g,f} = \overline{\mu_{f,g}}$. Also, since each $\mu_{f,f}$ has mass $\|f\|^2$, we see that the inner product $\langle f, g \rangle$ can be recovered from the spectral measure $\mu_{f,g}$:

(10.10) $$\langle f, g \rangle = \int_{\mathbf{R}} d\mu_{f,g}(x).$$

Exercise 10.2.3. With the above notation and assumptions, establish the bound $\|\mu_{f,g}\|_{TV} \ll \|f\|\|g\|$ for all $f, g \in H$. In particular, for any bounded Borel-measurable function $m \colon \mathbf{R} \to \mathbf{C}$, there exists a unique bounded operator $m(L) \colon H \to H$ such that

(10.11) $$\langle m(L)f, g \rangle = \int_{\mathbf{R}} m(x) \, d\mu_{f,g}(x).$$

Thus, for instance $m(L)$ is the identity when $m = 1$, and $m(L) = R(z)$ when $m(x) = \frac{1}{x-z}$.

We have just created a map $m \mapsto m(L)$ from the bounded Borel-measurable functions $B(\mathbf{R} \to \mathbf{C})$ on $\mathbf{R}$, to the bounded operators $B(H \to H)$ on H. Now we verify some basic properties of this map.

Exercise 10.2.4 (Bounded functional calculus). Let $L \colon D \to H$ be a self-adjoint densely defined operator, and let $m \mapsto m(L)$ be as above.

(i) Show that the map $m \mapsto m(L)$ is *-*linear*. In particular, if m is real-valued, then $m(L)$ is self-adjoint.

(ii) For any $f, g \in H$ and any strictly complex z, show that $d\mu_{R(z)f,g}(x) = \frac{1}{x-z} d\mu_{f,g}(x)$. (*Hint:* Use the resolvent identity.)

10.2. Self-adjointness and spectral measure

(iii) For any $f, g \in H$ and any $m \in B(\mathbf{R} \to \mathbf{C})$, show that $d\mu_{m(L)f,g} = m \, d\mu_{f,g}$.

(iv) Show that the map $m \mapsto m(L)$ is a $*$-homomorphism. In particular, $m(L)$ and $m'(L)$ commute for all $m, m' \in B(\mathbf{R} \to \mathbf{C})$.

(v) Show that $\|m(L)\|_{op} \ll \sup_{x \in \mathbf{R}} |m(x)|$ for all $m \in B(\mathbf{R} \to \mathbf{C})$. Improve the $\ll$ to $\leq$. (*Hint:* To get this improvement, use the TT^* identity $\|T\|_{op} = \|TT^*\|_{op}^{1/2}$ and the *tensor power trick*, see [**Ta2008**, §1.9].)

(vi) For any Borel subset E of $\mathbf{R}$, show that $\mu(E) := 1_E(L)$ is an orthogonal projection of H. Show that μ is a countably additive *projection-valued measure*, thus $\sum_{n=1}^{\infty} \mu(E_n) = \mu(\bigcup_{n=1}^{\infty} E_n)$ for any sequence of disjoint Borel E_n, where the convergence is in the strong operator topology.

(vii) For any bounded Borel set E, show that the image of $\mu(E)$ lies in D.

(viii) Show that for any $f \in D$ and $g \in H$, one has $d\mu_{Lf,g}(x) = x \, d\mu_{f,g}(x)$.

(ix) Let D_c be the union of the images of $\mu([-N, N])$ for $N = 1, 2, \ldots$. Show that D_c is a dense subspace of D (and hence of H), and that L maps D_c to D_c.

(x) Show that D is the space of all functions $f \in H$ such that
$$\int_{\mathbf{R}} |x|^2 \, d\mu_{f,f}(x) < \infty.$$
Conclude in particular that $m(L)$ maps D to D for all $m \in B(\mathbf{R} \to \mathbf{C})$.

(xi) If $m, m' \in B(\mathbf{R} \to \mathbf{C})$ is such that $m'(x) = xm(x)$ for all $x \in \mathbf{R}$, show that $m(L)$ takes values in D, that $m'(L)f = Lm(L)f$ for all $f \in H$, and $m'(L)f = m(L)Lf$ for all $f \in D$.

(xii) Let $\sigma(L)$ be the space of all z such that $L - z: D \to H$ is not invertible. Show that $\sigma(L)$ is a closed set which is the union of the supports of the $\mu_{f,g}$ as f, g range over H. In particular, $\sigma(L) \subset \mathbf{R}$, and $\sigma(L) \subset [0, +\infty)$ when L is positive. Show that $\mu(\sigma(L)) = \mu(\mathbf{R})$ is the identity map.

(xiii) Extend the bounded functional calculus from the bounded Borel measurable functions $B(\mathbf{R} \to \mathbf{C})$ on $\mathbf{R}$ to the bounded measurable functions $B(\sigma(L) \to \mathbf{C})$ on the spectrum $\sigma(L)$ (i.e., show that the previous statements (i)–(xi) continue to hold after $\mathbf{R}$ is replaced by $\sigma(L)$ throughout).

The above collection of facts (or various subcollections thereof) is often referred to as the *spectral theorem*. It is stated for self-adjoint operators, but one can of course generalise the spectral theorem to essentially self-adjoint operators by applying the spectral theorem to the closure. (One has to replace the domain D of L by the domain $\overline{D}$ of the closure $\overline{L}$, of course, when doing so.)

Exercise 10.2.5 (Spectral measure and eigenfunctions). Let the notation be as in the preceding exercise, and let $\lambda \in \mathbf{R}$. Show that the space $\{f \in D : Lf = \lambda f\}$ is the range of the projection $\mu(\{\lambda\})$. In particular, L has an eigenfunction at λ if and only if $\mu(\{\lambda\})$ is nontrivial.

We have seen how the existence of resolvents gives us a bounded functional calculus (i.e., the conclusions in the above exercise). Conversely, if a symmetric densely defined closed operator L has a bounded functional calculus, one can define the resolvents $R(z)$ simply as $m_z(L)$, where $m_z(x) := \frac{1}{x-z}$. Thus we see that the existence of a bounded functional calculus is equivalent to the existence of resolvents, which by the previous discussion is equivalent to self-adjointness.

Using the bounded functional calculus, one cannot only recover the resolvents $R(z)$, but can now also build Schrödinger propagators e^{itL}, and when L is positive definite one can also build heat operators e^{tL} for $t > 0$ and wave operators $e^{it\sqrt{L}}$ for $t \in \mathbf{R}$ (and also define resolvents $R(-k^2)$ for negative choices $-k^2$ of the spectral parameter). We will study these operators more in the next section.

Exercise 10.2.6 (Locally bounded functional calculus). Let $L : D \to H$ be a densely defined self-adjoint operator, and let $B_{\mathrm{loc}}(\sigma(L) \to \mathbf{C})$ be the space of Borel-measurable functions from $\sigma(L)$ to $\mathbf{C}$ which are bounded on every bounded set.

   (i) Show that for every $m \in B_{\mathrm{loc}}(\sigma(L) \to \mathbf{C})$ there is a unique linear operator $m(L) : D_c \to D_c$ such that
$$d\mu_{m(L)f,g}(x) = m(x)\, d\mu_{f,g}(x)$$
   whenever $f, g \in D_c$.

   (ii) Show that the map $m \mapsto m(L)$ is a $*$-homomorphism.

   (iii) Show that when m is actually bounded (rather than merely locally bounded), then this definition of $m(L)$ agrees with that in the preceding exercise (after restricting from H to D_c).

   (iv) Show that $L = \iota(L)$, where ι is the identity map $\iota(x) := x$.

(v) State and prove a rigorous version of the formal assertion that
$$L = \int_{\mathbf{R}} x d\mu(x)$$
and, more generally,
$$m(L) = \int_{\mathbf{R}} m(x) d\mu(x).$$

Now we use the spectral theorem to place self-adjoint operators in a normal form. Let us say that two operators $L\colon D \to H$ and $L'\colon D' \to H'$ are *unitarily equivalent* if there is a unitary map $U\colon H \to H'$ with $U(D) = D'$ and $L' = U^{-1}LU$. It is easy to see that all the constructions given above (such as the bounded functional calculus) are preserved by unitary equivalence.

If $L\colon D \to H$ is a densely defined self-adjoint operator, define an *invariant subspace* to be a subspace V of H such that $m(L)V \subset V$ for all $m \in B(\sigma(L) \to \mathbf{C})$. We say that a vector $f \in H$ is a *cyclic vector* for H if the set $\{m(L)f : m \in B(\sigma(L) \to \mathbf{C})\}$ is dense in H.

Exercise 10.2.7.

(i) Show that if V is an invariant subspace of H, then so is $V^\perp$; furthermore, the orthogonal projections to V and $V^\perp$ commute with $m(L)$ for every $m \in B(\sigma(L) \to \mathbf{C})$.

(ii) Show that if V is a invariant subspace of H, then the restriction $L\!\mid_V\colon D \cap V \to V$ of L to V is a densely defined self-adjoint operator on V with $\sigma(L\!\mid_V) \subset \sigma(L)$. Furthermore, one has $m(L\!\mid_V) = m(L)\!\mid_V$ for all $m \in B(\sigma(L) \to \mathbf{C})$.

(iii) Show that H decomposes as a direct sum $\bigoplus_{\alpha \in A} H_\alpha$, where the index set A is at most countable, and each H_α is a closed invariant subspace of H (with the H_α mutually orthogonal), such that each H_α has a cyclic vector f_α. (*Hint:* Use Zorn's lemma and the separability of H.)

(iv) If H has a cyclic vector f_0, show that L is unitarily conjugate to a multiplication operator $f(x) \mapsto xf(x)$ on $L^2(\mathbf{R}, \nu)$ for some non-negative Radon measure ν, defined on the domain $\{f \in L^2(\mathbf{R}, \nu) : xf \in L^2(\mathbf{R}, \nu)\}$.

(v) For general H, show that L is unitarily conjugate to a multiplication operator $f(x) \mapsto g(x)f(x)$ on $L^2(X, \nu)$ for some measure space (X, ν) with a countably generated σ-algebra and some measurable $g\colon X \to \mathbf{R}$, defined on the domain $\{f \in L^2(X, \nu) : gf \in L^2(X, \nu)\}$.

The above exercise gives a satisfactory concrete description of a self-adjoint operator (up to unitary equivalence) as a multiplication operator on some measure space $L^2(X, \nu)$, although we caution that this equivalence is not canonical (there is some flexibility in the choice of the underlying measure space (X, ν) and multiplier g, as well as the unitary conjugation map).

Exercise 10.2.8. Let $L\colon D \to H$ be a self-adjoint densely defined operator.

(i) Show that L is positive if and only if $\sigma(L) \subset [0, +\infty)$.
(ii) Show that L is bounded if and only if $\sigma(L)$ is bounded. Furthermore, in this case we have $\|L\|_{op} = \sup_{x \in \sigma(L)} |x|$.
(iii) Show that H is trivial if and only if $\sigma(L)$ is empty.

10.3. Self-adjointness and flows

Now we relate self-adjointness to a variety of flows, beginning with the heat flow.

Exercise 10.3.1. Let $L\colon D \to H$ be a self-adjoint positive densely defined operator, and for each $t \geq 0$, let $S(t)\colon H \to H$ be the heat operator $S(t) := e^{-tL}$.

(i) Show that for each $t \geq 0$, $S(t)$ is a bounded self-adjoint operator of norm at most 1, that the map $t \mapsto S(t)$ is continuous in the strong operator topology, and such that $S(0) = 1$ and $S(t)S(t') = S(t+t')$ for all $t, t' \geq 0$. (The latter two properties are asserting that $t \mapsto S(t)$ is a *one-parameter semigroup*.)
(ii) Show that for any $f \in D$, $\frac{S(t)f - f}{t}$ converges to $-Lf$ as $t \to 0^+$.
(iii) Conversely, if $f \in H$ is not in D, show that $\frac{S(t)f - f}{t}$ does not converge as $t \to 0^+$.

We remark that the above exercise can be viewed as a special case of the *Hille-Yoshida theorem*.

We now establish a converse to the above statement:

Theorem 10.3.1. *Let H be a separable Hilbert space, and suppose one has a family $t \mapsto S(t)$ of bounded self-adjoint operators of norm at most 1 for each $t \geq 0$, which is continuous in the strong operator topology, and such that $S(0) = 1$ and $S(t)S(t') = S(t+t')$ for all $t, t' \geq 0$. Then there exists a unique self-adjoint positive densely defined operator $L\colon D \to H$ such that $S(t) = e^{-tL}$ for all $t \geq 0$.*

We leave the proof of this result to the exercise below. The basic idea is to somehow use the identity
$$(x+1)^{-1} = \int_0^\infty e^{-tx} e^{-t} \, dt$$

10.3. Self-adjointness and flows

which suggests that

$$(L+1)^{-1} = \int_0^\infty S(t)e^{-t}\,dt$$

which should allow one to recover L from the $S(t)$.

Exercise 10.3.2. Let the notation be as in the above theorem.

(i) Establish the uniqueness claim. (*Hint:* Use Exercise 10.3.1.)

(ii) Let R be the operator

$$R := \int_0^\infty S(t)e^{-t}\,dt.$$

Show that R is well defined, bounded, positive semidefinite, and self-adjoint, with operator norm at most one, and that R commutes with all the $S(t)$.

(iii) Show that the spectrum of R lies in $[0,1]$, but that R has no eigenvalue at 0.

(iv) Show that there exists a densely defined self-adjoint operator $R^{-1}\colon D \to H$ such that RR^{-1} is the identity on D and $R^{-1}R$ is the identity on H.

(v) Show that $L := R^{-1} - 1$ is densely defined self-adjoint and positive definite, and commutes with all the $S(t)$.

(vi) Show that for all $v \in H$ and $t \geq 0$, one has $\frac{d}{dt}S(t)Rf = -LS(t)Rf$ and $\frac{d}{dt}e^{-tL}Rf = -Le^{-tL}Rf$, where the derivatives are in the classical limiting Newton quotient sense (in the strong topology of H).

(vii) Conclude the proof of Theorem 10.3.1. (*Hint:* Show that $\frac{d}{dt}\|S(t)Rf - e^{-tL}Rf\|^2$ is nonpositive.)

The above exercise gives an important way to establish essential self-adjointness, namely by solving the heat equation (10.1):

Exercise 10.3.3. Let $L\colon D \to H$ be a densely defined symmetric positive definite operator. Suppose that for every $f \in D$ there exists a continuously differentiable solution $u\colon [0, +\infty) \to D$ to (10.1). Show that L is essentially self-adjoint. (*Hint:* By investigating $\frac{d}{dt}\|u(t)\|^2$, establish the uniqueness of solutions to the heat equation, which allows one to define linear contractions $S(t)$. To establish self-adjointness of the $S(t)$, take an inner product of a solution to the heat equation against a time-reversed solution to the heat equation, and differentiate that inner product in time. Now apply the preceding exercises to obtain a self-adjoint extension $L'\colon D' \to H$ of L. To show that L' is the closure of L, it suffices to show that D is dense in D' with the inner product $\langle f, g \rangle + \langle L'f, g \rangle$. But if D is not dense in D', then it has a nontrivial orthogonal complement; apply $S(t)$ to this complement to show that D also has a nontrivial orthogonal complement in H, a contradiction.)

Remark 10.3.2. When applying the above criterion for essential self-adjointness, one usually cannot use the space C_c^∞ of compactly supported smooth functions as the dense subspace, because this space is usually not preserved by the heat flow. However, in practice one can get around this by enlarging the class, for instance to the class of *Schwartz functions*.

Now we obtain analogous results for the Schrödinger propagators e^{itL}. We begin with the analogue of Exercise 10.3.1:

Exercise 10.3.4. Let $L\colon D \to H$ be a self-adjoint densely defined operator, and for each $t \in \mathbf{R}$, let $U(t)\colon H \to H$ be the Schrödinger operator $S(t) := e^{itL}$.

(i) Show that for each $t \in \mathbf{R}$, $U(t)$ is a unitary operator, that the map $t \mapsto U(t)$ is continuous in the strong operator topology, and such that $U(0) = 1$ and $U(t)U(t') = U(t+t')$ for all $t, t' \geq 0$.

(ii) Show that for any $f \in D$, $\frac{U(t)f - f}{t}$ converges to iLf as $t \to 0$.

(iii) Conversely, if $f \in H$ is not in D, show that $\frac{U(t)f - f}{t}$ does not converge as $t \to 0$.

Now we can give the converse, known as *Stone's theorem on one-parameter unitary groups*:

Theorem 10.3.3 (Stone's theorem). *Let H be a separable Hilbert space, and suppose one has a family $t \mapsto U(t)$ such that $U(t)$ is unitary for each $t \in \mathbf{R}$, which is continuous in the strong operator topology, and such that $U(0) = 1$ and $U(t)U(t') = U(t+t')$ for all $t, t' \in \mathbf{R}$. Then there exists a unique self-adjoint densely defined operator $L\colon D \to H$ such that $U(t) = e^{itL}$ for all $t \in \mathbf{R}$.*

We outline a proof of this theorem in an exercise below, based on using the group $U(t)$ to build spectral measure.

Exercise 10.3.5. Let the notation be as in the above theorem.

(i) Establish the uniqueness component of Stone's theorem.

(ii) Show that for any $f \in H$, there is a nonnegative Radon measure $\mu_{f,f}$ of total mass $\|f\|^2$ such that

$$\langle U(t)f, f \rangle = \int_{\mathbf{R}} e^{itx}\, d\mu_{f,f}(x)$$

for all $t \in \mathbf{R}$. (*Hint:* Use *Bochner's theorem*, Proposition 2.3.5.)

10.3. Self-adjointness and flows

(iii) Show that for any $f, g \in H$ there is a unique complex measure $\mu_{f,g}$ such that
$$\langle U(t)f, g \rangle = \int_{\mathbf{R}} e^{itx}\, d\mu_{f,g}(x)$$
for all $t \in \mathbf{R}$. Show that $\mu_{f,g}$ is sesquilinear in f, g with $\mu_{g,f} = \overline{\mu_{f,g}}$.

(iv) Show that for any $m \in B(\mathbf{R} \to \mathbf{C})$, there is a unique bounded operator $m(L) \colon H \to H$ such that
$$\langle m(L)f, g \rangle = \int_{\mathbf{R}} m(x)\, d\mu_{f,g}(x)$$
for all $f, g \in H$.

(v) Show that $d\mu_{m(L)f,g}(x) = m(x)\, d\mu_{f,g}(x)$ for all $m \in B(\mathbf{R} \to \mathbf{C})$ and $f, g \in H$.

(vi) Show that the map $m \mapsto m(L)$ is a $*$-homomorphism from $B(\mathbf{R} \to \mathbf{C})$ to $B(H \to H)$.

(vii) Show that there is a projection-valued measure μ with $\mu(\mathbf{R}) = 1$, such that for every f, g, the complex measure $\mu_{f,g} := \langle \mu f, g \rangle$ is such that
$$\langle U(t)f, g \rangle = \int_{\mathbf{R}} e^{itx}\, d\mu_{f,g}(x)$$
for all $t \in \mathbf{R}$.

(viii) Show that there exists a self-adjoint densely defined operator $L \colon D \to H$ whose spectral measure is μ.

(ix) Conclude the proof of Stone's theorem.

Exercise 10.3.6. Let $L \colon D \to H$ be a densely defined symmetric operator. Suppose that for every $f \in D$ there exists a continuously differentiable solution $u \colon \mathbf{R} \to D$ to (10.2). Show that L is essentially self-adjoint.

We can now see a clear link between essential self-adjointness and completeness, at least in the case of scalar first-order differential operators:

Exercise 10.3.7. Let M be a smooth manifold with a smooth measure μ, and let X be a smooth vector field on M which is divergence-free with respect to the measure μ. Suppose that the vector field X is *complete* in the sense that for any $x_0 \in M$, there exists a global smooth solution $x \colon \mathbf{R} \to M$ to the ODE $\frac{d}{dt}x(t) = X(x(t))$ with initial data $x(0) = x_0$. Show that the first-order differential operator $i\nabla_X \colon C_c(M) \to L^2(M)$ is essentially self-adjoint.

Extend the above result to nondivergence-free vector fields, after replacing ∇_X with $\nabla_X + \frac{1}{2}\operatorname{div}(X)$.

Remark 10.3.4. The requirement of completeness is basically necessary; one can still have essential self-adjointness if there is a measure zero set of initial data x_0 for which the trajectories of X are incomplete, but once a positive measure set of trajectories become incomplete, the propagators $\exp(t\nabla_X)$ do not make sense globally, and so self-adjointness should fail.

Finally, we turn to the relationship between self-adjointness and the wave equation, which is a more complicated variant of the relationship between self-adjointness and the Schrödinger equation. More precisely, we will show the following version of Exercise 10.3.6:

Theorem 10.3.5. *Let $L\colon D \to H$ be a densely defined positive symmetric operator. Suppose for every $f, g \in D$, there exists a twice continuously differentiable (in D) solution $u\colon \mathbf{R} \to D$ to (10.3). Then L is essentially self-adjoint.*

(One could also obtain wave equation analogues of Exercise 10.3.4 or Theorem 10.3.3, but these are somewhat messy to state, and we will not do so here.)

We now prove this theorem. To simplify the exposition, we will assume that L is strictly positive, in the sense that $\langle Lf, f \rangle > 0$ for all nonzero $f \in D$, and leave the general case as an exercise.

We introduce a new inner product $\langle , \rangle_{\dot{H}^1}$ on D by the formula

$$\langle f, g \rangle_{\dot{H}^1} := \langle Lf, g \rangle.$$

By hypothesis, this is a Hermitian inner product on D. We then define an inner product $\langle , \rangle_E$ on $D \times D$ by the formula

$$\langle (u_0, u_1), (v_0, v_1) \rangle_E := \langle u_0, v_0 \rangle_{\dot{H}^1} + \langle u_1, v_1 \rangle,$$

then this is a Hermitian inner product on $D \times D$. We define the *energy space* $\mathcal{E}$ to be the completion of $D \times D$ with respect to this inner product; we can factor this as $\mathcal{E} = \dot{H}^1 \times H$, where $\dot{H}^1$ is the completion of D using the $\dot{H}^1$ inner product.

Suppose $u\colon \mathbf{R} \to D$ is a twice continuously differentiable solution to (10.3) for some $f, g \in D$. Then if we define the energy

$$E(t) := \frac{1}{2} \langle (u, \partial_t u), (u, \partial_t u) \rangle_E,$$

then one easily computes using (10.3) that $\partial_t E(t) = 0$, and so

$$E(t) = E(0) = \frac{1}{2} \langle (f, g), (f, g) \rangle_E.$$

In particular, if $f = g = 0$, then $u = 0$, and so twice continuously differentiable solutions to (10.3) are unique. This allows us to define wave operators $W(t)\colon D \times D \to D \times D$ by defining $W(t)(f, g) := (u(t), \partial_t u(t))$. This is clearly linear, and from the energy identity we see that W is an isometry, and thus extends to an isometry on the energy space $\mathcal{E}$. From uniqueness

10.3. Self-adjointness and flows

we also see that $t \mapsto W(t)$ is a one-parameter group, i.e., a homomorphism that is continuous in the strong operator topology. In particular, the isometries $W(t)$ are invertible and are thus unitary. By Stone's theorem, there thus exists a densely defined self-adjoint operator $A \colon \tilde{D} \to \mathcal{E}$ on some dense subspace $\tilde{D}$ of $\mathcal{E}$ such that $W(t) = e^{itA}$ for all $t \in \mathbf{R}$.

If $(f, g) \in D \times D$, then from the twice differentiability of the solution u to the wave equation, we see that

$$\lim_{t \to 0} \frac{W(t)(f,g) - (f,g)}{t} = (g, -Lf).$$

From Exercise 10.2, we conclude that $D \times D \subset \tilde{D}$ and

(10.12) $$A(f, g) = (-ig, iLf)$$

for all $(f, g) \in D \times D$.

Now we need to pass from the self-adjointness of A to the essential self-adjointness of L. Suppose for contradiction that L was not essentially self-adjoint. Then $L + 1$ does not have dense image, and so there exists a nonzero $h \in H$ such that

$$\langle (L+1)f, h \rangle = 0$$

for all $f \in D$.

It will be convenient to work with a "band-limited" portion of $D \times D$, to get around the problem that A and L can leave this domain. Let $\phi \colon \mathbf{R} \to \mathbf{R}$ be a compactly supported even smooth function of total mass one. For any $R > 0$, define the *Littlewood-Paley projection*

$$P_{\leq R} := \int_{\mathbf{R}} R\phi(Rt) W(t) \, dt = \int_{\mathbf{R}} R\phi(Rt) e^{itA} \, dt = \hat{\phi}(A/R),$$

where $\hat{\phi}$ is the Schwartz function $\hat{\phi}(\xi) := \int_{\mathbf{R}} \phi(t) e^{it\xi} \, dt$. Then (by strong continuity of $W(t)$) these operators map $D \times D$ to $D \times D$, commutes with the entire functional calculus of A, and also maps $\mathcal{E}$ to $\tilde{D}$. Also, from functional calculus one sees that

$$AP_{\leq R} = i \int_{\mathbf{R}} R^2 \phi'(Rt) W(t) \, dt,$$

and in particular, $AP_{\leq R}$ maps $D \times D$ to $D \times D$ as well.

Let $\tau \colon \mathcal{E} \to \mathcal{E}$ be the reflection operator $\tau(f, g) := (f, -g)$. From time reversal of the wave equation, we have $W(t)\tau = \tau W(-t)$, and thus τ commutes with $P_{\leq R}$. In particular, we can write

$$P_{\leq R}(f, g) = (P_{\leq R}^0 f, P_{\leq R}^1 g)$$

for some operators $P_{\leq R}^0 \colon \dot{H}^1 \to \dot{H}^1$ and $P_{\leq R}^1 \colon H \to H$. Since $P_{\leq R}$ preserves $D \times D$, the operators $P_{\leq R}^0, P_{\leq R}^1$ preserve D.

For any $(f,g) \in D \times D$, we have
$$P_{\leq R} A(f,g) = A P_{\leq R}(f,g);$$
combining this with (10.12) we see that
$$P^0_{\leq R} g = P^1_{\leq R} g$$
and
$$P^1_{\leq R} L f = L P^0_{\leq R} f$$
for all $f, g \in D$. Also, since $A P_{\leq R}$ preserves $D \times D$, we see that $L P^0_{\leq R} = P^1_{\leq R} L$ preserves D.

We can then expand $A^2 P_{\leq R}(f,g)$ as
$$A P_{\leq R} A(f,g) = A P_{\leq R}(-ig, iLf)$$
$$= A(-i P^1_{\leq R} g, i L P^0_{\leq R} f)$$
$$= (L P^0_{\leq R} f, L P^1_{\leq R} g)$$
and thus
$$(A^2 + 1) P_{\leq R}(f,g) = ((L+1) P^0_{\leq R} f, (L+1) P^1_{\leq R} g).$$
In particular,
$$\langle (A^2+1) P_{\leq R}(f,g), (0,h) \rangle_{\mathcal{E}} = 0$$
for all $(f,g) \in D \times D$. By duality (noting that $(A^2+1) P_{\leq R}$ is a bounded real function of A) we conclude that
$$(A^2+1) P_{\leq R}(0,h) = 0.$$
But by sending $R \to \infty$ and using the spectral theorem, this implies from monotone convergence that the spectral measure $\mu_{(0,h),(0,h)}$ is zero, and thus $(0,h)$ vanishes in $\mathcal{E}$, a contradiction. This establishes the essential self-adjointness of L.

Exercise 10.3.8. Establish Theorem 10.3.5 without the assumption that L is strictly positive. (*Hint:* Use the Cauchy-Schwarz inequality to show that strict positivity is equivalent to the absence of eigenfunctions with eigenvalue zero, and then quotient out such eigenfunctions.)

10.4. Essential self-adjointness of the Laplace-Beltrami operator

We now discuss how one can use the above criteria to establish essential self-adjointness of a Laplace-Beltrami operator $-\Delta_g$ on a smooth complete Riemannian manifold (M, g), viewed as a densely defined symmetric positive operator on the dense subspace $C^\infty_c(M)$ of $L^2(M)$. This result was first established in [**Ga1951**], [**Ro1960**].

10.4. Essential self-adjointness of the Laplace-Beltrami operator

To do this, we have to solve a PDE—either the Helmholtz equation (10.4) (for some z not on the positive real axis), the heat equation (10.1), the Schrödinger equation (10.3.4), or the wave equation (10.3).

The Schrödinger formalism is quite suggestive. From a semiclassical perspective, the Schrödinger equation associated to the Laplace-Beltrami operator $-\Delta_g$ should be viewed as a quantum version of the classical flow associated to the corresponding Hamiltonian $g^{ij}(x)\xi_i\xi_j$, i.e., geodesic flow. From Exercise 10.3.7 we know that the generator of Hamiltonian flows (normalised by i) are essentially self-adjoint when they are complete (note from *Liouville's theorem* that such generators are automatically divergence-free with respect to Liouville measure), which suggests that the Schrödinger operator $-\Delta_g$ should be also. Unfortunately, this is not a rigorous argument, and is difficult to make it so due to the nature of the time-dependent Schrödinger equation, which has infinite speed of propagation, and no dissipative properties. In practice, we therefore establish esssential self-adjointness by solving one of the other equations.

To solve any of these equations, it is difficult to solve any of them by an exact formula (unless the manifold M is extremely symmetric); but one can proceed by solving them *approximately*, and using perturbation theory to eliminate the error. This method works[1] as long as the error created by the approximate solution is both sufficiently small and sufficiently smooth that perturbative techniques (such as Neumann series, or the inverse function theorem) become applicable.

For instance, suppose one is trying to solve the Helmholtz equation

$$(-\Delta_g + k^2)u = f$$

for some large real k, and some $f \in L^2(M)$. For the sake of concreteness let us take M to be three-dimensional. If M was a Euclidean space $\mathbf{R}^3$, then we have an explicit formula

$$u(x) = \frac{1}{4\pi} \int_{\mathbf{R}^3} \frac{e^{-k|x-y|}}{|x-y|} f(y)\, dy$$

for the solution; note that the exponential decay of $e^{-k|x-y|}$ will keep u in L^2 as well. Inspired by this, we can try to solve the Helmholtz equation in curved space by proposing as an approximate solution

$$u(x) = \frac{1}{4\pi} \int_M \frac{e^{-kd(x,y)}}{d(x,y)} f(y)\, dg(y).$$

---

[1] This type of method, for instance, is used to deduce essential self-adjointness of various perturbations $-\Delta + V$ of the Laplace-Beltrami operator from the essential self-adjointness of the original operator $-\Delta$.

This is a little problematic because $d(x,y)$ develops singularities after a certain point, but if one has a uniform lower bound on the injectivity radius (here we are implicitly using the hypothesis of completeness), and also uniform bounds on the curvature and its derivatives, one can truncate this approximate solution to the region where $d(x,y)$ is small, and obtain an approximate solution u whose error
$$e := (-\Delta_g + k^2)u - f$$
can be made to be smaller in L^2 norm than that of f if the spectral parameter k is chosen large enough; we omit the details. From this, we can then solve the Helmholtz equation by Neumann series and thus establish essential self-adjointness. A similar method also works (under the same hypotheses on M) to construct approximate heat kernels, giving another way to establish essential self-adjointness of the Laplace-Beltrami operator.

What about when there is no uniform bound on the geometry? In that case, it is best to work with the wave equation formalism, because the finite speed of propagation property of that equation allows one to localise to compact portions of the manifold for which uniform bounds on the geometry are automatic. Indeed, to solve the wave equation in $C_c^\infty(M)$ for some fixed period of time, one can use finite speed of propagation, combined with completeness of the manifold, to work in a compact subset of this manifold, which by suitable alteration of the metric beyond the support of the solution, one can view as a subset of a compact complete manifold. On such manifolds, we already know essential self-adjointness by the previous arguments, so we may solve the wave equation in that setting (one can also solve the wave equation using methods from microlocal analysis, if desired), and then take repeated advantage of finite speed of propagation (which can be proven rigorously by energy methods) to glue together these local solutions to obtain a global solution; we omit the details.

In all of these cases, a somewhat nontrivial application of PDE theory is required. Unfortunately, this seems to be inevitable; at some point one must somehow use the hypothesis of completeness of the underlying manifold, and PDE methods are the only known way to connect that hypothesis to the dynamics of the Laplace-Beltrami operator.

Chapter 11

Notes on Lie algebras

A *abstract finite-dimensional complex Lie algebra*, or *Lie algebra* for short, is a finite-dimensional complex vector space $\mathfrak{g}$ together with an anti-symmetric bilinear form $[,] = [,]_\mathfrak{g} \colon \mathfrak{g} \times \mathfrak{g} \to \mathfrak{g}$ that obeys the *Jacobi identity*

$$[[x,y],z] + [[y,z],x] + [[z,x],y] = 0 \tag{11.1}$$

for all $x, y, z \in \mathfrak{g}$; by anti-symmetry one can also rewrite the Jacobi identity as

$$[x,[y,z]] = [[x,y],z] + [y,[x,z]]. \tag{11.2}$$

We will usually omit the subscript from the Lie bracket $[,]_\mathfrak{g}$ when this will not cause ambiguity. A *homomorphism* $\phi \colon \mathfrak{g} \to \mathfrak{h}$ between two Lie algebras $\mathfrak{g}, \mathfrak{h}$ is a linear map that respects the Lie bracket, thus $\phi([x,y]_\mathfrak{g}) = [\phi(x), \phi(y)]_\mathfrak{h}$ for all $x, y \in \mathfrak{g}$. As with many other classes of mathematical objects, the class of Lie algebras together with their homomorphisms then form a *category*. One can of course also consider Lie algebras in infinite dimension or over other fields, but we will restrict attention throughout these notes to the finite-dimensional complex case. The trivial, zero-dimensional Lie algebra is denoted 0; Lie algebras of positive dimension will be called *nontrivial*.

Lie algebras come up in many contexts in mathematics, in particular, arising as the tangent space of complex *Lie groups*. It is thus very profitable to think of Lie algebras as being the infinitesimal component of a Lie group, and in particular, almost all of the notation and concepts that are applicable to Lie groups (e.g., nilpotence, solvability, extensions, etc.) have infinitesimal counterparts in the category of Lie algebras (often with exactly the same terminology). See [**Ta2013**, §1.2] for more discussion about the connection between Lie algebras and Lie groups (that article was focused over the reals instead of the complexes, but much of the discussion carries over to the complex case).

A particular example of a Lie algebra is the general linear Lie algebra $\mathfrak{gl}(V)$ of linear transformations $x\colon V \to V$ on a finite-dimensional complex vector space (or *vector space* for short) V, with the commutator Lie bracket $[x, y] := xy - yx$; one easily verifies that this is indeed an abstract Lie algebra. We will define a *concrete* Lie algebra to be a Lie algebra that is a subalgebra of $\mathfrak{gl}(V)$ for some vector space V, and similarly define a *representation* of a Lie algebra $\mathfrak{g}$ to be a homomorphism $\rho\colon \mathfrak{g} \to \mathfrak{h}$ into a concrete Lie algebra $\mathfrak{h}$. It is a deep theorem of Ado (discussed in [**Ta2013**, §2.3]) that every abstract Lie algebra is in fact isomorphic to a concrete one (or equivalently, that every abstract Lie algebra has a faithful representation), but we will not need or prove this fact here.

Even without Ado's theorem, though, the structure of abstract Lie algebras is very well understood. As with objects in many other algebraic categories, a basic way to understand a Lie algebra $\mathfrak{g}$ is to factor it into two simpler algebras $\mathfrak{h}, \mathfrak{k}$ via a *short exact sequence*[1]

(11.3) $$0 \to \mathfrak{h} \to \mathfrak{g} \to \mathfrak{k} \to 0,$$

thus one has an injective homomorphism from $\mathfrak{h}$ to $\mathfrak{g}$ and a surjective homomorphism from $\mathfrak{g}$ to $\mathfrak{k}$ such that the image of the former homomorphism is the kernel of the latter. Given such a sequence, one can (nonuniquely) identify $\mathfrak{g}$ with the vector space $\mathfrak{h} \times \mathfrak{k}$ equipped with a Lie bracket of the form

(11.4) $$[(t,x),(s,y)]_\mathfrak{g} = ([t,s]_\mathfrak{h} + A(t,y) - A(s,x) + B(x,y), [x,y]_\mathfrak{k})$$

for some bilinear maps $A\colon \mathfrak{h} \times \mathfrak{k} \to \mathfrak{h}$ and $B\colon \mathfrak{k} \times \mathfrak{k} \to \mathfrak{h}$ that obey some Jacobi-type identities which we will not record here. Understanding exactly what maps A, B are possible here (up to coordinate change) can be a difficult task (and is one of the key objectives of *Lie algebra cohomology*), but in principle at least, the problem of understanding $\mathfrak{g}$ can be reduced to that of understanding its factors $\mathfrak{k}, \mathfrak{h}$. To emphasise this, I will (perhaps idiosyncratically) express the existence of a short exact sequence (11.3) by the ATLAS-type notation [**CoCuNoPaWi1985**]

(11.5) $$\mathfrak{g} = \mathfrak{h}.\mathfrak{k},$$

although one should caution that for given $\mathfrak{h}$ and $\mathfrak{k}$, there can be multiple nonisomorphic $\mathfrak{g}$ that can form a short exact sequence with $\mathfrak{h}, \mathfrak{k}$, so that $\mathfrak{h}.\mathfrak{k}$ is not a uniquely defined combination of $\mathfrak{h}$ and $\mathfrak{k}$; one could emphasise this by writing $\mathfrak{h}._{A,B}\mathfrak{k}$ instead of $\mathfrak{h}.\mathfrak{k}$, though we will not do so here. We will

---

[1] To be pedantic, a short exact sequence in a general category requires these homomorphisms to be *monomorphisms* and *epimorphisms* respectively, but in the category of Lie algebras these turn out to reduce to the more familiar concepts of injectivity and surjectivity respectively.

refer to $\mathfrak{g}$ as an *extension* of $\mathfrak{k}$ by $\mathfrak{h}$, and read the notation (11.5) as " $\mathfrak{g}$ is $\mathfrak{h}$-by-$\mathfrak{k}$"; confusingly, these two notations reverse the subject and object of "by", but unfortunately both notations are well entrenched in the literature. We caution that the operation . is not commutative, and it is only partly associative: every Lie algebra of the form $\mathfrak{k}.(\mathfrak{h}.\mathfrak{l})$ is also of the form $(\mathfrak{k}.\mathfrak{h}).\mathfrak{l}$, but the converse is not true (see [**Ta2013b**, §2.4] for some related discussion). As we are working in the infinitesimal world of Lie algebras (which have an additive group operation) rather than Lie groups (in which the group operation is usually written multiplicatively), it may help to think of $\mathfrak{h}.\mathfrak{k}$ as a (twisted) "sum" of $\mathfrak{h}$ and $\mathfrak{k}$ rather than a "product"; for instance, we have $\mathfrak{g} = 0.\mathfrak{g}$ and $\mathfrak{g} = \mathfrak{g}.0$, and also $\dim \mathfrak{h}.\mathfrak{k} = \dim \mathfrak{h} + \dim \mathfrak{k}$.

Special examples of extensions $\mathfrak{h}.\mathfrak{k}$ of $\mathfrak{k}$ by $\mathfrak{h}$ include the *direct sum* (or *direct product*) $\mathfrak{h} \oplus \mathfrak{k}$ (also denoted $\mathfrak{h} \times \mathfrak{k}$), which is given by the construction (11.4) with A and B both vanishing, and the *split extension* (or *semidirect product*) $\mathfrak{h} : \mathfrak{k} = \mathfrak{h} :_\rho \mathfrak{k}$ (also denoted $\mathfrak{h} \ltimes \mathfrak{k} = \mathfrak{h} \ltimes_\rho \mathfrak{k}$), which is given by the construction (11.4) with B vanishing and the bilinear map $A \colon \mathfrak{h} \times \mathfrak{k} \to \mathfrak{h}$ taking the form

$$A(t,x) = \rho(x)(t)$$

for some representation $\rho \colon \mathfrak{k} \to \operatorname{Der} \mathfrak{h}$ of $\mathfrak{k}$ in the concrete Lie algebra[2] of *derivations* $\operatorname{Der} \mathfrak{h} \subset \mathfrak{gl}(\mathfrak{h})$ of $\mathfrak{h}$, that is to say the algebra of linear maps $D \colon \mathfrak{h} \to \mathfrak{h}$ that obey the *Leibniz rule*

$$D[s,t]_\mathfrak{h} = [Ds,t]_\mathfrak{h} + [s,Dt]_\mathfrak{h}$$

for all $s,t \in \mathfrak{h}$.

There are two general ways to factor a Lie algebra $\mathfrak{g}$ as an extension $\mathfrak{h}.\mathfrak{k}$ of a smaller Lie algebra $\mathfrak{k}$ by another smaller Lie algebra $\mathfrak{h}$. One is to locate a *Lie algebra ideal* (or *ideal* for short) $\mathfrak{h}$ in $\mathfrak{g}$, thus $[\mathfrak{h}, \mathfrak{g}] \subset \mathfrak{h}$, where $[\mathfrak{h}, \mathfrak{g}]$ denotes the Lie algebra generated by $\{[x,y] : x \in \mathfrak{h}, y \in \mathfrak{g}\}$, and then take $\mathfrak{k}$ to be the quotient space $\mathfrak{g}/\mathfrak{h}$ in the usual manner; one can check that $\mathfrak{h}, \mathfrak{k}$ are also Lie algebras and that we do indeed have a short exact sequence

$$\mathfrak{g} = \mathfrak{h}.(\mathfrak{g}/\mathfrak{h}).$$

Conversely, whenever one has a factorisation $\mathfrak{g} = \mathfrak{h}.\mathfrak{k}$, one can identify $\mathfrak{h}$ with an ideal in $\mathfrak{g}$, and $\mathfrak{k}$ with the quotient of $\mathfrak{g}$ by $\mathfrak{h}$.

The other general way to obtain such a factorisation is to start with a homomorphism $\rho \colon \mathfrak{g} \to \mathfrak{m}$ of $\mathfrak{g}$ into another Lie algebra $\mathfrak{m}$, take $\mathfrak{k}$ to be the

---

[2] The derivation algebra $\operatorname{Der} \mathfrak{g}$ of a Lie algebra $\mathfrak{g}$ is analogous to the *automorphism group* $\operatorname{Aut}(G)$ of a Lie group G, with the two concepts being intertwined by the tangent space functor $G \mapsto \mathfrak{g}$ from Lie groups to Lie algebras (i.e., the derivation algebra is the infinitesimal version of the automorphism group). Of course, this functor also intertwines the Lie algebra and Lie group versions of most of the other concepts discussed here, such as extensions, semidirect products, etc.

image $\rho(\mathfrak{g})$ of $\mathfrak{g}$, and $\mathfrak{h}$ to be the kernel $\ker\rho := \{x \in \mathfrak{g} : \rho(x) = 0\}$. Again, it is easy to see that this does indeed create a short exact sequence:

$$\mathfrak{g} = \ker\rho.\rho(\mathfrak{g}).$$

Conversely, whenever one has a factorisation $\mathfrak{g} = \mathfrak{h}.\mathfrak{k}$, one can identify $\mathfrak{k}$ with the image of $\mathfrak{g}$ under some homomorphism, and $\mathfrak{h}$ with the kernel of that homomorphism. Note that if a representation $\rho \colon \mathfrak{g} \to \mathfrak{m}$ is *faithful* (i.e., injective), then the kernel is trivial and $\mathfrak{g}$ is isomorphic to $\rho(\mathfrak{g})$.

Now we consider some examples of factoring some class of Lie algebras into simpler Lie algebras. The easiest examples of Lie algebras to understand are the *abelian* Lie algebras $\mathfrak{g}$, in which the Lie bracket identically vanishes. Every one-dimensional Lie algebra is automatically abelian, and thus isomorphic to the scalar algebra $\mathbf{C}$. Conversely, by using an arbitrary linear basis of $\mathfrak{g}$, we see that an abelian Lie algebra is isomorphic to the direct sum of one-dimensional algebras. Thus, a Lie algebra is abelian if and only if it is isomorphic to the direct sum of finitely many copies of $\mathbf{C}$.

Now consider a Lie algebra $\mathfrak{g}$ that is not necessarily abelian. We then form the *derived algebra* $[\mathfrak{g}, \mathfrak{g}]$; this algebra is trivial if and only if $\mathfrak{g}$ is abelian. It is easy to see that $[\mathfrak{h}, \mathfrak{k}]$ is an ideal whenever $\mathfrak{h}, \mathfrak{k}$ are ideals, so, in particular, the derived algebra $[\mathfrak{g}, \mathfrak{g}]$ is an ideal and we thus have the short exact sequence

$$\mathfrak{g} = [\mathfrak{g}, \mathfrak{g}].(\mathfrak{g}/[\mathfrak{g}, \mathfrak{g}]).$$

The algebra $\mathfrak{g}/[\mathfrak{g}, \mathfrak{g}]$ is the maximal abelian quotient of $\mathfrak{g}$, and is known as the *abelianisation* of $\mathfrak{g}$. If it is trivial, we call the Lie algebra *perfect*. If instead it is nontrivial, then the derived algebra has strictly smaller dimension than $\mathfrak{g}$. From this, it is natural to associate two series to any Lie algebra $\mathfrak{g}$, the *lower central series*

$$\mathfrak{g}_1 = \mathfrak{g}; \mathfrak{g}_2 := [\mathfrak{g}, \mathfrak{g}_1]; \mathfrak{g}_3 := [\mathfrak{g}, \mathfrak{g}_2]; \ldots$$

and the *derived series*

$$\mathfrak{g}^{(1)} := \mathfrak{g}; \mathfrak{g}^{(2)} := [\mathfrak{g}^{(1)}, \mathfrak{g}^{(1)}]; \mathfrak{g}^{(3)} := [\mathfrak{g}^{(2)}, \mathfrak{g}^{(2)}]; \ldots.$$

By induction we see that these are both decreasing series of ideals of $\mathfrak{g}$, with the derived series being slightly smaller ($\mathfrak{g}^{(k)} \subseteq \mathfrak{g}_k$ for all k). We say that a Lie algebra is *nilpotent* if its lower central series is eventually trivial, and *solvable* if its derived series eventually becomes trivial. Thus, abelian Lie algebras are nilpotent, and nilpotent Lie algebras are solvable, but the converses are not necessarily true. For instance, in the general linear group $\mathfrak{gl}_n = \mathfrak{gl}(\mathbf{C}^n)$, which can be identified with the Lie algebra of $n \times n$ complex matrices, the subalgebra $\mathfrak{n}$ of strictly upper triangular matrices is nilpotent (but not abelian for $n \geq 3$), while the subalgebra $\mathfrak{n}$ of upper triangular matrices is solvable (but not nilpotent for $n \geq 2$). It is also clear that any subalgebra of a nilpotent algebra is nilpotent, and similarly for solvable or abelian algebras.

From the above discussion we see that a Lie algebra is solvable if and only if it can be represented by a tower of abelian extensions, thus
$$\mathfrak{g} = \mathfrak{a}_1.(\mathfrak{a}_2.\ldots(\mathfrak{a}_{k-1}.\mathfrak{a}_k)\ldots)$$
for some abelian $\mathfrak{a}_1, \ldots, \mathfrak{a}_k$. Similarly, a Lie algebra $\mathfrak{g}$ is nilpotent if it is expressible as a tower of *central* extensions (so that in all the extensions $\mathfrak{h}.\mathfrak{k}$ in the above factorisation, $\mathfrak{h}$ is central in $\mathfrak{h}.\mathfrak{k}$, where we say that $\mathfrak{h}$ is central in $\mathfrak{g}$ if $[\mathfrak{h}, \mathfrak{g}] = 0$). We also see that an extension $\mathfrak{h}.\mathfrak{k}$ is solvable if and only if both factors $\mathfrak{h}, \mathfrak{k}$ are solvable. Splitting abelian algebras into cyclic (i.e., one-dimensional) ones, we thus see that a finite-dimensional Lie algebra is solvable if and only if it is *polycylic*, i.e., it can be represented by a tower of cyclic extensions.

For our next fundamental example of using short exact sequences to split a general Lie algebra into simpler objects, we observe that every abstract Lie algebra $\mathfrak{g}$ has an *adjoint representation* $\mathrm{ad}\colon \mathfrak{g} \to \mathrm{ad}\,\mathfrak{g} \subset \mathfrak{gl}(\mathfrak{g})$, where for each $x \in \mathfrak{g}$, $\mathrm{ad}\,x \in \mathfrak{gl}(\mathfrak{g})$ is the linear map $(\mathrm{ad}\,x)(y) := [x, y]$; one easily verifies that this is indeed a representation (indeed, (11.2) is equivalent to the assertion that $\mathrm{ad}[x, y] = [\mathrm{ad}\,x, \mathrm{ad}\,y]$ for all $x, y \in \mathfrak{g}$). The kernel of this representation is the *centre* $Z(\mathfrak{g}) := \{x \in \mathfrak{g} : [x, \mathfrak{g}] = 0\}$, which is the maximal central subalgebra of $\mathfrak{g}$. We thus have the short exact sequence

(11.6) $$\mathfrak{g} = Z(\mathfrak{g}).\,\mathrm{ad}\,g$$

which, among other things, shows that every abstract Lie algebra is a central extension of a concrete Lie algebra (which can serve as a cheap substitute for Ado's theorem, mentioned earlier).

For our next fundamental decomposition of Lie algebras, we need some more definitions. A Lie algebra $\mathfrak{g}$ is simple if it is nonabelian and has no ideals other than 0 and $\mathfrak{g}$; thus simple Lie algebras cannot be factored $\mathfrak{g} = \mathfrak{h}.\mathfrak{k}$ into strictly smaller algebras $\mathfrak{h}, \mathfrak{k}$. In particular, simple Lie algebras are automatically perfect and centreless. We have the following fundamental theorem:

Theorem 11.0.1 (Equivalent definitions of semisimplicity)**.** *Let $\mathfrak{g}$ be a Lie algebra. Then the following are equivalent:*

(i) *$\mathfrak{g}$ does not contain any nontrivial solvable ideal.*

(ii) *$\mathfrak{g}$ does not contain any nontrivial abelian ideal.*

(iii) *The Killing form $K\colon \mathfrak{g} \times \mathfrak{g} \to \mathbf{C}$, defined as the bilinear form $K(x, y) := \mathrm{tr}_{\mathfrak{g}}((\mathrm{ad}\,x)(\mathrm{ad}\,y))$, is nondegenerate on $\mathfrak{g}$.*

(iv) *$\mathfrak{g}$ is isomorphic to the direct sum of finitely many nonabelian simple Lie algebras.*

We review the proof of this theorem later in this chapter. A Lie algebra obeying any (and hence all) of the properties (i)–(iv) is known as a *semisimple* Lie algebra. The statement (iv) is usually taken as the *definition* of semisimplicity; the equivalence of (iv) and (i) is a special case of *Weyl's complete reducibility theorem* (see Theorem 11.9.1), and the equivalence of (iv) and (iii) is known as the *Cartan semisimplicity criterion*. (The equivalence of (i) and (ii) is easy.)

If $\mathfrak{h}$ and $\mathfrak{k}$ are solvable ideals of a Lie algebra $\mathfrak{g}$, then it is not difficult to see that the vector sum $\mathfrak{h}+\mathfrak{k}$ is also a solvable ideal (because on quotienting by $\mathfrak{h}$ we see that the derived series of $\mathfrak{h}+\mathfrak{k}$ must eventually fall inside $\mathfrak{h}$, and thence must eventually become trivial by the solvability of $\mathfrak{h}$). As our Lie algebras are finite-dimensional, we conclude that $\mathfrak{g}$ has a unique maximal solvable ideal, known as the *radical* $\operatorname{rad} \mathfrak{g}$ of $\mathfrak{g}$. The quotient $\mathfrak{g}/\operatorname{rad}\mathfrak{g}$ is then a Lie algebra with trivial radical, and is thus semisimple by the above theorem, giving the *Levi decomposition*

$$\mathfrak{g} = \operatorname{rad}\mathfrak{g}.(\mathfrak{g}/\operatorname{rad}\mathfrak{g})$$

expressing an arbitrary Lie algebra as an extension of a semisimple Lie algebra $\mathfrak{g}/\operatorname{rad}\mathfrak{g}$ by a solvable algebra $\operatorname{rad}\mathfrak{g}$ (and it is not hard to see that this is the only possible such extension up to isomorphism). Indeed, a deep theorem of Levi allows one to upgrade this decomposition to a split extension

$$\mathfrak{g} = \operatorname{rad}\mathfrak{g} : (\mathfrak{g}/\operatorname{rad}\mathfrak{g})$$

although we will not need or prove this result here.

In view of the above decompositions, we see that we can factor any Lie algebra (using a suitable combination of direct sums and extensions) into a finite number of simple Lie algebras and the scalar algebra $\mathbf{C}$. In principle, this means that one can understand an arbitrary Lie algebra once one understands all the simple Lie algebras (which, being defined over $\mathbf{C}$, are somewhat confusingly referred to as *simple complex Lie algebras* in the literature). Amazingly, this latter class of algebras are completely classified:

Theorem 11.0.2 (Classification of simple Lie algebras). *Up to isomorphism, every simple Lie algebra is of one of the following forms:*

   (i) $A_n = \mathfrak{sl}_{n+1}$ *for some* $n \geq 1$.

   (ii) $B_n = \mathfrak{so}_{2n+1}$ *for some* $n \geq 2$.

   (iii) $C_n = \mathfrak{sp}_{2n}$ *for some* $n \geq 3$.

   (iv) $D_n = \mathfrak{so}_{2n}$ *for some* $n \geq 4$.

   (v) $E_6, E_7,$ *or* E_8.

   (vi) F_4.

   (vii) G_2.

(*The precise definition of the* classical Lie algebras A_n, B_n, C_n, D_n *and the* exceptional Lie algebras E_6, E_7, E_8, F_4, G_2 *will be recalled later.*)

One can extend the families A_n, B_n, C_n, D_n of classical Lie algebras a little bit to smaller values of n, but the resulting algebras are either isomorphic to other algebras on this list, or cease to be simple; see [**Ta2013c**, §2.1] for further discussion.)

This classification is a basic starting point for the classification of many other related objects, including Lie algebras and Lie groups over more general fields (e.g., the reals $\mathbf{R}$), as well as finite simple groups. Being so fundamental to the subject, this classification is covered in almost every basic textbook in Lie algebras; I will review it in this chapter. The material here is all drawn from standard reference texts (I relied particularly on [**FuHa**]). In fact it seems remarkably hard to deviate from the standard routes given in the literature to the classification.

11.1. Abelian representations

One of the key strategies in the classification of a Lie algebra $\mathfrak{g}$ is to work with representations of $\mathfrak{g}$, particularly the adjoint representation ad: $\mathfrak{g} \to \operatorname{ad} \mathfrak{g}$, and then restrict such representations to various simpler subalgebras $\mathfrak{h}$ of $\mathfrak{g}$, for which the representation theory is well understood. In particular, one aims to exploit the representation theory of *abelian* algebras (and to a lesser extent, nilpotent and solvable algebras), as well as the fundamental example of the two-dimensional *special linear Lie algebra* $\mathfrak{sl}_2$, which is the smallest and easiest to understand of the simple Lie algebras, and plays an absolutely crucial role in exploring and then classifying all the other simple Lie algebras.

We begin this program by recording the representation theory of abelian Lie algebras. We begin with representations $\rho \colon \mathbf{C} \to \mathfrak{gl}(V)$ of the one-dimensional algebra $\mathbf{C}$. Setting $x := \rho(1)$, this is essentially the representation theory of a single linear transformation $x \colon V \to V$. Here, the theory is given by the *Jordan decomposition*. First, for each complex number $\lambda \in \mathbf{C}$, we can define the *generalised eigenspace*

$$V_\lambda^x := \{v \in V : (x - \lambda)^n v = 0 \text{ for some } n\}.$$

One easily verifies that the V_λ^x are all linearly independent x-invariant subspaces of V, and in particular, that there are only finitely many λ (the *spectrum* $\sigma(x)$ of x) for which V_λ^x is nontrivial. If one quotients out all the generalised eigenspaces, one can check that the quotiented transformation x no longer has any spectrum, which contradicts the *fundamental theorem of algebra* applied to the *characteristic polynomial* of this quotiented transformation (or, if is more analytically inclined, one could apply *Liouville's theorem* to the *resolvent operators* to obtain the required contradiction). Thus the generalised eigenspaces span V:

$$V = \bigoplus_{\lambda \in \sigma(x)} V_\lambda^x.$$

On each space V_λ^x, the operator $x - \lambda$ only has spectrum at zero, and thus (again from the fundamental theorem of algebra) has nontrivial kernel; similarly for any x-invariant subspace of V_λ^x, such as the range $(x-\lambda)V_\lambda^x$ of $x-\lambda$. Iterating this observation we conclude that $x - \lambda$ is a *nilpotent operator* on V_λ^x, thus $(x - \lambda)^n = 0$ for some n. If we then write x_{ss} to be the direct sum of the scalar multiplication operators λ on each generalised eigenspace V_λ^x, and x_n to be the direct sum of the operators $x - \lambda$ on these spaces, we have obtained the *Jordan decomposition* (or *Jordan-Chevalley decomposition*)

$$x = x_{ss} + x_n$$

where the operator $x_{ss} \colon V \to V$ is *semisimple* in the sense that it is a diagonalisable linear transformation on V (or equivalently, all generalised eigenspaces are actually eigenspaces), and x_n is nilpotent. Furthermore, as we may use polynomial interpolation to find a polynomial $P \colon \mathbf{C} \to \mathbf{C}$ such that $P(z) - \lambda$ vanishes to arbitrarily high order at $z = \lambda$ for each $\lambda \in \sigma(V)$ (and also $P(0) = 0$), we see that x_{ss} (and hence x_n) can be expressed as polynomials in x with zero constant coefficient; this fact will be important later. In particular, x_{ss} and x_n commute.

Conversely, given an arbitrary linear transformation $x \colon V \to V$, the Jordan-Chevalley decomposition is the unique decomposition into commuting semisimple and nilpotent elements. Indeed, if we have an alternate decomposition $x = x'_{ss} + x'_n$ into a semisimple element x'_{ss} commuting with a nilpotent element x'_n, then the generalised eigenspaces of x must be preserved by both x'_{ss} and x'_n, and so without loss of generality we may assume that there is just a single generalised eigenspace $V = V_\lambda^x$; subtracting λ we may then assume that $\lambda = 0$, but then x is nilpotent, and so $x'_{ss} = x - x'_n$ is also nilpotent; but the only transformation which is both semisimple and nilpotent is the zero transformation, and the claim follows.

From the Jordan-Chevalley decomposition it is not difficult to then place x in *Jordan normal form* by selecting a suitable basis for V; see, e.g.. [**Ta2008**, §1.13]. But in contrast to the Jordan-Chevalley decomposition, the basis is not unique in general, and we will not explicitly use the Jordan normal form in the rest of this chapter.

Given an abstract complex vector space V, there is in general no canonical notion of complex conjugation on V, or of linear transformations $x \colon V \to V$. However, we can define the conjugate $\overline{x}$ of any *semisimple* transformation $x \colon V \to V$, defined as the direct sum of $\overline{\lambda}$ on each eigenspace V_λ^x of x.

In particular, we can define the conjugate $\overline{x_{ss}}\colon V \to V$ of the semisimple component x_{ss} of an arbitrary linear transformation x_{ss}, which will be the direct sum of $\overline{\lambda}$ on each *generalised* eigenspace V_λ^x of x. The significance of this transformation lies in the observation that the product $\overline{x_{ss}}x$ has trace $|\lambda|^2 \dim V_\lambda^x$ on each generalised eigenspace (since nilpotent operators have zero trace), and in particular, we see that

$$(11.7) \qquad \operatorname{tr}(\overline{x_{ss}}x) = 0$$

if and only if the spectrum consists only of zero, or equivalently that x is nilpotent. Thus (11.7) provides a test for nilpotency, which will be turn out to be quite useful later in this chapter. (Note that this trick relies very much on the special structure of $\mathbf{C}$, in particular, the fact that it has characteristic zero.)

In the above arguments we have used the basic fact that if two operators $x\colon V \to V$ and $y\colon V \to V$ commute, then the generalised eigenspaces of one operator are preserved by the other. Iterating this fact, we can now start understanding the representations $\rho\colon \mathfrak{h} \to \mathfrak{gl}(V)$ of an abelian Lie algebra. Namely, there is a finite set $\sigma(\rho) \subset \mathfrak{h}^*$ of linear functionals (or homomorphisms) $\lambda\colon \mathfrak{h} \to \mathbf{C}$ on $\mathfrak{h}$ (i.e., elements of the dual space $\mathfrak{h}^*$) for which the generalised eigenspaces

$$V_\lambda^{\mathfrak{h}} := \{v \in V : (\rho(\mathfrak{h}) - \lambda)^n v = 0 \text{ for some } n\}$$

are nontrivial and $\mathfrak{h}$-invariant, and we have the decomposition

$$V = \bigoplus_{\lambda \in \sigma(x)} V_\lambda^{\mathfrak{h}}.$$

Here we use $(\rho(\mathfrak{h}) - \lambda)^n v = 0$ as shorthand for writing $(x_1 - \lambda(x_1)) \ldots (\rho(x_n) - \lambda(x_n))v = 0$ for all $x_1, \ldots, x_n \in \mathfrak{h}$. An important special case arises when the action of $\mathfrak{h}$ is semisimple in the sense that $\rho(x)$ is semisimple for all $x \in \mathfrak{h}$. Then all the generalised eigenspaces are just eigenspaces (or *weight spaces*), thus

$$\rho(x)v = \lambda(x)v$$

for all $v \in V_\lambda^{\mathfrak{h}}$ and $x \in \mathfrak{h}$. When this occurs we call v a *weight vector* with weight λ.

11.2. Engel's theorem and Lie's theorem

In the introduction we gave the two basic examples of nilpotent and solvable Lie algebras, namely the strictly upper triangular and upper triangular matrices. The theorems of Engel and Lie assert, roughly speaking, that these examples (and subalgebras thereof) are essentially the only type of solvable and nilpotent Lie algebras that can exist, at least in the concrete setting of subalgebras of $\mathfrak{gl}(V)$. Among other things, these theorems greatly clarify the representation theory of nilpotent and solvable Lie algebras.

We begin with Engel's theorem.

Theorem 11.2.1 (Engel's theorem). *Let $\mathfrak{g} \subset \mathfrak{gl}(V)$ be a concrete Lie algebra such that every element x of $\mathfrak{g}$ is nilpotent as a linear transformation on V.*

(i) *If V is nontrivial, then there is a nonzero element v of V which is annihilated by every element of $\mathfrak{g}$.*

(ii) *There is a basis of V for which all elements of $\mathfrak{g}$ are strictly upper triangular. In particular, $\mathfrak{g}$ is nilpotent.*

Proof. We begin with (i). We induct on the dimension of $\mathfrak{g}$. The claim is trivial for dimensions 0 and 1, so suppose that $\mathfrak{g}$ has dimension greater than 1, and that the claim is already proven for smaller dimensions.

Let $\mathfrak{h}$ be a maximal proper subalgebra of $\mathfrak{g}$, then $\mathfrak{h}$ has dimension strictly between zero and $\dim \mathfrak{g}$ (since all one-dimensional subspaces are proper subalgebras). Observe that for every $x \in \mathfrak{h}$, $\operatorname{ad} x$ acts on both the vector spaces $\mathfrak{g}$ and $\mathfrak{h}$ and thus also on the quotient space $\mathfrak{g}/\mathfrak{h}$. As x is nilpotent, all of these actions are nilpotent also. In particular, by induction hypothesis, there is $v \in \mathfrak{g}/\mathfrak{h}$ which is annihilated by $\operatorname{ad} x$ for all $x \in \mathfrak{h}$. Let w be a representative of v in $\mathfrak{g}$, then $[w, \mathfrak{h}] \subset \mathfrak{h}$, and so $\operatorname{span}(w, \mathfrak{h})$ is a subalgebra and is thus all of $\mathfrak{g}$.

By induction hypothesis again, the space W of vectors in V annihilated by $\mathfrak{h}$ is nontrivial; as $[w, \mathfrak{h}] \subset \mathfrak{h}$, it is preserved by w. As w is nilpotent, there is a nontrivial element of W annihilated by w and hence by $\mathfrak{g}$, as required.

Now we prove (ii). We induct on the dimension of V. The case of dimension zero is trivial, so suppose V has dimension at least one, and the claim has already been proven for dimension $\dim(V) - 1$. By (i), we may find a nontrivial vector v annihilated by $\mathfrak{g}$, and so we may project $\mathfrak{g}$ down to $\mathfrak{gl}(V/\operatorname{span}(v))$. By the induction hypothesis, there is a basis for $V/\operatorname{span}(v)$ on which the projection of any element of $\mathfrak{g}$ is strictly upper-triangular; pulling this basis back to V and adjoining v, we obtain the claim. $\square$

As a corollary of this theorem and the short exact sequence (11.6) we see that an abstract Lie algebra $\mathfrak{g}$ is nilpotent iff $\operatorname{ad} \mathfrak{g}$ is nilpotent iff $\operatorname{ad} x$ is nilpotent in $\mathfrak{g}$ for every $x \in \mathfrak{g}$ (i.e., every element of $\mathfrak{g}$ is *ad-nilpotent*).

Engel's theorem is in fact valid over every field. The analogous theorem of Lie for solvable algebras, however, relies much more strongly on the specific properties of the complex field $\mathbf{C}$.

Theorem 11.2.2 (Lie's theorem). *Let $\mathfrak{g} \subset \mathfrak{gl}(V)$ be a solvable concrete Lie algebra.*

(i) If V is nontrivial, there exists a nonzero element v of V which is an eigenvector for every element of $\mathfrak{g}$.

(ii) There is a basis for V such that every element of $\mathfrak{g}$ is upper triangular.

Note that if one specialises Lie's theorem to abelian $\mathfrak{g}$ then one essentially recovers the abelian theory of the previous section.

Proof. We prove (i). As before we induct on the dimension of $\mathfrak{g}$. The dimension zero case is trivial, so suppose that $\mathfrak{g}$ has dimension at least one and that the claim has been proven for smaller dimensions.

Let $\mathfrak{h}$ be a codimension one subalgebra of $\mathfrak{g}$; such an algebra can be formed by taking a codimension one subspace of the abelianisation $\mathfrak{g}/[\mathfrak{g},\mathfrak{g}]$ (which has dimension at least one, else $\mathfrak{g}$ will not be solvable) and then pulling back to $\mathfrak{g}$. Note that $\mathfrak{h}$ is automatically an ideal.

By induction, there is a nonzero element v of V such that every element of $\mathfrak{h}$ has v as an eigenvector, thus we have

$$xv = \lambda(x)v$$

for all $x \in \mathfrak{h}$ and some linear functional $\lambda \colon \mathfrak{h} \to \mathbf{C}$. If we then set $W = V_\lambda^{\mathfrak{h}}$ to be the simultaneous eigenspace

$$W := \{w \in V : xv = \lambda(x)v \text{ for all } x \in \mathfrak{h}\},$$

then W is a nontrivial subspace of V.

Let y be an element of $\mathfrak{g}$ that is not in $\mathfrak{h}$, and let $w \in W$. Consider the space spanned by the orbit $w, yw, y^2w, \ldots$. By finite dimensionality, this space has a basis $w, yw, y^2w, \ldots, y^{n-1}w$ for some n. By induction and definition of W, we see that every $x \in \mathfrak{h}$ acts on this space by an upper-triangular matrix with diagonal entries $\lambda(x)$ in this basis. Of course, y acts on this space as well, and so $[x,y]$ has trace zero on this space, thus $n\lambda([x,y]) = 0$ and so $\lambda([x,y]) = 0$ (here we use the characteristic zero nature of $\mathbf{C}$). From this we see that y fixes W. If we let v' be an eigenvector of y on W (which exists from the Jordan decomposition of y), we conclude that v' is a simultaneous eigenvector of $\mathfrak{g}$ as required.

The claim (ii) follows from (i) much as in Engel's theorem. $\square$

11.3. Characterising semisimplicity

The objective of this section will be to prove Theorem 11.0.1.

Let $\mathfrak{g} \subset \mathfrak{gl}(V)$ be a concrete Lie algebra, and x an element of $\mathfrak{g}$. Then the components $x_{ss}, x_n \colon V \to V$ of x need not lie in $\mathfrak{g}$. However, they behave "as if" they lie in $\mathfrak{g}$ for the purposes of taking Lie brackets, in the following sense:

Lemma 11.3.1. *Let $\mathfrak{g} \subset \mathfrak{gl}(V)$ and let $x \in \mathfrak{g}$ have Jordan decomposition $x = x_{ss} + x_n$. Then $[x_{ss}, \mathfrak{g}] \subset [\mathfrak{g}, \mathfrak{g}]$, $[\overline{x_{ss}}, \mathfrak{g}] \subset [\mathfrak{g}, \mathfrak{g}]$ and $[x_n, \mathfrak{g}] \subset [\mathfrak{g}, \mathfrak{g}]$.*

Proof. As x_{ss} and x_n are semisimple and nilpotent on V and commute with each other, $\operatorname{ad} x_{ss}$ and $\operatorname{ad} x_n$ are semisimple and nilpotent on $\mathfrak{gl}(V)$ and also commute with each other (this can, for instance, by using Lie's theorem (or the Jordan normal form) to place x in upper triangular form and compute everything explicitly). Thus $\operatorname{ad} x = \operatorname{ad} x_{ss} + \operatorname{ad} x_n$ is the Jordan-Chevalley decomposition of $\operatorname{ad} x$, and in particular, $\operatorname{ad} x_{ss} = Q(\operatorname{ad} x)$ for some polynomial Q with zero constant coefficient. Since $\operatorname{ad} x$ maps $\mathfrak{g}$ to the subalgebra $[\mathfrak{g}, \mathfrak{g}]$, we conclude that $\operatorname{ad} x_{ss} = Q(\operatorname{ad} x)$ does also, thus $[x_{ss}, \mathfrak{g}] \subset [\mathfrak{g}, \mathfrak{g}]$ as required. Similarly for $\overline{x_{ss}}$ and x_n (note that $\operatorname{ad} \overline{x_{ss}} = \overline{\operatorname{ad} x_{ss}}$). $\square$

We can now use this (together with Engel's theorem and the test (11.7) for nilpotency) to obtain a part of Theorem 11.0.1:

Proposition 11.3.2. *Let $\mathfrak{g}$ be a simple Lie algebra. Then the Killing form K is nondegenerate.*

Proof. As $\mathfrak{g}$ is simple, its centre $Z(\mathfrak{g})$ is trivial, so by (11.6) $\mathfrak{g}$ is isomorphic to $\operatorname{ad} \mathfrak{g}$. In particular we may assume that $\mathfrak{g}$ is a concrete Lie algebra, thus $\mathfrak{g} \subset \mathfrak{gl}(V)$ for some vector space V.

Suppose for contradiction that K is degenerate. Using the skew-adjointness identity
$$K([z, x], y) = -K(x, [z, y])$$
for all $x, y, z \in \mathfrak{g}$ (which comes from the cyclic properties of trace), we see that the kernel $\{x \in \mathfrak{g} : K(x, y) = 0 \text{ for all } y \in \mathfrak{g}\}$ is a nontrivial ideal of $\mathfrak{g}$, and is thus all of $\mathfrak{g}$ as $\mathfrak{g}$ is simple. Thus $K(x, y) = 0$ for all $x, y \in \mathfrak{g}$.

Now let $x, y, z \in \mathfrak{g}$. By Lemma 11.3.1, $\overline{x_{ss}}$ acts by Lie bracket on $\mathfrak{g}$ and so one can define $\operatorname{ad} \overline{x_{ss}} \in \mathfrak{gl}(\mathfrak{g})$. We now consider the quantity
$$\operatorname{tr}_{\mathfrak{g}}(\operatorname{ad} \overline{x_{ss}})(\operatorname{ad}[y, z]).$$
We can rearrange this as
$$\operatorname{tr}_{\mathfrak{g}}(\operatorname{ad}[y, \overline{x_{ss}}])(\operatorname{ad} z).$$
By Lemma 11.3.1, $[y, \overline{x_{ss}}] \in \mathfrak{g}$, so this is equal to
$$K([y, \overline{x_{ss}}], z) = 0,$$
and so
$$\operatorname{tr}_{\mathfrak{g}}(\operatorname{ad} \overline{x_{ss}})(\operatorname{ad} w) = 0$$

11.3. Characterising semisimplicity

for all $w \in [\mathfrak{g}, \mathfrak{g}]$. On the other hand, $[\mathfrak{g}, \mathfrak{g}]$ is an ideal of $\mathfrak{g}$; as $\mathfrak{g}$ is simple, we must thus have $\mathfrak{g} = [\mathfrak{g}, \mathfrak{g}]$ (i.e., $\mathfrak{g}$ is perfect). As $x \in \mathfrak{g}$, we conclude that

$$\operatorname{tr}_{\mathfrak{g}}(\operatorname{ad} \overline{x_{ss}})(\operatorname{ad} x) = 0.$$

From (11.7) we conclude that $\operatorname{ad} x$ is nilpotent for every x. By Engel's theorem, this implies that $\operatorname{ad} \mathfrak{g}$, and hence $\mathfrak{g}$, is nilpotent; but $\mathfrak{g}$ is simple, giving the desired contradiction. □

Corollary 11.3.3. *Let $\mathfrak{h}$ be a simple ideal of a Lie algebra $\mathfrak{g}$. Then $\mathfrak{h}$ is complemented by another ideal $\mathfrak{k}$ of $\mathfrak{g}$ (thus $\mathfrak{h} \cap \mathfrak{k} = \{0\}$ and $\mathfrak{h} + \mathfrak{k} = \mathfrak{g}$), with $\mathfrak{g}$ isomorphic to the direct sum $\mathfrak{h} \oplus \mathfrak{k}$.*

Proof. The adjoint action of $\mathfrak{g}$ restricts to the ideal $\mathfrak{h}$ and gives a restricted Killing form

$$K_{\mathfrak{h}}(x, y) := \operatorname{tr}_{\mathfrak{h}}((\operatorname{ad} x)(\operatorname{ad} y)).$$

By Proposition 11.3.2, this bilinear form is nondegenerate on $\mathfrak{h}$, so the orthogonal complement

$$\mathfrak{k} := \mathfrak{h}^{\perp} = \{x \in \mathfrak{g} : K_{\mathfrak{h}}(x, y) = 0 \text{ for all } y \in \mathfrak{h}\}$$

is a complementary subspace to $\mathfrak{h}$. It can be verified to also be an ideal. Since $[\mathfrak{h}, \mathfrak{k}]$ lies in both $\mathfrak{h}$ and $\mathfrak{k}$, we see that $[\mathfrak{h}, \mathfrak{k}] = 0$, and so $\mathfrak{g}$ is isomorphic to $\mathfrak{h} \oplus \mathfrak{k}$ as claimed. □

Now we can prove Theorem 11.0.1. We first observe that (i) trivially implies (ii); conversely, if $\mathfrak{g}$ has a nontrivial solvable ideal $\mathfrak{h}$, then every element of the derived series of $\mathfrak{h}$ is also an ideal of $\mathfrak{g}$, and in particular, $\mathfrak{g}$ will have a nontrivial abelian ideal. Thus (i) and (ii) are equivalent.

Now we show that (i) implies (iv), which we do by induction on the dimension of $\mathfrak{g}$. Of course we may assume $\mathfrak{g}$ is nontrivial. Let $\mathfrak{h}$ be a nontrivial ideal of $\mathfrak{g}$ of minimal dimension. If $\mathfrak{h} = \mathfrak{g}$ then $\mathfrak{g}$ is simple (note that it cannot be abelian as $\mathfrak{g}$ is nontrivial and semisimple) and we are done. If $\mathfrak{h}$ is strictly smaller than $\mathfrak{g}$, then it also has no nontrivial solvable ideals (because the radical of $\mathfrak{h}$ is a *characteristic* subalgebra of $\mathfrak{h}$ (that is, it is invariant with respect to automorphisms of $\mathfrak{h}$) and is thus an ideal in $\mathfrak{g}$) and so by induction is isomorphic to the direct sum of simple Lie algebras; as $\mathfrak{h}$ was minimal, we conclude that $\mathfrak{h}$ is itself simple. By Corollary 11.3.3, $\mathfrak{g}$ then splits as the direct sum of $\mathfrak{h}$ and a semisimple Lie algebra of strictly smaller dimension, and the claim follows from the induction hypothesis.

From Proposition 11.3.2 we see that (iv) implies (iii), so to finish the proof of Theorem 11.0.1 it suffices to show that (iii) implies (ii). Indeed, if $\mathfrak{g}$ has a nontrivial abelian ideal $\mathfrak{h}$, then for any $x \in \mathfrak{g}$ and $y \in \mathfrak{h}$, $\operatorname{ad} x \operatorname{ad} y$ annihilates $\mathfrak{h}$ and also has range in $\mathfrak{h}$, hence has trace zero, so $\mathfrak{h}$ is K-orthogonal to $\mathfrak{g}$, giving the degeneracy of the Killing form.

Remark 11.3.4. Similar methods also give the *Cartan solvability criterion*: a Lie algebra $\mathfrak{g}$ is solvable if and only if $\mathfrak{g}$ is orthogonal to $[\mathfrak{g}, \mathfrak{g}]$ with respect to the Killing form. Indeed, the "only if" part follows easily from Lie's theorem, while for the "if" part one can adapt the proof of Proposition 11.3.2 to show that if $\mathfrak{g}$ is orthogonal to $[\mathfrak{g}, \mathfrak{g}]$, then every element of $\operatorname{ad}[\mathfrak{g}, \mathfrak{g}]$ is nilpotent, hence by Engel's theorem $\operatorname{ad}[\mathfrak{g}, \mathfrak{g}]$ is nilpotent, and so from the short exact sequence (11.6) we see that $[\mathfrak{g}, \mathfrak{g}]$ is nilpotent, and hence $\mathfrak{g}$ is solvable.

Remark 11.3.5. The decomposition of a semisimple Lie algebra as the direct sum of simple Lie algebras is unique up to isomorphism and permutation. Indeed, suppose that $\bigoplus_{i=1}^{n} \mathfrak{g}_i$ is isomorphic to $\bigoplus_{j=1}^{n'} \mathfrak{g}'_j$ for some simple $\mathfrak{g}_i, \mathfrak{g}'_j$. We project each $\mathfrak{g}'_j$ to $\mathfrak{g}_i$ and observe from simplicity that these projections must either be zero or isomorphisms (cf. *Schur's lemma*). For fixed i, there must be at least one j for which the projection is an isomorphism (otherwise $\bigoplus_{j=1}^{n'} \mathfrak{g}'_j$ could not generate all of $\bigoplus_{i=1}^{n} \mathfrak{g}_i$); on the other hand, as any two $\mathfrak{g}'_j$ commute with each other in the direct sum, and $\mathfrak{g}_i$ is nonabelian, there is at most one j for which the projection is an isomorphism. This gives the required identification of the $\mathfrak{g}_i$ and $\mathfrak{g}'_j$ up to isomorphism and permutation.

Remark 11.3.6. One can also establish complete reducibility by using the *Weyl unitary trick*, in which one first creates a real compact Lie group whose Lie algebra is a real form of the complex Lie algebra being studied, and then uses the complete reducibility of actions of compact groups. This also gives an alternate way to establish Theorem 11.9.1 in Section 11.9.

Semisimple Lie algebras have a number of important nondegeneracy properties. For instance, they have no nontrivial outer automorphisms (at the infinitesimal level, at least):

Lemma 11.3.7 (Semisimple derivations are inner)**.** *Let $\mathfrak{g}$ be a semisimple Lie algebra. Then every derivation $D \in \operatorname{Der} \mathfrak{g}$ on $\mathfrak{g}$ is inner, thus $D = \operatorname{ad} x$ for some $x \in \mathfrak{g}$.*

Proof. From the identity $[D, \operatorname{ad} x] = \operatorname{ad} Dx$ we see that $\operatorname{ad} \mathfrak{g}$ is an ideal in $\operatorname{Der} \mathfrak{g}$. The trace form $(D_1, D_2) \mapsto \operatorname{tr}(D_1 D_2)$ on $\operatorname{Der} \mathfrak{g}$ restricts to the Killing form on $\operatorname{ad}(\mathfrak{g})$, which is nondegenerate.

Suppose for contradiction that $\operatorname{ad}(\mathfrak{g})$ is not all of $\operatorname{Der}(\mathfrak{g})$, then there is a nontrivial derivation D which is trace-form orthogonal to $\operatorname{ad}(\mathfrak{g})$, thus D is trace-orthogonal to $[\operatorname{ad} x, \operatorname{ad} y]$ for all $x, y \in \mathfrak{g}$, so that $[D, \operatorname{ad} x] = \operatorname{ad} Dx$ is trace-orthogonal to $\operatorname{ad} y$ for all $x, y \in \mathfrak{g}$. As K is nondegenerate, we conclude that $Dx = 0$ for all x, and so D is trivial, a contradiction. $\square$

11.3. Characterising semisimplicity

This fact, combined with the complete reducibility of $\mathfrak{g}$-modules (a fact which we will prove in an appendix) implies that the Jordan decomposition preserves concrete semisimple Lie algebras:

Corollary 11.3.8. *Let* $\mathfrak{g} \subset \mathfrak{gl}(V)$ *be a concrete semisimple Lie algebra, and let* $x \in \mathfrak{g}$. *Then* $x_{ss}, x_n, \overline{x}_{ss}$ *also lie in* $\mathfrak{g}$.

Proof. By Theorem 11.0.1, $\mathfrak{g}$ is the direct sum of commuting simple algebras. It is easy to see that if x, y commute then the Jordan decomposition of $x+y$ arises from the sum of the Jordan decompositions of x and y separately, so we may assume without loss of generality that $\mathfrak{g}$ is simple.

Observe that if V splits as the direct sum $V = V_1 \oplus V_2$ of two $\mathfrak{g}$-invariant subspaces (so that $\mathfrak{g}$ can be viewed as a subalgebra of $\mathfrak{gl}(V_1) \oplus \mathfrak{gl}(V_2)$, and the elements of x can be viewed as being block-diagonal in a suitable basis of V_1, V_2), then the claim for V follows from that of V_1 and V_2. So by an induction on dimension, it suffices to establish the claim under the hypothesis that V is *indecomposable*, in that it cannot be expressed as the direct sum of two nontrivial invariant subspaces.

In Section 11.9 we will show that every invariant subspace W of V is *complemented* in that one can write $V = W \oplus W'$ for some invariant subspace W'. Assuming this fact, it suffices to establish the claim in the case that V is *irreducible*, in the sense that it contains no proper invariant subspaces.

By Lemma 11.3.3, the operation $y \mapsto [x_{ss}, y]$ is a derivation on $\mathfrak{g}$, thus there exists $a \in \mathfrak{g}$ such that $[x_{ss}, y] = (\operatorname{ad} a)y$ for all $y \in \mathfrak{g}$, thus $x_{ss} - a \in \mathfrak{gl}(V)$ centralises $\mathfrak{g}$. By *Schur's lemma* and the hypothesis of irreducibility, we conclude that $x_{ss} - a$ is a multiple of a constant λ. On the other hand, every element of $\mathfrak{g}$ has trace zero since $\mathfrak{g} = [\mathfrak{g}, \mathfrak{g}]$; in particular, a and x have trace zero, and so $x_{ss} - a$ has trace zero. But this trace is just $\lambda \dim W$, so we conclude that $x_{ss} - a$ has zero as its only generalised eigenvalue and is thus nilpotent. $\square$

This allows us to make the Jordan decomposition universal for semisimple algebras:

Lemma 11.3.9 (Semisimple Jordan decomposition). *Let* $\mathfrak{g}$ *be a semisimple Lie algebra, and let* $x \in \mathfrak{g}$. *Then we have a unique decomposition* $x = x_{ss} + x_n$ *in* $\mathfrak{g}$ *such that* $\rho(x_{ss}) = (\rho(x))_{ss}$ *and* $\rho(x_n) = (\rho(x))_n$ *for every representation* ρ *of* $\mathfrak{g}$.

Proof. As the adjoint representation is faithful we may assume without loss of generality that $\mathfrak{g}$ is a concrete algebra, thus $\mathfrak{g} \subset \mathfrak{gl}(V)$. The uniqueness is then clear by taking ρ to be the identity. To obtain existence, we take x_{ss}, x_n to be the concrete Jordan decomposition. We need to verify

$\rho(x_{ss}) = (\rho(x))_{ss}$ and $\rho(x_n) = (\rho(x))_n$ for any representation $\rho\colon \mathfrak{g} \to \mathfrak{m}$. The adjoint actions of $\rho(x_{ss})$ and $\rho(x_n)$ on $\rho(\mathfrak{g})$ commute and are semisimple and nilpotent respectively and so

$$\operatorname{ad}\rho(x_{ss}) = (\operatorname{ad}\rho(x))_{ss}; \quad \operatorname{ad}\rho(x_n) = (\operatorname{ad}\rho(x))_n$$

in $\operatorname{ad}\rho(g)$ (cf. the proof of Lemma 11.3.1). A similar argument (applying Corollary 11.3.8 to $\rho(g)$, which is isomorphic to a quotient of $\mathfrak{g}$ and is thus semisimple, to keep $\rho(x)_{ss}, \rho(x)_n$ in $\rho(\mathfrak{g})$) gives

$$\operatorname{ad}(\rho(x)_{ss}) = (\operatorname{ad}\rho(x))_{ss}; \quad \operatorname{ad}(\rho(x)_n) = (\operatorname{ad}\rho(x))_n.$$

Since the adjoint representation of the semisimple algebra $\rho(\mathfrak{g})$ is faithful, the claim follows. $\square$

One can also show that x_{ss}, x_n commute with each other and with the centraliser $C(x) := \{y \in \mathfrak{g} : [x, y] = 0\}$ of x by using the faithful nature of the adjoint representation for semisimple algebras, though we will not need these facts here. Using this lemma we have a well-defined notion of an element x of a semisimple algebra $\mathfrak{g}$ being semisimple (resp. nilpotent), namely that $x = x_{ss}$ or $x = x_n$. Lemma 11.3.9 then implies that any representation of a semisimple element of $\mathfrak{g}$ is again semisimple, and any representation of a nilpotent element of $\mathfrak{g}$ is again nilpotent. This apparently innocuous statement relies heavily on the semisimple nature of $\mathfrak{g}$; note for instance that the representation

$$t \mapsto \begin{pmatrix} 0 & t \\ 0 & 0 \end{pmatrix}$$

of the nonsemisimple algebra $\mathbf{C} \equiv \mathfrak{gl}_1$ into $\mathfrak{gl}_2$ takes semisimple elements to nilpotent ones.

11.4. Cartan subalgebras

While simple Lie algebras do not have any nontrivial ideals, they do have some very useful subalgebras known as *Cartan subalgebras* which will eventually turn out to be abelian and which can be used to dramatically clarify the structure of the rest of the algebra.

We need some definitions. An element x of $\mathfrak{g}$ is said to be *regular* if its generalised null space

$$\mathfrak{g}_0^x := \{y \in \mathfrak{g} : (\operatorname{ad} x)^n y = 0 \text{ for some } n\}$$

has minimal dimension. A *Cartan subalgebra* of $\mathfrak{g}$ is a nilpotent subalgebra $\mathfrak{h}$ of $\mathfrak{g}$ which is its own *normaliser*, thus $N(\mathfrak{h}) := \{x \in \mathfrak{g} : [x, \mathfrak{h}] \subset \mathfrak{h}\}$ is equal to $\mathfrak{h}$. From the polynomial nature of the Lie algebra operations (and the Noetherian nature of algebraic geometry) we see that the regular elements of $\}$ are generic (i.e., they form a nonempty Zariski-open subset of $\mathfrak{g}$).

11.4. Cartan subalgebras

Example 11.4.1. In $\mathfrak{gl}(V)$, the regular elements consist of the semisimple elements with distinct eigenvalues. Fixing a basis for V, the space of elements of $\mathfrak{gl}(V)$ that are diagonalised by that basis form a Cartan subalgebra of $\mathfrak{gl}(V)$.

Cartan algebras always exist, and can be constructed as generalised null spaces of regular elements:

Proposition 11.4.2 (Existence of Cartan subalgebras). *Let $\mathfrak{g}$ be an abstract Lie algebra. If $x \in \mathfrak{g}$ is regular, then the generalised null space $\mathfrak{h} := \mathfrak{g}_0^x$ of x is a Cartan subalgebra.*

Proof. Suppose that $\mathfrak{h}$ is not nilpotent, then by Engel's theorem the adjoint action of at least one element of $\mathfrak{h}$ on $\mathfrak{h}$ is not nilpotent. By the polynomial nature of the Lie algebra operations, we conclude that the adjoint action of a generic element of $\mathfrak{h}$ on $\mathfrak{h}$ is not nilpotent.

The action of x on $\mathfrak{g}/\mathfrak{h}$ is nonsingular, so the action of generic elements of $\mathfrak{h}$ on $\mathfrak{g}/\mathfrak{h}$ is also nonsingular. Thus we can find $y \in \mathfrak{h}$ such that $\operatorname{ad} y$ is not nilpotent on $\mathfrak{h}$ and not singular on $\mathfrak{g}/\mathfrak{h}$. From this we see that $\mathfrak{g}_0^y$ is a proper subspace of $\mathfrak{g}_0^x$, contradicting the regularity of x. Thus $\mathfrak{h}$ is nilpotent.

Finally, we show that $\mathfrak{h}$ is its own normaliser. Suppose that $y \in \mathfrak{g}$ normalises $\mathfrak{h}$, then $(\operatorname{ad} x) y \in \mathfrak{h}$. But $\mathfrak{h}$ is the generalised null space of $\operatorname{ad} x$, and so $y \in \mathfrak{h}$ as required. □

Furthermore, all Cartan algebras arise as generalised null spaces:

Proposition 11.4.3 (Cartans are null spaces). *Let $\mathfrak{g}$ be an abstract Lie algebra, and let $\mathfrak{h}$ be a Cartan subalgebra. Let*

$$\mathfrak{g}_0^{\mathfrak{h}} = \{x \in \mathfrak{g} : (\operatorname{ad} \mathfrak{h})^n x = 0 \text{ for some } n\}$$

be the generalised null space of $\mathfrak{h}$. Then $\mathfrak{g}_0^{\mathfrak{h}} = \mathfrak{h}$. Furthermore, for generic $x \in \mathfrak{h}$, one has

$$\mathfrak{h} = \mathfrak{g}_0^x.$$

Proof. As $\mathfrak{h}$ is nilpotent, we certainly have $\mathfrak{h} \subset \mathfrak{g}_0^{\mathfrak{h}}$. Now, for any $x \in \mathfrak{h}$, $\operatorname{ad} x$ acts nilpotently on both $\mathfrak{g}_0^{\mathfrak{h}}$ and $\mathfrak{h}$ and hence on $\mathfrak{g}_0^{\mathfrak{h}}/\mathfrak{h}$. By Engel's theorem, we can thus find $y \in \mathfrak{g}_0^{\mathfrak{h}}/\mathfrak{h}$ that is annihilated by the adjoint action of $\mathfrak{h}$; pulling back to $\mathfrak{g}_0^{\mathfrak{h}}$, we conclude that the normaliser of $\mathfrak{h}$ is strictly larger than $\mathfrak{h}$, contradicting the hypothesis that $\mathfrak{h}$ is a Cartan subalgebra. This shows that $\mathfrak{g}_0^{\mathfrak{h}} = \mathfrak{h}$.

Now let $x \in \mathfrak{h}$ be generic, then $\mathfrak{g}_0^x$ has minimal dimension amongst $x \in \mathfrak{h}$. Let $y \in \mathfrak{h}$ be arbitrary. Then for any scalar t, $\operatorname{ad}(x + ty)$ acts on $\mathfrak{g}$ and on $\mathfrak{h}$ and hence on $\mathfrak{g}/\mathfrak{h}$. This action is invertible when $t = 0$, and hence is also invertible for generic t; thus for generic t, $\mathfrak{g}_0^{x+ty} \subset \mathfrak{g}_0^x$. By minimality we conclude that $\mathfrak{g}_0^{x+ty} = \mathfrak{g}_0^x$, so $\mathrm{ad}(x+ty)$ is nilpotent on $\mathfrak{g}_0^x$ for generic t, and thus for all t. In particular, $\mathrm{ad}(x+y)$ is nilpotent on $\mathfrak{g}_0^x$ for any $y \in \mathfrak{h}$, thus $\mathfrak{g}_0^x \subset \mathfrak{g}_0^{\mathfrak{h}} = \mathfrak{h}$. Since $\mathfrak{h} \subset \mathfrak{g}_0^x$, we obtain $\mathfrak{h} = \mathfrak{g}_0^x$ as required. □

Corollary 11.4.4 (Cartans are conjugate). *Let $\mathfrak{g}$ be a Lie algebra, and let $\mathfrak{h}$ be a Cartan algebra. Then for generic $x \in \mathfrak{g}$, $\mathfrak{h}$ is conjugate to $\mathfrak{g}_0^x$ by an inner automorphism of $\mathfrak{g}$ (i.e., an element of the algebraic group generated by $\exp(\mathrm{ad}\, y)$ for $y \in \mathfrak{g}$). In particular, any two Cartan subalgebras are conjugate to each other by an inner automorphism.*

Proof. Let S be the set of $x' \in \mathfrak{h}$ with $\mathfrak{h} = \mathfrak{g}_0^{x'}$, then x' is a Zariski open dense subset of $\mathfrak{h}$ by Proposition 11.4.3. Then let T be the collection of $x \in \mathfrak{g}$ that are conjugate to an $x' \in S$, then T is a algebraically constructible subset of $\mathfrak{g}$. For $x' \in S$, observe that $(\mathrm{ad}\, x')(\mathfrak{g})$ and $\mathfrak{h}$ span $\mathfrak{g}$, since $\mathfrak{h} = \mathfrak{g}_0^{x'}$, and so by the *inverse function theorem*, a (topological) neighbourhood of x' is contained in T. This implies that T is Zariski dense, and the claim follows. □

In the case of semisimple algebras, the Cartan structure is particularly clean:

Proposition 11.4.5. *Let $\mathfrak{g}$ be a semisimple Lie algebra. Then every Cartan subalgebra $\mathfrak{h}$ is abelian, and K is nondegenerate on $\mathfrak{h}$.*

The dimension of the Cartan algebra of a semisimple Lie algebra is known as the *rank* of the algebra.

Proof. The nilpotent algebra $\mathfrak{h}$ acts via the adjoint action on $\mathfrak{g}$, and by Lie's theorem this action can be made upper triangular. From this it is not difficult to obtain a decomposition

$$\mathfrak{g} = \bigoplus_{\lambda \in \sigma(\mathfrak{h})} \mathfrak{g}_\lambda^{\mathfrak{h}}$$

for some finite set $\sigma(\mathfrak{h}) \subset \mathfrak{h}^*$, where $\mathfrak{g}_\lambda^{\mathfrak{h}}$ are the generalised eigenspaces

$$\mathfrak{g}_\lambda^{\mathfrak{h}} = \{x \in \mathfrak{g} : (\mathrm{ad}\,\mathfrak{h} - \lambda)^k x = 0 \text{ for some } k\}.$$

From the Jacobi identity (11.2) we see that $[\mathfrak{g}_\lambda^{\mathfrak{h}}, \mathfrak{g}_\mu^{\mathfrak{h}}] \subset \mathfrak{g}_{\lambda+\mu}^{\mathfrak{h}}$. Among other things, this shows that $\mathfrak{g}_\lambda^{\mathfrak{h}}$ has ad-trace zero for any nonzero λ, and hence $\mathfrak{g}_\lambda^{\mathfrak{h}}, \mathfrak{g}_\mu^{\mathfrak{h}}$ are K-orthogonal if $\lambda + \mu \neq 0$. In particular, $\mathfrak{g}_0^{\mathfrak{h}}$ is K-orthogonal to $\bigoplus_{\lambda \neq 0} \mathfrak{g}_\lambda^{\mathfrak{h}}$. By Theorem 11.0.1, K is nondegenerate on $\mathfrak{g}$, and thus also nondegenerate on $\mathfrak{g}_0^{\mathfrak{h}}$; by Proposition 11.4.3, K is thus nondegenerate on 𝔥. But by Lie's theorem, we can find a basis for which 𝔥 consists of upper-triangular matrices in the adjoint representation of 𝔤, so that [𝔥, 𝔥] is strictly upper-triangular and thus K-orthogonal to 𝔥. As K is nondegenerate on 𝔥, this forces [𝔥, 𝔥] to be abelian, as required. □

We now use the semisimple Jordan decomposition (Lemma 11.3.9) to obtain a further nondegeneracy property of the Cartan subalgebras of semisimple algebras:

Proposition 11.4.6. *Let 𝔤 be a semisimple Lie algebra. Then every Cartan subalgebra 𝔥 consists entirely of semisimple elements.*

Proof. Let $x \in 𝔥$, then (by the abelian nature of 𝔥) $\operatorname{ad} x$ annihilates 𝔥; as $\operatorname{ad} x_n$ is a polynomial in $\operatorname{ad} x$ with zero constant coefficient, $\operatorname{ad} x_n$ annihilates 𝔥 as well; thus x_n normalises 𝔥 and thus also lies in 𝔥 as 𝔥 is Cartan. If $y \in 𝔥$, then y commutes with x_n and so $\operatorname{ad} y$ commutes with $\operatorname{ad} x_n$. As the latter is nilpotent, we conclude that $\operatorname{ad} x_n \operatorname{ad} y$ is nilpotent and thus has trace zero. Thus x_n is K-orthogonal to 𝔥 and thus vanishes since the Killing form is nondegenerate on 𝔥. Thus every element of 𝔥 is semisimple as required. □

11.5. $\mathfrak{sl}_2$ representations

To proceed further, we now need to perform some computations on a very specific Lie algebra, the special linear algebra $\mathfrak{sl}_2$ of 2×2 complex matrices with zero trace. This is a three-dimensional concrete Lie algebra, spanned by the three generators

$$H := \begin{pmatrix} 1 & 0 \\ 0 & -1 \end{pmatrix}; X := \begin{pmatrix} 0 & 1 \\ 0 & 0 \end{pmatrix}; Y := \begin{pmatrix} 0 & 0 \\ 1 & 0 \end{pmatrix}$$

which obey the commutation relations

(11.8) $\qquad [H, X] = 2X; [H, Y] = -2Y; [X, Y] = H.$

Conversely, any abstract three-dimensional Lie algebra generated by H, X, Y with relations (11.8) is clearly isomorphic to $\mathfrak{sl}_2$. One can check that this is a simple Lie algebra, with the one-dimensional space generated by H being a Cartan subalgebra.

Now we classify by hand the representations $\rho \colon \mathfrak{sl}_2 \to \mathfrak{gl}(V)$ of $\mathfrak{sl}_2$. Observe that $\mathfrak{sl}_2$ acts infinitesimally on $\mathbf{R}^2$ by the differential operators (or vector fields)

$$H \to x\partial_x - y\partial_y; \quad X \to x\partial_y; \quad Y \to y\partial_x.$$

In particular, we see that for each natural number n, the space P_n of homogeneous polynomials in two variables x, y of degree n has a representation $\sigma_n \colon \mathfrak{sl}_2 \to \mathfrak{gl}(P_n)$; if we give this space the basis $e_{2i-n} := x^i y^{n-i}$ for

$i = 0, \ldots, n$, the action is then described by the formulae

(11.9) $\quad \sigma_n(H)e_j = je_j; \quad \sigma_n(X)e_j = \dfrac{n-j}{2}e_{j+2}; \quad \sigma_n(Y)e_j = \dfrac{n+j}{2}e_{j-2}$

for $j = n, n-2, \ldots, -n+2, n$. From these formulae it is also easy to see that these representations are irreducible in the sense that the P_n have no nontrivial $\mathfrak{sl}_2$-invariant subspaces.

Conversely, these representations (and their direct sums) describe (up to isomorphism) all of the representations of $\mathfrak{sl}_2$:

Theorem 11.5.1 (Representations of $\mathfrak{sl}_2$). *Any representation $\rho\colon \mathfrak{sl}_2 \to \mathfrak{gl}(V)$ is isomorphic to the direct sum of finitely many of the representations $\sigma_n\colon \mathfrak{sl}_2 \to \mathfrak{gl}(P_n)$.*

Here of course the direct sum $\rho_1 \oplus \rho_2 \colon \mathfrak{g} \to \mathfrak{gl}(V_1 \oplus V_2)$ of two representations $\rho_1\colon \mathfrak{g} \to \mathfrak{gl}(V_1)$, $\rho_2\colon \mathfrak{g} \to \mathfrak{gl}(V_2)$ is defined as $\rho_1 \oplus \rho_2(x) := (\rho_1(x), \rho_2(x))$, and two representations $\rho_1\colon \mathfrak{g} \to \mathfrak{gl}(V_1)$, $\rho_2\colon \mathfrak{g} \to \mathfrak{gl}(V_2)$ are isomorphic if there is an invertible linear map $\phi\colon V_1 \to V_2$ such that $\phi \circ \rho_1(x) = \rho_2(x) \circ \phi$ for all $x \in \mathfrak{g}$.

Proof. By induction we may assume that V is nontrivial, the claim has already been proven for any smaller dimensional spaces than V.

As H is semisimple, $\rho(H)$ is semisimple by Lemma 11.3.9, and so we can split V into the direct sum

$$V = \bigoplus_{\lambda \in \sigma(H)} V_\lambda^H$$

of eigenspaces of H for some finite $\sigma(H) \subset \mathbf{C}$.

From (11.8) we have the raising law

$$\rho(X)V_\lambda^H \subset V_{\lambda+2}^H$$

and the lowering law

$$\rho(Y)V_\lambda^H \subset V_{\lambda-2}^H.$$

As $\sigma(H)$ is finite, we may find a "highest weight" $\lambda \in \sigma(H)$ with the property that $\lambda + 2 \notin \sigma(H)$, thus $\rho(X)$ annihilates V_λ^H by the raising law. We will use the basic strategy of starting from the highest weight space and applying lowering operators to discover one of the irreducible components of the representation.

From (11.8) one has

$$\rho(X)\rho(Y) = \rho(Y)\rho(X) + \rho(H)$$

and so from induction and the lowering law we see that

(11.10) $\quad \rho(X)\rho(Y)^{k+1}v = (\lambda + (\lambda - 2) + \ldots + (\lambda - 2k))\rho(Y)^k v$

for all natural numbers k and all $v \in V_\lambda^H$. If $\lambda + (\lambda - 2) + \ldots + (\lambda - 2k)$ is never zero, this creates an infinite sequence $V_\lambda^H, V_{\lambda-2}^H, V_{\lambda-4}^H, \ldots$ of nontrivial eigenspaces, which is absurd, so we have $\lambda + (\lambda - 2) + \ldots + (\lambda - 2n) = 0$ for some natural number n, thus $\lambda = n$. If we let

$$W := \bigoplus_{k=0}^{n} \rho(Y)^k V_n^H,$$

then we see that W is invariant under H, X, and Y, and thus $\mathfrak{g}$-invariant; also if for each $\lambda \in \sigma(H)$ we let $\tilde{V}_\lambda^H$ be the set of all $v \in V_\lambda^H$ such that $\rho(X)^k v$ is never a nonzero element of V_n^H then we see that

$$\tilde{W} := \bigoplus_{\lambda \in \sigma(H)} \tilde{V}_\lambda^H$$

is also $\mathfrak{g}$-invariant, and furthermore that W and $\tilde{W}$ are complementary subspaces in V. Applying the induction hypothesis, we are done unless $W = V$, but then by splitting V_n^H into one-dimensional spaces and applying the lowering operators, we see that we reduce to the case that V_n^H is one-dimensional. But if one then lets e_n be a generator of V_n^H and recursively defines $e_{n-2}, e_{n-4}, \ldots, e_{-n}$ by

$$\rho(Y) e_j = \frac{n+j}{2} e_{j-2},$$

one then checks using (11.10) that ρ is isomorphic to σ_n, and the claim follows. $\square$

Remark 11.5.2. Theorem 11.5.1 shows that all representations of $\mathfrak{sl}_2$ are *completely reducible* in that they can be decomposed as the direct sum of irreducible representations. In fact, all representations of semisimple Lie algebras are completely reducible; this can be proven by a variant of the above arguments (in combination with the analysis of weights given below), and can also be proven by the unitary trick, or by analysing the action of *Casimir elements* of the universal enveloping algebra of $\mathfrak{g}$, as done in Section 11.9.

11.6. Root spaces

Now we use the $\mathfrak{sl}_2$ theory to analyse more general semisimple algebras.

Let $\mathfrak{g}$ be a semisimple Lie algebra, and let $\mathfrak{h}$ be a Cartan algebra, then by Proposition 11.4.5 $\mathfrak{h}$ is abelian and acts in a semisimple fashion on $\mathfrak{g}$, and by Proposition 11.4.3 $\mathfrak{h}$ is its own null space $\mathfrak{g}_0^{\mathfrak{h}}$ in the weight decomposition of $\mathfrak{g}$, thus we have the *Cartan decomposition*

$$\mathfrak{g} = \mathfrak{h} \oplus \bigoplus_{\alpha \in \Phi} \mathfrak{g}_\alpha^{\mathfrak{h}}$$

as vector spaces (not as Lie algebras) where Φ is a finite subset of $\mathfrak{h}^*\backslash\{0\}$ (known as the set of *roots*) and $\mathfrak{g}_\alpha^\mathfrak{h}$ is the nontrivial eigenspace

(11.11) $$\mathfrak{g}_\alpha^\mathfrak{h} = \{x \in \mathfrak{g} : [y, x] = \alpha(y)x \text{ for all } y \in \mathfrak{h}\}.$$

Example 11.6.1. A key example to keep in mind is when $\mathfrak{g} = \mathfrak{sl}_n$ is the Lie algebra of $n \times n$ matrices of trace zero. An explicit computation using the Killing form and Theorem 11.0.1 shows that this algebra is semisimple; in fact it is simple, but we will not show this yet. The space $\mathfrak{h}$ of diagonal matrices of trace zero can then be verified to be a Cartan algebra; it can be identified with the space $\mathbf{C}_0^n$ of complex n-tuples summing to zero, and using the usual Hermitian inner product on $\mathbf{C}^n$ we can also identify $\mathfrak{h}^*$ with $\mathbf{C}_0^n$. The roots are then of the form $e_i - e_j$ for distinct $1 \leq i, j \leq n$, where $e_1, \ldots, e_n$ is the standard basis for $\mathbf{C}^n$, with $\mathfrak{g}_{e_i-e_j}^\mathfrak{h}$ being the one-dimensional space of matrices that are vanishing except possibly at the (i, j) coefficient.

From the Jacobi identity (11.2) we see that the Lie bracket acts additively on the weights, thus

(11.12) $$[\mathfrak{g}_\alpha^\mathfrak{h}, \mathfrak{g}_\beta^\mathfrak{h}] \subset \mathfrak{g}_{\alpha+\beta}^\mathfrak{h}$$

for all $\alpha, \beta \in \mathfrak{h}^*$. Taking traces, we conclude that

$$K(\mathfrak{g}_\alpha^\mathfrak{h}, \mathfrak{g}_\beta^\mathfrak{h}) = 0$$

whenever $\alpha + \beta \neq 0$. As K is nondegenerate, we conclude that if $\mathfrak{g}_\alpha^\mathfrak{h}$ is nontrivial, then $\mathfrak{g}_{-\alpha}^\mathfrak{h}$ must also be nontrivial, thus Φ is symmetric around the origin.

We also claim that Φ spans $\mathfrak{h}^*$ as a vector space. For if this were not the case, then there would be a nontrivial $x \in \mathfrak{h}$ that is annihilated by Φ, which by (11.11) implies that $\operatorname{ad} x$ annihilates all of the $\mathfrak{g}_\alpha^\mathfrak{h}$ and is thus central, contradicting the semisimplicity of $\mathfrak{g}$.

From Proposition 11.4.5, K is nondegenerate on $\mathfrak{h}$. Thus, for each root $\alpha \in \Phi$, there is a corresponding nonzero element t_α of $\mathfrak{h}$ such that $K(t_\alpha, x) = \alpha(x)$ for all $x \in \mathfrak{h}$. If we let $x \in \mathfrak{g}_\alpha^\mathfrak{h}, y \in \mathfrak{g}_{-\alpha}^\mathfrak{h}$ and $z \in \mathfrak{g}_0^\mathfrak{h} = \mathfrak{h}$, we have

$$K([x, y], z) = K(y, [z, x])$$
$$= K(y, \alpha(z)x)$$
$$= \alpha(K(x, y)z)$$
$$= K(K(x, y)t_\alpha, z)$$

and thus by the nondegeneracy of K on $\mathfrak{h}$ we obtain the useful formula

(11.13) $$[x, y] = K(x, y)t_\alpha$$

for $x \in \mathfrak{g}_\alpha^\mathfrak{h}$ and $y \in \mathfrak{g}_{-\alpha}^\mathfrak{h}$.

11.6. Root spaces

As K is nondegenerate, we can find $X = X_\alpha \in \mathfrak{g}_\alpha^\mathfrak{h}$ and $Y = Y_\alpha \in \mathfrak{g}_{-\alpha}^\mathfrak{h}$ with $K(X,Y) \neq 0$ (which can be found as K is nondegenerate). We divide into two cases depending on whether $\alpha(t_\alpha)$ vanishes or not. If $\alpha(t_\alpha)$ vanishes, then $[X,Y]$ is nontrivial but commutes with X and Y, and so $\operatorname{ad} X, \operatorname{ad} Y$ generate a solvable algebra. By Lie's theorem, this algebra is upper-triangular in some basis, and so $\operatorname{ad}[X,Y]$ is nilpotent, hence $\operatorname{ad} t_\alpha$ is nilpotent; but by Proposition 11.4.6 $\operatorname{ad} t_\alpha$ is also semisimple, contradicting the nonzero nature of t_α (and the semisimple nature of $\mathfrak{g}$). Thus $\alpha(t_\alpha)$ is nonvanishing. If we then scale X, Y so that $[X, Y] = H$, where $H = H_\alpha$ is the *coroot* of α, defined as the element of $\mathfrak{h}$ given by the formula

$$H := \frac{2}{\alpha(t_\alpha)} t_\alpha$$

so that

(11.14) $$\alpha(H) = 2,$$

then X, Y, H obey the relations (11.8) and thus generate a copy of $\mathfrak{sl}_2$, rather than a solvable algebra. The representation theory of $\mathfrak{sl}_2$ can then be applied to the space

(11.15) $$\bigcup_{n \in S_\alpha} \mathfrak{g}_{n\alpha/2}^\mathfrak{h},$$

where $S_\alpha := \{n \in \mathbf{R} : n\alpha/2 \in \Phi \cup \{0\}\}$. By (11.12), this space is invariant with respect to x and y and hence to the copy of $\mathfrak{sl}_2$, and by (11.11), (11.14) each $\mathfrak{g}_{n\alpha/2}^\mathfrak{h}$ is the weight space of H of weight n for each $n \in S$. By Theorem 11.5.1, we conclude that the set S consists of integers. On the other hand, from (11.13) we see that any copy of the representation σ_n with n a positive even integer must have its 0 weight space contained in the span of t_α, and so there is only one such representation in (11.15). As X, Y, H already give a copy of σ_2 in (11.15), there are no other copies of σ_n with n positive even, thus we have that $\mathfrak{g}_\alpha^\mathfrak{h}$ is one-dimensional and that the only even multiples of $\alpha/2$ in Φ are $\pm\alpha$. In particular, $2\alpha \notin \Phi$ whenever $\alpha \in \Phi$, which also implies that $\alpha/2 \notin \Phi$ whenever $\alpha \in \Phi$. Returning to Theorem 11.5.1, we conclude that the set S_α contains no odd integers, and so α and $-\alpha$ are the only multiples of α in Φ.

Next, let β be any nonzero element of $\mathfrak{h}^*$ orthogonal to α with respect to the inner product $\langle,\rangle$ of $\mathfrak{h}^*$ that is dual to the restriction of the Killing form to $\mathfrak{h}$, and consider the space

(11.16) $$\bigcup_{n \in S_{\alpha,\beta}} \mathfrak{g}_{\beta+n\alpha/2}^\mathfrak{h}$$

where

$$S_{\alpha,\beta} := \{n \in \mathbf{R} : \beta + n\alpha/2 \in \Phi\}.$$

By (11.12), this is again an $\mathfrak{sl}_2$-invariant space, and by (11.11), (11.14) each $\mathfrak{g}^{\mathfrak{h}}_{\beta+n\alpha/2}$ is the weight space of H of weight n. From Theorem 11.5.1 we see that $S_{\alpha,\beta}$ is an arithmetic progression $\{-m, -m+2, \ldots, m-2, m\}$ of spacing 2; in particular, $S_{\alpha,\beta}$ is symmetric around the origin and consists only of integers. This implies that the set Φ is symmetric with respect to reflection across the hyperplane that is orthogonal to α, and also implies that
$$2\frac{\langle \alpha, \beta \rangle}{\langle \alpha, \alpha \rangle} \in \mathbf{Z}$$
for all roots $\alpha, \beta \in \Phi$.

We summarise the various geometric properties of Φ as follows:

Proposition 11.6.2 (Root systems). *Let $\mathfrak{g}$ be a semisimple Lie algebra, let $\mathfrak{h}$ be a Cartan subalgebra, and let $\langle, \rangle$ be the inner product on $\mathfrak{h}^*$ that is dual to the Killing form restricted to $\mathfrak{h}$. Let $\Phi \subset \mathfrak{h}^*$ be the set of roots. Then:*

(i) *Φ does not contain zero.*

(ii) *If α is a root, then Φ is symmetric with respect to the reflection operation $s_\alpha \colon \mathfrak{h}^* \to \mathfrak{h}^*$ across the hyperplane orthogonal to α; in particular, $-\alpha$ is also a root.*

(iii) *If α is a root, then no multiple of α other than $\pm\alpha$ are roots.*

(iv) *If α, β are roots, then $\frac{\langle \alpha, \beta \rangle}{\langle \alpha, \alpha \rangle}$ is an integer or half-integer. Equivalently, $s_\alpha(\beta) = \beta + m\alpha$ for some integer m.*

(v) *Φ spans $\mathfrak{h}^*$.*

A set of vectors Φ obeying the above axioms (i)–(v) is known as a *root system* on $\mathfrak{h}^*$ (viewed as a finite-dimensional complex Hilbert space with the inner product $\langle, \rangle$).

Remark 11.6.3. A short calculation reveals the remarkable fact that if Φ is a root system, then the associated system of coroots $\{H_\alpha : \alpha \in \Phi\}$ is also a root system. This is one of the starting points for the deep phenomenon of *Langlands duality*, which we will not discuss here.

When $\mathfrak{g}$ is simple, one can impose a useful additional axiom on Φ. Say that a root system Φ is *irreducible* if Φ cannot be covered by the union $V \cup W$ of two orthogonal proper subspaces of $\mathfrak{h}^*$.

Lemma 11.6.4. *If $\mathfrak{g}$ is a simple Lie algebra, then the root system of Φ is irreducible.*

Proof. If Φ can be covered by two orthogonal subspaces $V \cup W$, then if we consider the subspace of $\mathfrak{g}$
$$V \oplus \bigoplus_{\alpha \in \Phi \cap V} \mathfrak{g}^{\mathfrak{h}}_\alpha$$

where we use the inner product $\langle,\rangle$ to identify $\mathfrak{h}^*$ with $\mathfrak{h}$ and thus V with a subspace of $\mathfrak{h}$ (thus, for instance, this identifies α with t_α), then one can check using (11.12) and (11.13) that this is a proper ideal of $\mathfrak{g}$, contradicting simplicity. $\square$

It is easy to see that every root system is expressible as the union of irreducible root systems (on orthogonal subspaces of $\mathfrak{h}^*$). As it turns out, the irreducible root systems are completely classified, with the complete list of root systems (up to isomorphism) being described in terms of the *Dynkin diagrams* $A_n, B_n, C_n, D_n, E_6, E_7, E_8, F_4, G_2$ briefly mentioned in Theorem 11.0.2. We will now turn to this classification in the next section, and then use root systems to recover the Lie algebra.

11.7. Classification of root systems

In this section we classify all the irreducible root systems Φ on a finite-dimensional complex Hilbert space $\mathfrak{h}^*$, up to Hilbert space isometry. Of course, we may take $\mathfrak{h}^*$ to be a standard complex Hilbert space $\mathbf{C}^n$ without loss of generality. The arguments here are purely elementary, proceeding purely from the root system axioms rather than from any Lie algebra theory.

Actually, we can quickly pass from the complex setting to the real setting. By axiom (v), Φ contains a basis $\alpha_1, \ldots, \alpha_n$ of $\mathbf{C}^n$; by axiom (iv), the inner products between these basis vectors are real, as are the inner products between any other root and a basis root. From this we see that Φ lies in the *real* vector space spanned by the basis roots, so by a change of basis we may assume without loss of generality that $\Phi \subset \mathbf{R}^n$.

Henceforth Φ is assumed to lie in $\mathbf{R}^n$. From two applications of (iv) we see that for any two roots α, β, the expression

$$\frac{\langle \alpha, \beta \rangle}{\langle \alpha, \alpha \rangle} \frac{\langle \alpha, \beta \rangle}{\langle \beta, \beta \rangle}$$

lies in $\frac{1}{4}\mathbf{Z}$; but it is also equal to $\cos^2 \angle(\alpha, \beta)$, and hence

$$\cos^2 \angle(\alpha, \beta) \in \{0, \frac{1}{4}, \frac{1}{2}, \frac{3}{4}, 1\}$$

for all roots α, β. Analysing these cases further using (iv) again, we conclude that there are only a restricted range of options for a pair of roots α, β:

Lemma 11.7.1. *Let α, β be roots. Then one of the following occurs:*

- (0) *β and α are orthogonal.*
- (1/4) *α, β have the same length and subtend an angle of $\pi/3$ or $2\pi/3$.*
- (1/2) *α has $\sqrt{2}$ times the length of β or vice versa, and α, β subtend an angle of $\pi/4$ or $3\pi/4$.*

(3/4) α has $\sqrt{3}$ times the length of β or vice versa, and α, β subtend an angle of $\pi/6$ or $5\pi/6$.

(1) $\beta = \pm \alpha$.

We next record a useful corollary of Lemma 11.7.1 (and axiom (ii)):

Corollary 11.7.2. *Let α, β be roots. If α, β subtend an acute angle, then $\alpha - \beta$ and $\beta - \alpha$ are also roots, but $\alpha + \beta$ is not a root. Equivalently, if α, β subtend an obtuse angle, then $\alpha + \beta$ is a root, but $\alpha - \beta$ and $\beta - \alpha$ are not roots.*

This follows from a routine case analysis and is omitted.

We can leverage Corollary 11.7.2 as follows. Call an element h of $\mathbf{R}^n$ *regular* if it is not orthogonal to any root, thus generic elements of $\mathbf{R}^n$ are regular. Given a regular element h, let $\Phi_h^+ := \{\alpha \in \Phi : \langle \alpha, h \rangle > 0\}$ denote the roots α which are *h-positive* in the sense that their inner product with h is positive; thus Φ is partitioned into Φ_h^+ and $-\Phi_h^+$. We will abbreviate h-positive as *positive* if h is understood from context. Call a positive root $\alpha \in \Phi_h^+$ a *h-simple root* (or *simple root* for short) if it cannot be written as the sum of two positive roots. Clearly every positive root is then a linear combination of simple roots with natural number coefficients. By Corollary 11.7.2, two simple roots cannot subtend an acute angle, and so any two distinct simple roots subtend a right or obtuse angle.

Example 11.7.3. Using the root system $\{e_i - e_j : 1 \leq i, j \leq n; i \neq j\}$ of $\mathfrak{sl}_n$ discussed previously, if one takes h to be any vector in $\mathbf{C}_0^n$ with decreasing coefficients, then the positive roots are those roots $e_i - e_j$ with $i < j$, and the simple roots are the roots $e_i - e_{i+1}$ for $1 \leq i < n$.

Define an *admissible configuration* to be a collection of unit vectors in $\mathbf{R}^n$ in an open half-space $\{v : \langle v, h \rangle > 0\}$ with the property that any two vectors in this collection form an angle of $\pi/2, 2\pi/3, 3\pi/4$, or $5\pi/6$, and call the configuration *irreducible* if it cannot be decomposed into two nonempty orthogonal subsets. From Lemma 11.7.1 and the above discussion we see that the unit vectors $\alpha/\|\alpha\|$ associated to the simple roots are an admissible configuration. They are also irreducible, for if the simple roots partition into two orthogonal sets then it is not hard to show (using Corollary 11.7.2) that all positive roots lie in the span of one of these two sets, contradicting irreducibility of the root system.

We can say quite a bit about admissible configurations; the fact that the vectors in the system always subtend right or obtuse angles, combined with the half-space restriction, is quite limiting (basically because this information can be in violation of inequalities such as the *Bessel inequality*, or the

11.7. Classification of root systems

positive (semi-)definiteness $\|\sum_i c_i v_i\|^2 \geq 0$ of the *Gram matrix*). We begin with an assertion of linear independence:

Lemma 11.7.4. *If $v_1, \ldots, v_n$ is an admissible configuration, then it is linearly independent.*

Among other things, this shows that the number of simple roots of a semisimple Lie algebra is equal to the rank of that algebra.

Proof. Suppose this is not the case, then one has a nontrivial linear constraint
$$\sum_{i \in A} c_i v_i = \sum_{j \in B} c_j v_j$$
for some positive c_i, c_j and disjoint $A, B \subset \{1, \ldots, n\}$. But as any two vectors in an admissible configuration subtend a right or obtuse angle, $\langle \sum_{i \in A} c_i v_i, \sum_{j \in B} c_j v_j \rangle \leq 0$, and thus $\sum_{i \in A} c_i v_i = \sum_{j \in B} c_j v_j = 0$. But this is not possible as all the v_i lie in an open half-space. $\square$

Define the *Coxeter diagram* of an admissible configuration $v_1, \ldots, v_n$ to be the graph with vertices $v_1, \ldots, v_n$, and with any two vertices v_i, v_j connected by an edge of multiplicity $4\cos^2 \angle v_i, v_j$, thus two vertices are unconnected if they are orthogonal, connected with a single edge if they subtend an angle of $2\pi/3$, a double edge if they subtend an angle of $3\pi/4$, and a triple edge if they subtend an angle of $5\pi/6$. The irreducibility of a configuration is equivalent to the connectedness of a Coxeter diagram. Note that the Coxeter diagram describes all the inner products between the v_i and thus describes the v_i up to an orthogonal transformation (as can be seen for instance by applying the Gram-Schmidt process).

Lemma 11.7.5. *The Coxeter diagram of an admissible configuration does not contain a cycle; in other words, it is acyclic. In particular, the Coxeter diagram of an irreducible admissible configuration is a tree.*

Proof. Suppose for contradiction that the Coxeter diagram contains a cycle $v_1, \ldots, v_n$, we see that $\langle v_i, v_{i+1} \rangle \leq -\frac{1}{2}$ for $i = 1, \ldots, n$ (with the convention $v_{n+1} = v_1$) and $\langle v_i, v_j \rangle \leq 0$ for all other i. This implies that $\|\sum_{i=1}^n v_i\|^2 \leq 0$, which contradicts the linear independence of the v_i. $\square$

Lemma 11.7.6. *Any vertex in the Coxeter diagram has degree at most three (counting multiplicity).*

Proof. Let v_0 be a vertex which is adjacent to some other vertices $v_1, \ldots, v_d$, which are then an orthonormal system. By Bessel's inequality (and linear

independence) one has
$$\sum_{i=1}^{d} \langle v_0, v_i \rangle^2 < 1.$$
But from construction of the Coxeter diagram we have $\langle v_0, v_i \rangle^2 = -\frac{m_i}{4}$ for each i, where $m_i \in \{1, 2, 3\}$ is the multiplicity of the edge connecting v_0 and v_i. The claim follows. □

We can also contract simple edges:

Lemma 11.7.7. *If $v_1, \ldots, v_n$ is an admissible configuration with v_i, v_j joined by a single edge, then the configuration formed from $v_1, \ldots, v_n$ by replacing v_i, v_j with the single vertex $v_i + v_j$ is again an admissible configuration, with the resulting Coxeter diagram formed from the original Coxeter diagram by deleting the edge between v_i and v_j and then identifying v_i, v_j together.*

This follows easily from acyclicity and direct computation.

By Lemma 11.7.6 and Lemma 11.7.7, the Coxeter diagram can never form a vertex of degree three no matter how many simple edges are contracted. From this we can easily show that connected Coxeter diagrams must have one of the following shapes:

(A_n) n vertices joined in a chain of simple edges;

(BCF_n) n vertices joined in a chain of edges, one of which is a double edge and all others are simple edges;

(DE_n) three chains of simple edges emenating from a common vertex (forming a "Y" shape), connecting n vertices in all;

(G_2) Two vertices joined by a triple edge.

We can cut down the BCF_n and DE_n cases further:

Lemma 11.7.8. *The Coxeter diagram of an admissible configuration cannot contain as a subgraph:*

(a) *a chain of four edges, with one of the interior edges a double edge;*

(b) *three chains of two simple edges each, emenating from a common vertex;*

(c) *three chains of simple edges of length $1, 2, 5$ respectively, emenating from a common vertex.*

Proof. To exclude (a), suppose for contradiction that we have two chains (u_1, u_2) and (v_1, v_2, v_3) of simple edges, with u_2, v_3 joined by a double edge. Writing $U := \frac{1}{\sqrt{3}}(u_1 + 2u_2)$ and $V := \frac{1}{\sqrt{6}}(v_1 + 2v_2 + 3v_3)$, one computes that U, V are unit vectors with inner product $\langle U, V \rangle = -1$, implying that U, V are parallel, contradicting linear independence.

11.7. Classification of root systems

To exclude (b), suppose that we have three chains (u_1, u_2, x), (v_1, v_2, x), (w_1, w_2, x) of simple edges joined at x. Then the vectors $U := \frac{1}{\sqrt{3}}(u_1 + 2u_2)$, $V := \frac{1}{\sqrt{3}}(v_1 + 2v_2)$, $W := \frac{1}{\sqrt{3}}(w_1 + 2w_2)$ are an orthonormal system that each have an inner product of $-1/\sqrt{3}$ each with x. Comparing this with Bessel's inequality we conclude that x lies in the span of U, V, W, contradicting linear independence.

Finally, to exclude (c), suppose we have three chains (u_1, x), (v_1, v_2, x), $(w_1, w_2, w_3, w_4, w_5, x)$ of simple edges joined at x. Writing $U := u_1$, $V := \frac{1}{\sqrt{3}}(v_1 + 2v_2)$, $W := \frac{1}{\sqrt{15}}(w_1 + 2w_2 + 3w_3 + 4w_4 + 5w_5)$, we compute that U, V, W are an orthonormal system that have inner products of $-1/2, -1/\sqrt{3}, -\frac{5}{\sqrt{60}}$ respectively with x. As $\frac{1}{4} + \frac{1}{3} + \frac{25}{60} = 1$, this forces x to lie in the span of U, V, W, again contradicting linear independence. □

We remark that one could also obtain the required contradictions in the above proof by verifying in all three cases that the *Gram matrix* of the subconfiguration has determinant zero.

Corollary 11.7.9. *The Coxeter diagram of an irreducible admissible configuration must take one of the following forms:*

(A_n) *n vertices joined in a chain of simple edges for some $n \geq 1$;*

(BC_n) *n vertices joined in a chain of edges for some $n \geq 2$, with one boundary edge being a double edge and all other edges simple;*

(D_n) *three chains of simple edges of length $1, 1, n-3$ respectively for some $n \geq 4$, emanating from a single vertex;*

(E_n) *three chains of simple edges of length $1, 2, n-4$ respectively for some $n = 6, 7, 8$, emanating from a single vertex;*

(F_4) *tour vertices joined in a chain of edges, with the middle edge being a double edge and the other two edges simple;*

(G_2) *two vertices joined by a triple edge.*

Now we return to root systems. Fixing a regular h, we define the *Dynkin diagram* to be the Coxeter diagram associated to the (unit vectors of the) simple roots, except that we orient the double or triple edges to point from the longer root to the shorter root. (Note from Lemma 11.7.1 that we know exactly what the ratio between lengths is in these cases; in particular, the Dynkin diagram describes the root system up to a unitary transformation and dilation.) We conclude

Corollary 11.7.10. *The Dynkin diagram of an irreducible root system must take one of the following forms:*

(A_n) *n vertices joined in a chain of simple edges for some $n \geq 1$;*

$$
\begin{array}{l}
A_r \\
B_r \\
C_r \\
D_r \\
E_r \\
F_4 \\
G_2
\end{array}
$$

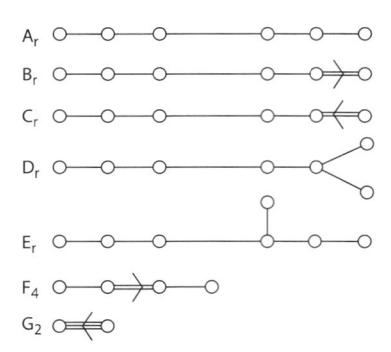

Figure 1. The Dynkin diagrams.

- (B_n) n vertices joined in a chain of edges for some $n \geq 2$, with one boundary edge being a double edge (pointing outward) and all other edges simple;
- (C_n) n vertices joined in a chain of edges for some $n \geq 3$, with one boundary edge being a double edge (pointing inward) and all other edges simple;
- (D_n) three chains of simple edges of length $1, 1, n-3$ respectively for some $n \geq 4$, emanating from a single vertex;
- (E_n) three chains of simple edges of length $1, 2, n-4$ respectively for some $n = 6, 7, 8$, emanating from a single vertex;
- (F_4) four vertices joined in a chain of edges, with the middle edge being a double (oriented) edge and the other two edges simple;
- (G_2) two vertices joined by a triple (oriented) edge.

This describes (up to isomorphism and dilation) the simple roots:

- (A_n) The simple roots take the form $e_i - e_{i+1}$ for $1 \leq i \leq n+1$ in the space $\mathbf{C}_0^{n+1}$ of vectors whose coefficients sum to zero.
- (B_n) The simple roots take the form $e_i - e_{i+1}$ for $1 \leq i \leq n-1$ and also e_n in $\mathbf{C}^n$.
- (C_n) The simple roots take the form $e_i - e_{i+1}$ for $1 \leq i \leq n-1$ and also $2e_n$ in $\mathbf{C}^n$.
- (D_n) The simple roots take the form $e_i - e_{i+1}$ for $1 \leq i \leq n-1$ and also $e_{n-1} + e_n$ in $\mathbf{C}^n$.
- (E_8) The simple roots take the form $e_i - e_{i+1}$ for $1 \leq i \leq 6$ and also $e_6 + e_7$ and $-\frac{1}{2}\sum_{i=1}^{8} e_i$ in $\mathbf{C}^8$.
- (E_6, E_7) This system is obtained from E_8 by deleting the first one or two simple roots (and cutting down $\mathbf{C}^8$ appropriately).

11.7. Classification of root systems

(F_4) The simple roots take the form $e_i - e_{i+1}$ for $1 \leq i \leq 2$ and also e_3 and $-\frac{1}{2}\sum_{i=1}^{4} e_i$ in $\mathbf{C}^4$.

(G_2) The simple roots take the form $e_1 - e_2$, $e_3 - 2e_2 + e_1$ in $\mathbf{C}_0^3$.

Remark 11.7.11. A slightly different way to reach the classification is to replace the Dynkin diagram by the *extended Dynkin diagram* in which one also adds the maximal negative root in addition to the simple roots; this breaks the linear independence, but one can then label each vertex by the coefficient in the linear combination needed to make the roots sum to zero, and one can then analyse these multiplicities to classify the possible diagrams and thence the root systems.

Now we show how the simple roots can be used to recover the entire root system. Define the *Weyl group* W to be the group generated by all the reflections s_α coming from all the roots α; as the roots span $\mathbf{R}^n$ and obey axiom (ii), the Weyl group acts faithfully on the finite set Φ and is thus itself finite.

Lemma 11.7.12. *Let h be regular, and let h' be any element of $\mathbf{R}^n$. Then there exists $w \in W$ such that $\langle w(h'), \alpha \rangle \geq 0$ for all h-simple roots α (or equivalently, for all h-positive roots α). In particular, if h' is regular, then $\Phi^+_{w(h')} = \Phi^+_h$, so that all h-simple roots are $w(h')$-simple and vice versa.*

Furthermore, every root can be mapped by an element of W to an h-simple root.

Finally, W is generated by the reflections s_α coming from the h-simple roots α.

Proof. Let α be a simple root. The action of the reflection s_α maps α to $-\alpha$, and maps all other simple roots β to $\beta + m\alpha$ for some nonnegative m (since α, β subtend a right or obtuse angle). In particular, we see that s_α maps all positive roots other than α to positive roots, and hence (as s_α is an involution)
$$s_\alpha(\Phi^+_h) = \Phi^+_h \cup \{-\alpha\} \setminus \{\alpha\}.$$
In particular, if we define $\rho := \frac{1}{2}\sum_{\beta \in \Phi^+_h} \beta$, then
$$(11.17) \qquad s_\alpha(\rho) = \rho - \alpha$$
for all simple roots α.

Let W_h be the subgroup of W generated by the s_α for the simple roots α, and choose $w \in W_h$ to maximise $\langle w(h'), \rho \rangle$. Then from (11.17) we have $\langle w(h'), \alpha \rangle \geq 0$, giving the first claim. Since every root α is h'-simple for some regular h' (by selecting h' to very nearly be orthogonal to α), we conclude that every root can be mapped by an element of W_h to a h-simple root in h, giving the second claim. Thus for any root β, s_β is conjugate in W_h to a reflection s_α for a h-simple root α, so s_β lies in W_h and so $W = W_h$, giving the final claim. $\square$

Remark 11.7.13. The set of all h' for which $\Phi^+_{h'} = \Phi^+_h$ is known as the *Weyl chamber* associated to h; this is an open polyhedral cone in $\mathbf{R}^n$, and the above lemma shows that it is the interior of a fundamental domain of the action of the Weyl group. In the case of the special linear group, the standard Weyl chamber (in $\mathbf{R}^n_0$ now instead of $\mathbf{R}^n$) would be the set of vectors $h' \in \mathbf{R}^n_0$ with decreasing coefficients.

From the above lemma we can reconstruct the root system from the simple roots by using the reflections s_α associated to the simple roots to generate the Weyl group W, and then applying the Weyl group to the simple roots to recover all the roots. Note that the lemma also shows that the set of h-simple roots and h'-simple roots are isomorphic for any regular h, h', so that the Dynkin diagram is indeed independent (up to isomorphism) of the choice of regular element h as claimed earlier. Thus we have in principle described the irreducible root systems (up to isomorphism) as coming from the Dynkin diagrams $A_n, B_n, C_n, D_n, E_6, E_7, E_8, F_4, G_2$; see for instance [**FuHa**] for explicit descriptions of all of these. With these explicit descriptions one can verify that all of these systems are indeed irreducible root systems.

11.8. Chevalley bases

Now that we have described root systems, we use them to reconstruct Lie algebras. We first begin with an abstract uniqueness result that shows that a simple Lie algebra is determined up to isomorphism by its root system.

Theorem 11.8.1 (Root system uniquely determines a simple Lie algebra). *Let $\mathfrak{g}, \tilde{\mathfrak{g}}$ be simple Lie algebras with Cartan subalgebras $\mathfrak{h}, \tilde{\mathfrak{h}}$ and root systems $\Phi \subset \mathfrak{h}^*$, $\tilde{\Phi} \subset \tilde{\mathfrak{h}}^*$. Suppose that one can identify $\mathfrak{h}$ with $\tilde{\mathfrak{h}}$ as vector spaces in such a way that the root systems agree: $\Phi = \tilde{\Phi}$. Then the identification between $\mathfrak{h}$ and $\tilde{\mathfrak{h}}$ can be extended to an identification of $\mathfrak{g}$ and $\tilde{\mathfrak{g}}$ as Lie algebras.*

Proof. First we note from (11.11) and the identification $\Phi = \tilde{\Phi}$ that the Killing forms on $\mathfrak{h}$ and $\tilde{\mathfrak{h}}$ agree, so we will identify $\mathfrak{h}, \tilde{\mathfrak{h}}$ as Hilbert spaces, not just as vector spaces.

The strategy will be to exploit a Lie algebra version of the *Goursat lemma* (or the *Schur lemma*), finding a sufficiently "nondegenerate" subalgebra $\mathfrak{k}$ of $\mathfrak{g} \oplus \tilde{\mathfrak{g}}$ and using the simple nature of $\mathfrak{g}$ and $\tilde{\mathfrak{g}}$ to show that this subalgebra is the graph of an isomorphism from $\mathfrak{g}$ to $\tilde{\mathfrak{g}}$. This strategy will follow the same general strategy used in Theorem 11.5.1, namely to start

11.8. Chevalley bases

with a "highest weight" space and apply lowering operators to discover the required graph.

We turn to the details. Pick a regular element h of $\mathfrak{h} = \tilde{\mathfrak{h}}$, so that one has a notion of a positive root. For every simple root α, we select nonzero elements X_α, Y_α, of $\mathfrak{g}_\alpha^\mathfrak{h}, \mathfrak{g}_{-\alpha}^\mathfrak{h}$ respectively such that

(11.18) $$[X_\alpha, Y_\alpha] = H_\alpha$$

where H_α is the coroot of α; similarly select $\tilde{X}_\alpha, \tilde{Y}_\alpha$ in $\tilde{\mathfrak{g}}_\alpha^\mathfrak{h}, \tilde{\mathfrak{g}}_{-\alpha}^\mathfrak{h}$, and set $X'_\alpha := X_\alpha \oplus \tilde{X}_\alpha$ and $Y'_\alpha := Y_\alpha \oplus \tilde{Y}_\alpha$. Let $\mathfrak{k}$ be the subalgebra of $\mathfrak{g} \oplus \mathfrak{g}'$ generated by the X'_α and Y'_α. It is not hard to see that the X_α, Y_α generate $\mathfrak{g}$ as a Lie algebra, so $\mathfrak{k}$ surjects onto $\mathfrak{g}$; similarly $\mathfrak{k}$ surjects onto $\mathfrak{g}'$.

Let β be a maximal root, that is to say a root such that $\beta + \alpha$ is not a root for any positive α; such a root always exists. (It is in fact unique, though we will not need this fact here.) Then we have one-dimensional spaces $\mathfrak{g}_\beta^\mathfrak{h}$ and $\tilde{\mathfrak{g}}_\beta^\mathfrak{h}$, and thus a two-dimensional subspace $\mathfrak{g}_\beta^\mathfrak{h} \oplus \tilde{\mathfrak{g}}_\beta^\mathfrak{h}$ in $\mathfrak{g} \oplus \tilde{\mathfrak{g}}$. Inside this subspace, we select a one-dimensional subspace L which is not equal to $\mathfrak{g}_\beta^\mathfrak{h} \oplus 0$ or $0 \times \tilde{\mathfrak{g}}_\beta^\mathfrak{h}$; in particular, L is not contained in $\mathfrak{g} \oplus 0$ or $0 \oplus \tilde{\mathfrak{g}}$.

Let $\mathfrak{l}$ be the subspace of $\mathfrak{g} \oplus \mathfrak{g}'$ generated by L and the adjoint action of the lowering operators Y'_α, thus it is spanned by elements of the form

(11.19) $$\mathrm{ad}\, Y'_{\alpha_1} \ldots \mathrm{ad}\, Y'_{\alpha_k} x$$

for simple roots $\alpha_1, \ldots, \alpha_k$ and $x \in L$. Then $\mathfrak{l}$ contains L and is thus not contained in $\mathfrak{g} \oplus 0, 0 \oplus \tilde{\mathfrak{g}}$; because (11.19) only involves lowering operators, we also see that L does not contain any other element of $\mathfrak{g}_\beta^\mathfrak{h} \oplus \tilde{\mathfrak{g}}_\beta^\mathfrak{h}$ other than $\mathfrak{l}$. In particular, L is not all of $\mathfrak{g} \oplus \tilde{\mathfrak{g}}$.

Clearly $\mathfrak{l}$ is closed under the adjoint action of the lowering operators Y'_α. We claim that it is also closed under the adjoint action of the raising operators X'_α. To see this, first observe that X'_α, Y'_γ commute when α, γ are distinct simple roots, because $\alpha - \beta$ cannot be a root (since this would make one of α, γ nonsimple). Next, from (11.18) we see that $\mathrm{ad}\, X'_\alpha \mathrm{ad}\, Y'_\alpha$ acts as a scalar on any element of the form (11.19), while from the maximality of β we see that $\mathrm{ad}\, X'_\alpha$ annihilates x. From this the claim easily follows.

As $\mathfrak{l}$ is closed under the adjoint action of both the X'_α and the Y'_α, we have $[\mathfrak{k}, \mathfrak{l}] \subset \mathfrak{l}$. Projecting onto $\mathfrak{g}$, we see that the projection of $\mathfrak{l}$ is an ideal of $\mathfrak{g}$, and is hence 0 or $\mathfrak{g}$ as $\mathfrak{g}$ is simple. As $\mathfrak{l}$ is not contained in $0 \oplus \tilde{\mathfrak{g}}$, we see that $\mathfrak{l}$ surjects onto $\mathfrak{g}$; similarly it surjects onto $\tilde{\mathfrak{g}}$. An analogous argument shows that the intersection of $\mathfrak{l}$ with $\mathfrak{g} \oplus 0$ is either 0 or $\mathfrak{g} \oplus 0$; the latter would force $\mathfrak{l} = \mathfrak{g} \oplus \tilde{\mathfrak{g}}$ by the surjective projection onto $\tilde{\mathfrak{g}}$, which was already ruled out. Thus $\mathfrak{l}$ has trivial intersection with $\mathfrak{g} \oplus 0$, and similarly with $0 \oplus \tilde{\mathfrak{g}}$, and is thus a graph. Such a graph cannot be an ideal of $\mathfrak{g} \oplus \tilde{\mathfrak{g}}$, so that $\mathfrak{k} \neq \mathfrak{g} \oplus \tilde{\mathfrak{g}}$. As $\mathfrak{k}$ was a subalgebra that surjected onto both $\mathfrak{g}$ and $\tilde{\mathfrak{g}}$, we conclude by arguing as before that $\mathfrak{k}$ is also a graph; as $\mathfrak{k}$ is a Lie algebra, the graph is that of a Lie algebra isomorphism. Since $[X'_\alpha, Y'_\alpha] = H_\alpha \oplus H_\alpha$, we see that $\mathfrak{k}$ restricts to the graph of the identity on $\mathfrak{h}$, and the claim follows. □

Remark 11.8.2. The above arguments show that every root can be obtained from the maximal root by iteratively subtracting off simple roots (while staying in $\Phi \cup \{0\}$), which among other things implies that the maximal root is unique. These facts can also be established directly from the axioms of a root system (or from the classification of root systems), but we will not do so here. By using Theorem 11.8.1, one can convert graph automorphisms of the Dynkin diagram (e.g., the automorphism sending the A_n Dynkin diagram to its inverse, or the *triality* automorphism that rotates the D_4 diagram) to automorphisms of the Lie algebra; these are important in the theory of twisted groups of Lie type, and more specifically the Steinberg groups and Suzuki-Ree groups; see Section 12.3.

Remark 11.8.3. In a converse direction, once one establishes that in an irreducible root system Φ that every root can be obtained from the maximal root by subtracting off simple roots (while staying in $\Phi \cup \{0\}$), this shows that any Lie algebra $\mathfrak{g}$ associated to this system is necessarily simple. Indeed, given any nontrivial ideal $\mathfrak{h}$ in $\mathfrak{g}$ and a nontrivial element x of $\mathfrak{h}$, one locates a minimal element of $\Phi \cup \{0\}$ in which x has a nontrivial component, then iteratively applies raising operators to then locate a nontrivial element of the root space of the maximal root in $\mathfrak{h}$; if one then applies lowering operators one recovers all the other root spaces, so that $\mathfrak{h} = \mathfrak{g}$.

Theorem 11.8.1, when combined with the results from previous sections, already gives Theorem 11.0.2, but without a fully explicit way to determine the Lie algebras $A_n, B_n, C_n, D_n, E_6, E_7, E_8, F_4, G_2$ listed in that theorem (or even to establish whether these systems exist at all). In the case of the *classical Lie algebras* A_n, B_n, C_n, D_n, one can explicitly describe these algebras in terms of the *special linear algebras* $\mathfrak{sl}_n$, *special orthogonal algebras* $\mathfrak{so}_n$, and *symplectic algebras* $\mathfrak{sp}_n$, but this does not give very much guidance as to how to explicitly describe the *exceptional Lie algebras* E_6, E_7, E_8, F_4, G_2. We now turn to the question of how to explicitly describe all the simple Lie algebras in a unified fashion.

Let $\mathfrak{g}$ be a simple Lie algebra, with Cartan algebra $\mathfrak{h}$. We view $\mathfrak{h}$ as a Hilbert space with the Killing form, and then identify this space with its dual $\mathfrak{h}^*$. Thus, for instance, the coroot H_α of a root $\alpha \in \mathfrak{h}^* \equiv \mathfrak{h}$ is now given by the simpler formula

$$(11.20) \qquad H_\alpha = \frac{2}{\langle \alpha, \alpha \rangle} \alpha.$$

11.8. Chevalley bases

Let $\Phi \subset \mathfrak{h}^* \equiv \mathfrak{h}$ be the root system, which is irreducible. As described in Section 11.6, we have the vector space decomposition

$$\mathfrak{g} \equiv \mathfrak{h} \oplus \bigoplus_{\alpha \in \Phi} \mathfrak{g}_\alpha^\mathfrak{h}$$

where the spaces $\mathfrak{g}_\alpha^\mathfrak{h}$ are one-dimensional, thus we can choose a generator E_α for each $\mathfrak{g}_\alpha^\mathfrak{h}$, though we have the freedom to multiply each E_α by a complex constant, which we will take advantage of to perform various normalisations. A basis for algebra $\mathfrak{h}$ together with the E_α then form a basis for $\mathfrak{g}$, known as a *Cartan-Weyl basis* for this Lie algebra. From (11.11), (11.20) we have

$$[H_\alpha, E_\beta] = A_{\alpha,\beta} E_\beta$$

where $A_{\alpha,\beta}$ is the quantity

$$A_{\alpha,\beta} := \frac{2\langle \alpha, \beta \rangle}{\langle \alpha, \alpha \rangle}$$

which is always an integer because Φ is a root system (indeed $A_{\alpha,\beta}$ takes values in $\{0, \pm 1, \pm 2, \pm 3\}$, and form an interesting matrix known as the *Cartan matrix*).

As discussed in Section 11.6, $[E_\alpha, E_{-\alpha}]$ is a multiple of the coroot H_α; by adjusting $E_\alpha, E_{-\alpha}$ for each pair $\{\alpha, -\alpha\}$ we may normalise things so that

(11.21) $$[E_\alpha, E_{-\alpha}] = H_\alpha$$

for all α (here we use the fact that $H_{-\alpha} = -H_\alpha$ to avoid inconsistency). Next, we see from (11.19) that

$$[E_\alpha, E_\beta] = 0$$

if $\alpha + \beta \notin \Phi \cup \{0\}$, and

(11.22) $$[E_\alpha, E_\beta] = N_{\alpha,\beta} E_{\alpha+\beta}$$

for some complex number $N_{\alpha,\beta}$ if $\alpha + \beta \in \Phi$. By considering the action of E_α on (11.16) using Theorem 11.5.1 one can verify that $N_{\alpha,\beta}$ is nonzero; however, its value is not yet fully determined because there is still residual freedom to normalise the E_α. Indeed, one has the freedom to multiply E_α by any nonzero complex scalar c_α as long as $c_{-\alpha} = c_\alpha^{-1}$ (to preserve the normalisation (11.21)), in which case the structure constant $N_{\alpha,\beta}$ gets transformed according to the law

(11.23) $$N_{\alpha,\beta} \mapsto \frac{c_\alpha c_\beta}{c_{\alpha+\beta}} N_{\alpha,\beta}.$$

However, observe that the combined structure constant $N_{\alpha,\beta} N_{-\alpha,-\beta}$ is unchanged by this rescaling; indeed, there is an explicit formula for this quantity:

Lemma 11.8.4. *For any roots α, β with $\alpha + \beta \in \Phi$, one has*
$$N_{\alpha,\beta} N_{-\alpha,-\beta} = (r+1)^2$$
where $\beta - r\alpha, \ldots, \beta, \ldots, \beta + q\alpha$ are the string of roots of the form $\beta + m\alpha$ for integer m.

This formula can be confirmed by an explicit computation using Theorem 11.5.1 (using, say, the standard basis for P_n to select $E_{\beta+m\alpha}$, which then fixes $E_{-\beta-m\alpha}$ by (11.21)); we omit the details.

On the other hand, we have the following clever renormalisation trick of Chevalley, exploiting the abstract isomorphism from Theorem 11.8.1:

Lemma 11.8.5 (Chevalley normalisation). *There exist choices of E_α such that*
$$N_{\alpha,\beta} = N_{-\alpha,-\beta}$$
for all roots α, β with $\alpha + \beta \in \Phi$.

Proof. We first select E_α arbitrarily, then we will have
$$N_{\alpha,\beta} = a_{\alpha,\beta} N_{-\alpha,-\beta}$$
for some nonzero $a_{\alpha,\beta}$ for all roots α, β. The plan is then to locate coefficients c_α so that the transformation (11.23) eliminates all of the $a_{\alpha,\beta}$ factors.

To do this, observe that we may identify $\mathfrak{h}$ with itself and Φ with itself via the negation map $x \mapsto -x$ for $x \in \mathfrak{h}$ and $\alpha \mapsto -\alpha$ for $\alpha \in \Phi$. From this and Theorem 11.8.1, we may find a Lie algebra isomorphism $\phi \colon \mathfrak{g} \to \mathfrak{g}$ that maps x to $-x$ on $\mathfrak{h}$, and thus maps $\mathfrak{g}_\alpha^\mathfrak{h}$ to $\mathfrak{g}_{-\alpha}^\mathfrak{h}$ for any root α. In particular, we have
$$\phi(E_\alpha) = b_\alpha E_{-\alpha}$$
for some nonzero coefficients b_α; from (11.21) we see in particular that
$$b_\alpha b_{-\alpha} = 1. \tag{11.24}$$
If we then apply ϕ to (11.22), we conclude that
$$b_\alpha b_\beta N_{-\alpha,-\beta} = b_{\alpha+\beta} N_{\alpha,\beta}$$
when $\alpha + \beta$ is a root, so that $a_{\alpha,\beta}$ takes the special form
$$a_{\alpha,\beta} = \frac{b_\alpha b_\beta}{b_{\alpha+\beta}}.$$
If we then select c_α so that
$$c_\alpha = b_\alpha c_{-\alpha}$$
for all roots α (this is possible thanks to (11.24)), then the transformation (11.23) eliminates $a_{\alpha,\beta}$ as desired. □

From the two lemmas above, we see that we can select a special Cartan-Weyl basis, known as a *Chevalley basis*, such that

$$[E_\alpha, E_\beta] = \pm(r+1)E_{\alpha+\beta} \tag{11.25}$$

whenever $\alpha + \beta$ is a root; in particular, the structure constants $N_{\alpha,\beta}$ are all integers, which is a crucial fact when one wishes to construct Lie algebras and *Chevalley groups* over fields of arbitrary characteristic. This comes very close to fully describing the Lie algebra structure associated to a given Dynkin diagram, except that one still has to select the signs $\pm$ in (11.25) so that one actually gets a Lie algebra (i.e., that the Jacobi identity (11.1) is obeyed). This turns out to be nontrivial; see[3] [**Ti1972**] for details. Among other things, this construction shows that every root system actually creates a Lie algebra (thus far we have only established uniqueness, not existence), though once one has the classification one could also build a Lie algebra explicitly for each Dynkin diagram by hand (in particular, one can build the simply laced classical Lie algebras A_n, D_n and the maximal simply laced exceptional algebra E_8, and construct the remaining Lie algebras by taking fixed points of suitable involutions; see, e.g., [**BoHaReSe2011**] for this approach).

11.9. Casimirs and complete reducibility

Finally, we supply a proof of the following fact, used in the proof of Corollary 11.3.8:

Theorem 11.9.1 (Weyl's complete reducibility theorem). *Let $\mathfrak{g} \subset \mathfrak{gl}(V)$ be a simple Lie algebra, and let W be a $\mathfrak{g}$-invariant subspace of V. Then there exists a complementary $\mathfrak{g}$-invariant subspace W' such that $V = W \oplus W'$.*

Among other things, Theorem 11.9.1 shows that every finite-dimensional linear representation of $\mathfrak{g}$ splits into the direct sum of irreducible representations, which explains the terminology. The claim is also true for semisimple Lie algebras $\mathfrak{g}$, but we will only need the simple case here, which allows for some minor simplifications to the argument.

The proof of this theorem requires a variant $B\colon \mathfrak{g} \to \times \mathfrak{g} \to \mathbf{C}$ of the Killing form associated to V, defined by the formula

$$B(x,y) := \operatorname{tr}(xy), \tag{11.26}$$

and a certain element of $\mathfrak{gl}(V)$ associated to this form known as the *Casimir operator*. We first need to establish a variant of Theorem 11.0.1:

---

[3]There are other approaches to demonstrate existence of a Lie algebra associated to a given root system; one popular one proceeds using the Chevalley-Serre relations, see, e.g., [**Se1966**]. There is still a certain amount of freedom to select the signs, but this ambiguity can be described precisely; see [**Ca1993**] for details.

Proposition 11.9.2. *With the hypotheses of Theorem* 11.9.1, B *is nondegenerate.*

Proof. This is a routine modification of Proposition 11.3.2 (one simply omits the use of the adjoint representation). □

Once one establishes nondegeneracy, one can then define the *Casimir operator* $C \in \mathfrak{gl}(V)$ by setting

$$C := \sum_{i=1}^{n} e_i f_i$$

whenever $e_1, \ldots, e_n$ is a basis of $\mathfrak{g}$ and $f_1, \ldots, f_n$ is its *dual basis*, thus $B(e_i, f_j) = \delta_{ij}$ where δ_{ij} is the Kronecker delta. It is easy to see that this definition does not depend on the choice of basis, which in turn (by infinitesimally conjugating both bases by an element x of the algebra $\mathfrak{g}$) implies that C commutes with every element x of $\mathfrak{g}$.

On the other hand, C does not vanish entirely. Indeed, taking traces and using (11.26) we see that

(11.27) $$\operatorname{tr}(C) = \dim(\mathfrak{g}).$$

This already gives an important special case of Theorem 11.9.1:

Proposition 11.9.3. *Theorem* 11.9.1 *is true when* W *has codimension one and is irreducible.*

Proof. The Lie algebra $\mathfrak{g}$ acts on the one-dimensional space V/W; since $\mathfrak{g} = [\mathfrak{g}, \mathfrak{g}]$ (from the simplicity hypothesis), we conclude that this action is trivial. In other words, each element of $\mathfrak{g}$ maps V to W, so the Casimir operator C does as well. In particular, the trace of C on V is the same as the trace of C on W. On the other hand, by Schur's lemma, C is a constant on W; applying (11.27), we conclude that this constant is nonzero. Thus C is nondegenerate on W, but is not full rank on V as it maps V to W. Thus it must have a one-dimensional null-space W' which is complementary to W. As C commutes with $\mathfrak{g}$, W' is $\mathfrak{g}$-invariant, and the claim follows. □

We can then remove the irreducibility hypothesis:

Proposition 11.9.4 (Whitehead's lemma). *Theorem* 11.9.1 *is true when* W *has codimension one.*

Proof. We induct on the dimension of W (or V). If W is irreducible then we are already done, so suppose that W has a proper invariant subspace U. Then W/U has codimension one in V/U, so by the induction hypothesis W/U is complemented by a one-dimensional invariant subspace Y of V/U, which lifts to an invariant subspace Z of V in which U has codimension one.

11.9. Casimirs and complete reducibility

By the induction hypothesis again, U is complemented by a one-dimensional invariant subspace W' in Z, and it is then easy to see that W' also complements W in V, and the claim follows. $\square$

Next, we remove the codimension one hypothesis instead:

Proposition 11.9.5. *Theorem* 11.9.1 *is true when W is irreducible.*

Proof. Let A be the space of linear maps $T\colon V \to W$ whose restriction to W is a constant multiple of the identity, and let B be the subalgebra of A whose restriction to W vanishes. Then A, B are $\mathfrak{g}$-invariant (using the Lie bracket action), and B has codimension one in A. Applying Proposition 11.9.4 (pushing $\mathfrak{g}$ forward to $\mathfrak{gl}(A)$, and treating the degenerate case when $\mathfrak{gl}(A)$ vanishes separately) we see that B is complemented by a one-dimensional invariant subspace B' of A. Thus there exist $T \in A$ that does not lie in B, and which commutes with every element of $\mathfrak{g}$. The kernel W' of T is then an invariant complement of W in V, and the claim follows. $\square$

Applying the induction argument used to prove Proposition 11.9.4, we now obtain Theorem 11.9.1 in full generality.

Chapter 12

Notes on groups of Lie type

In Chapter 11 we reviewed the structural theory of finite-dimensional complex Lie algebras (or *Lie algebras* for short), with a particular focus on those Lie algebras which were semisimple or simple. In particular, we discussed the Weyl complete reducibility theorem (asserting that semisimple Lie algebras are the direct sum of simple Lie algebras) and the classification of simple Lie algebras (with all such Lie algebras being (up to isomorphism) of the form A_n, B_n, C_n, D_n, E_6, E_7, E_8, F_4, or G_2).

Among other things, the structural theory of Lie algebras can then be used to build analogous structures in nearby areas of mathematics, such as *Lie groups* and Lie algebras over more general fields than the complex field $\mathbf{C}$ (leading in particular to the notion of a *Chevalley group*), as well as *finite simple groups of Lie type*, which form the bulk of the *classification of finite simple groups* (with the exception of the *alternating groups* and a finite number of *sporadic groups*).

In the case of complex Lie groups, it turns out that every simple Lie algebra $\mathfrak{g}$ is associated with a finite number of connected complex Lie groups, ranging from a "minimal" Lie group G_{ad} (the *adjoint form* of the Lie group) to a "maximal" Lie group $\tilde{G}$ (the *simply connected form* of the Lie group) that finitely covers G_{ad}, and occasionally also a number of intermediate forms which finitely cover G_{ad}, but are in turn finitely covered by $\tilde{G}$. For instance, $\mathfrak{sl}_n(\mathbf{C})$ is associated with the projective special linear group $\mathrm{PSL}_n(\mathbf{C}) = \mathrm{PGL}_n(\mathbf{C})$ as its adjoint form and the special linear group

$SL_n(\mathbf{C})$ as its simply connected form, and intermediate groups can be created by quotienting out $SL_n(\mathbf{C})$ by some subgroup of its centre (which is isomorphic to the nth roots of unity). The minimal form G_{ad} is *simple* in the group-theoretic sense of having no normal subgroups, but the other forms of the Lie group are merely *quasisimple*, although traditionally all of the forms of a Lie group associated to a simple Lie algebra are known as *simple Lie groups*.

Thanks to the work of Chevalley, a very similar story holds for over arbitrary fields k; given any Dynkin diagram, one can define a simple Lie algebra with that diagram over that field, and also one can find a finite number of groups over k (known as *Chevalley groups*) associated to that diagram, ranging from an adjoint group $G_{\text{ad},k}$ to a universal group $G_{u,k}$. Thus, for instance, one could construct the universal form $E_7(q)_u$ of the E_7 algebraic group over a finite field $\mathbf{F}_q$ of finite order. In the case that k is algebraically closed, the Chevalley groups are algebraic groups over k, in which case we refer to these groups as *forms* associated to the Dynkin diagram, with every form having an *isogeny* (the analogue of a finite cover for algebraic groups) to the adjoint form, and in turn receiving an isogeny from the universal form.

When one restricts the Chevalley group construction to adjoint groups over a finite field (e.g., $PSL_n(\mathbf{F}_q)$), one usually obtains a finite simple group (with a finite number of exceptions when the rank and the field are very small, and in some cases one also has to pass to a bounded index subgroup, such as the derived group, first). One could also use other versions of the Chevalley group than the adjoint group, but one then recovers the same finite simple group as before if one quotients out by the centre. This construction was then extended by Steinberg, Suzuki, and Ree by taking a Chevalley group over a finite field and then restricting to the fixed points of a certain automorphism of that group; after some additional minor modifications such as passing to a bounded index subgroup or quotienting out a bounded centre, this gives some additional finite simple groups of Lie type, including classical examples such as the projective special unitary groups $PSU_n(\mathbf{F}_{q^2})$, as well as some more exotic examples such as the *Suzuki groups* or the *Ree groups*.

In this chapter we review the constructions of these groups and their basic properties. The material here is standard, and was drawn from a number of sources, but primarily from [**Ca1993**], [**GoLySo1998**], [**FuHa**], [**St1967**].

12.1. Simple Lie groups over C

We begin with some discussion of Lie groups G over the complex numbers $\mathbf{C}$. We will restrict attention to the connected Lie groups, since more general

12.1. Simple Lie groups over C

Lie groups can be factored,
$$0 \to G^\circ \to G \to G/G^\circ \to 0$$
into an extension of an (essentially arbitrary) discrete group G/G° by the connected component G° (or, in the ATLAS notation of the previous chapter, $G = G^\circ.(G/G^\circ)$). One can interpret G° as the minimal open subgroup of G, thus a Lie group is connected if and only if there are no proper open subgroups.

To each Lie group G over $\mathbf{C}$ one can associate a complex Lie algebra $\mathfrak{g}$, which one can identify with the tangent space of G at the identity. This identification is, however, not injective; one can have nonisomorphic Lie groups with the same Lie algebra. For instance, the special linear group $\mathrm{SL}_2(\mathbf{C})$ and the projective special linear group $\mathrm{PSL}_2(\mathbf{C}) = \mathrm{SL}_2(\mathbf{C})/\{+1, -1\}$ have the same Lie algebra $\mathfrak{sl}_2(\mathbf{C})$; intuitively, the Lie algebra captures all the "local" information of the Lie group but not the "global" or "topological" information. (This statement can be made more precise using the *Baker-Campbell-Hausdorff formula*, discussed in [**Ta2013**, §1.2].) On the other hand, every connected Lie group G has a *universal cover* $\tilde{G}$ with the same Lie algebra (up to isomorphism) as G, which is a simply connected Lie group which projects onto G by a short exact sequence
$$0 \to \pi_1(G) \to \tilde{G} \to G \to 0$$
with $\pi_1(G)$ being (an isomorphic copy of) the (topological) *fundamental group* of G. Furthermore, two Lie groups have the same Lie algebra (up to isomorphism) if and only if their universal covers agree (up to isomorphism); this is essentially *Lie's second theorem*, discussed in [**Ta2013**, §1.2] (in the context of Lie groups and Lie algebras over the reals rather than the complex numbers, but the result holds over both fields). Conversely, every Lie algebra is the Lie algebra of some Lie group, and thus of some simply connected Lie group; this is essentially *Lie's third theorem*, also discussed in [**Ta2013**, §1.2]. Thus, the Lie groups associated to a given Lie algebra $\mathfrak{g}$ can all be viewed as quotients of a universal cover $\tilde{G}$ by a discrete normal subgroup Γ.

We can say a little more about the fundamental group $\pi_1(G)$. Observe that $\tilde{G}$ acts by conjugation on $\pi_1(G)$; however, $\pi_1(G)$ is discrete, and so the automorphism group of $\pi_1(G)$ is discrete also. Since $\tilde{G}$ is connected, we conclude that the action of $\tilde{G}$ on $\pi_1(G)$ is trivial; in other words, $\pi_1(G)$ is a *central* subgroup of G (and so $\tilde{G}$ is a central extension of G). In particular, the fundamental group $\pi_1(G)$ of a connected Lie group G is always abelian[1].

Not every subgroup of a Lie group is again a Lie group; for instance, the rational numbers $\mathbf{Q}$ are a subgroup of the one-dimensional complex Lie

---
[1] Of course, fundamental groups can be nonabelian for more general topological spaces; the key property of Lie groups that are being used here is that they are *H-spaces*.

group **C** but are clearly not a Lie group. However, a basic theorem of Cartan (proven in [**Ta2013**, §1.3]) says that any subgroup of a *real* Lie group which is *topologically* closed, is also a real Lie group. This theorem doesn't directly apply in the complex case (for instance **R** is a subgroup of the complex Lie group **C** but is only a real Lie group rather than a complex one), but it does say that a closed subgroup of a complex Lie group is a real Lie group, and if in addition one knows that the real tangent space of the subgroup at the origin is closed under complex multiplication then it becomes a complex Lie group again.

We expect properties about the Lie algebra $\mathfrak{g}$ to translate to analogous properties about the Lie group G. In the case of simple Lie algebras, we have the following:

Lemma 12.1.1. *Let G be a connected complex Lie group with Lie algebra $\mathfrak{g}$. Then the following are equivalent:*

(i) *$\mathfrak{g}$ is a simple Lie algebra.*

(ii) *G is nonabelian, and the only closed normal subgroups of G are discrete or all of G.*

(iii) *G is nonabelian, and the only normal subgroups of G are discrete or all of G.*

Proof. Suppose first that $\mathfrak{g}$ is simple (which implies that $\mathfrak{g}$, and hence G, is nonabelian), but G has a closed normal subgroup H which is not discrete or all of G, then by Cartan's theorem it is a real Lie group with positive dimension. Then the Lie algebra $\mathfrak{h}$ of H is a nontrivial real Lie algebra which is preserved by the adjoint action of $\mathfrak{g}$. If $\mathfrak{h} = \mathfrak{g}$ then H contains a neighbourhood of the identity in G and is thus all of G as G is connected, so $\mathfrak{h}$ is a proper subalgebra of $\mathfrak{g}$. Note that $[\mathfrak{h}, \mathfrak{g}]$ is a complex Lie algebra ideal of $\mathfrak{g}$, so by simplicity this ideal is trivial, thus $\mathfrak{h}$ lies in the centre of $\mathfrak{g}$, which is again trivial by simplicity, a contradiction.

If H is normal but not closed, one can adapt the above argument as follows. If H is central then it is discrete (because $\mathfrak{g}$ is centreless) so assume that H is not central, then it contains a nontrivial conjugacy class; after translation this means that H contains a curve through the identity whose derivative at the identity is a nonzero vector v in $\mathfrak{g}$. As $\mathfrak{g}$ is simple, $\mathfrak{g}$ is the minimal ideal generated by v, which implies that the orbit of v under the adjoint action of G spans $\mathfrak{g}$ as a linear space, thus there are a finite number of G-conjugates of v that form a basis for $\mathfrak{g}$. Lifting back up to G and using the inverse function theorem, we conclude that H contains an open neighbourhood of the identity and is thus all of G.

Now suppose that $\mathfrak{g}$ is not simple. If it has a nontrivial abelian ideal, then one can exponentiate this ideal and take closures to obtain a closed

12.1. Simple Lie groups over C   271

normal abelian subgroup of G, which is not all of G as G is nonabelian, and which is complex because the ideal is a complex vector space. So we may assume that no such ideal exists, which means from Theorem 11.0.1 that $\mathfrak{g}$ is semisimple and thus the direct sum $\mathfrak{g}_1 \oplus \ldots \oplus \mathfrak{g}_k$ of simple algebras for some $k \geq 2$. If we then take H to be the subgroup of G whose adjoint action on $\mathfrak{g}$ is the identity on $\mathfrak{g}_1$, then H is a closed subgroup of G, thus a real Lie group, and also a complex Lie group as the tangent space is $\mathfrak{g}_2 \oplus \ldots \oplus \mathfrak{g}_k$, giving a closed normal subgroup of intermediate dimension. □

In view of this lemma, we call a connected complex Lie group *simple* if it is nonabelian and the only closed normal subgroups of G are discrete or all of G. This differs slightly from the group-theoretic notion of simplicity, which asserts instead that the only normal subgroups of G (including the nonclosed normal subgroups) are *trivial* or all of G. However, these two notions are actually not that far apart from each other. First, given a simple Lie algebra $\mathfrak{g}$, one can form the *adjoint form* G_{ad} of the associated Lie group, defined as the closed subgroup of the general linear group $\mathrm{GL}(\mathfrak{g})$ on $\mathfrak{g}$ generated by the transformations $\mathrm{Ad}_x := \exp(\mathrm{ad}\,x)$ for $x \in \mathfrak{g}$. This is group is clearly connected. Because all such transformations are derivations on $\mathfrak{g}$, and derivations on a simple Lie algebra are inner (see Lemma 11.3.7), we see that the tangent space of this group is $\mathrm{ad}\,\mathfrak{g}$, which is isomorphic to $\mathfrak{g}$ as $\mathfrak{g}$ is simple (and thus centreless). In particular, G_{ad} is a complex Lie group whose Lie algebra is $\mathfrak{g}$. Furthermore, any other connected complex Lie group G with Lie algebra $\mathfrak{g}$ will map by a continuous homomorphism to G_{ad} by the conjugation action of G on $\mathfrak{g}$; this map is open near the origin, and so this homomorphism is surjective. Thus, G is a discrete cover of G_{ad}, much as $\tilde{G}$ is a discrete cover of G, and so all the Lie groups G with Lie algebra $\mathfrak{g}$ are sandwiched between the universal cover $\tilde{G}$ and the adjoint form G_{ad}. The same argument shows that G_{ad} itself has no nontrivial discrete normal subgroups, as one could then have nontrivial quotients of G_{ad} which still somehow cover G_{ad} by an inverse of the quotient map, which is absurd. Thus the adjoint form G_{ad} of the Lie group is simple in the group-theoretic sense, but none of the other forms are (since they can be quotiented down to G_{ad}). In particular, G_{ad} is centreless, so given any of the other covers G of G_{ad}, the kernel of the projection of G to G_{ad} is precisely $Z(G)$, thus $G_{\mathrm{ad}} \equiv G/Z(G)$ for any of the Lie group forms G.

Note that for any form G of the Lie group associated to the simple Lie algebra $\mathfrak{g}$, the commutator group $[G, G]$ contains a neighbourhood of the origin (as $\mathfrak{g}$ is *perfect*) and so is all of G. Thus we see that while any given form G of the Lie group is not necessarily simple in the group-theoretic sense, it is *quasisimple*, that is to say it is a perfect central extension of a simple group.

It is now of interest to understand the fundamental group $\pi(G_{\mathrm{ad}})$ of the adjoint form G_{ad}, as this measures the gap between $\tilde{G}$ and G_{ad} and will classify all the intermediate forms G of the Lie group associated to $\mathfrak{g}$ (as these all arise from quotienting $\tilde{G}$ by some subgroup of $\pi(G_{\mathrm{ad}})$). For this we have the following very useful tool:

Lemma 12.1.2 (Existence of compact form). *Let $\mathfrak{g}$ be a simple complex Lie algebra, and let $G_{\mathrm{ad}} \subset \mathrm{GL}(\mathfrak{g})$ be its adjoint form. Then there exists a compact subgroup G_c of G_{ad} with Lie algebra $i\mathfrak{g}_{\mathbf{R}}$, where $\mathfrak{g}_{\mathbf{R}}$ is a real Lie algebra that complexifies to $\mathfrak{g}$, thus $\mathfrak{g} = \mathfrak{g}_{\mathbf{R}} \oplus i\mathfrak{g}_{\mathbf{R}}$. Furthermore, every element A in G_{ad} has a unique polar decomposition $A = DU$, where $U \in G_c$ and $D \in \exp(\mathfrak{g}_{\mathbf{R}})$.*

Proof. Before we begin the proof, we give a (morally correct) example of the lemma: take $\mathfrak{g} = \mathfrak{sl}_n(\mathbf{C})$, and replace G_{ad} by $\mathrm{SL}_n(\mathbf{C})$ (this is not the adjoint form of $\mathfrak{g}$, but never mind this). Then the obvious choice of compact form is the special unitary group $G_c = \mathrm{SU}_n(\mathbf{C})$, which has as Lie algebra the real algebra $i\mathfrak{su}_n(\mathbf{C})$ of skew-adjoint transformations of trace zero. This suggests that we need a notion of "adjoint" $*\colon \mathfrak{g} \to \mathfrak{g}$ for more general Lie algebras $\mathfrak{g}$ in order to extract the skew-adjoint ones.

We now perform this construction. As discussed in Section 11.8, $\mathfrak{g}$ has a *Cartan-Weyl basis* consisting of vectors E_α for roots $\alpha \in \Phi$ as well as coroots H_α for simple roots α (with the H_β for other roots β then expressed as linear combinations of the simple coroots H_α, and where we have fixed some direction h in which to define the notions of positive and simple roots), obeying the relations

$$[H_\alpha, H_\beta] = 0,$$
$$[H_\alpha, E_\beta] = A_{\alpha,\beta} E_\beta,$$
$$[E_\alpha, E_{-\alpha}] = H_\alpha,$$
$$[E_\alpha, E_\beta] = N_{\alpha,\beta} E_{\alpha+\beta},$$

when $\alpha \neq -\beta$ and some integers $A_{\alpha,\beta}, N_{\alpha,\beta}$, with the convention that E_α vanishes when α is not a root. We can also arrange matters so that $N_{\alpha,\beta} = N_{-\alpha,-\beta}$; see Lemma 11.8.5. If we then define the adjoint map $*\colon \mathfrak{g} \to \mathfrak{g}$ to be the antilinear map that preserves all the coroots H_α, but maps E_α to $E_{-\alpha}$ for all α, one easily verifies that $*$ is an anti-homomorphism, so that $[X^*, Y^*] = -[X, Y]^*$ for all $X, Y \in \mathfrak{g}$. Furthermore, one can now make $\mathfrak{g}$ into a complex Hilbert space with the Hermitian form $\langle X, Y \rangle := K(X, Y^*)$ (with K being the Killing form), which one can verify using the Cartan-Weyl basis to be positive definite (indeed the Cartan-Weyl basis becomes an orthogonal basis with this Hermitian form). For any $X \in \mathfrak{g}$, one can also verify that

12.1. Simple Lie groups over C

the maps $\operatorname{ad} X\colon \mathfrak{g} \to \mathfrak{g}$ and $\operatorname{ad} X^*\colon \mathfrak{g} \to \mathfrak{g}$ are adjoints with respect to this Hermitian form.

If we now set $\mathfrak{g_R} := \{X \in \mathfrak{g} : X^* = X\}$ to be the self-adjoint elements of $\mathfrak{g}$, and G_c to be those elements of G_{ad} that are unitary with respect to the Hermitian form, we see that $\mathfrak{g_R}$ complexifies to $\mathfrak{g}$ and G_c is a compact group with real Lie algebra $i\mathfrak{g_R}$. Also, since $\operatorname{ad} X$ is the adjoint of $\operatorname{ad} X^*$, we see that G_{ad} is closed under the operation of taking adjoints.

Now we obtain the polar decomposition. If $A \in G_{\mathrm{ad}}$, then AA^* is a self-adjoint positive definite map on the Hilbert space $\mathfrak{g}$, which also lies in G_{ad} and thus respects the Lie bracket: $AA^*[X,Y] = [AA^*X, AA^*Y]$. By diagonalising AA^* and working with the structure constants of the Lie bracket in the eigenbasis of AA^* we conclude that all powers $(AA^*)^t$ for $t > 0$ also respect the Lie bracket; sending $t \to 0$ we conclude that $\log AA^*$ is a derivation of $\mathfrak{g}$, and thus inner, which implies that $(AA^*)^t \in G_{\mathrm{ad}}$ for all $t > 0$. In particular, the square root $D := (AA^*)^{1/2}$ lies in G_{ad}. Setting $U := D^{-1}A$ we obtain the required polar decomposition; the uniqueness can be obtained by observing that $DU = A$ implies $D = (AA^*)^{1/2}$. □

From the polar decomposition we see that G_{ad} can be contracted onto G_c (by deforming DU as $D^t U$ as t goes from 1 to 0). In particular, G_c is connected and has the same fundamental group as G_{ad}. On the other hand, the Hermitian form $\langle,\rangle$ restricts to a real positive definite form on the tangent space of G_c that is invariant with respect to the conjugation action of G_c, and thus defines a *Riemannian metric* on G_c. The definiteness of the Killing form then implies (after some computation) that this metric has strictly positive *sectional curvature* (and hence also strictly positive *Ricci curvature*), and so any cover of G_c also has a metric with Ricci and sectional curvatures uniformly bounded from below. Applying *Myers' theorem* (discussed in [**Ta2009b**, §2.10]), we conclude that any cover of G_c is necessarily compact also; this implies that the fundamental group of G_c, and hence of G_{ad}, is finite. Thus there are only finitely many different forms of G between G_{ad} and $\tilde{G}$, with the latter being a finite cover of the former. For instance, in the case of $\mathfrak{g} = \mathfrak{sl}_n(\mathbf{C})$ (i.e., the type A_{n-1} case), one can show that the adjoint form G_{ad} is isomorphic to $\mathrm{PSL}_n(\mathbf{C}) = \mathrm{PGL}_n(\mathbf{C})$ and the universal cover $\tilde{G}$ is isomorphic to $\mathrm{SL}_n(\mathbf{C})$, so that

$$\pi(G_{\mathrm{ad}}) \equiv Z(\tilde{G}) \equiv \mathbf{Z}/n\mathbf{Z}$$

(since the central elements of $\mathrm{SL}_n(\mathbf{C})$ come from the nth roots of unity), and all the intermediate forms of G then come from quotienting out $\mathrm{SL}_n(\mathbf{C})$ by some subgroup of the nth roots of unity. Actually, as it turns out, for all Lie algebras other than the A_n family, the fundamental group $\pi(G_{\mathrm{ad}}) \equiv Z(\tilde{G})$ is very small, having order at most 4; see below. For instance, in the orthogonal algebras $\mathfrak{so}_n(\mathbf{C})$ (coming from the B_r and D_r families) the adjoint form is $SO_n(\mathbf{C})$ and the universal cover is the *spin group* $\mathrm{Spin}_n(\mathbf{C})$, which is a double cover of $SO_n(\mathbf{C})$; in particular, there are no other models of the Lie groups associated to the B_r and D_r diagrams. This is in marked contrast with the case of abelian Lie groups, in which there is an infinity of Lie groups associated to a given abelian Lie algebra. For instance, with the one-dimensional Lie algebra $\mathbf{C}$, every lattice Γ in $\mathbf{C}$ gives a different Lie group $\mathbf{C}/\Gamma$ with the specified Lie algebra.

The compact form of the adjoint form G_{ad} of course lifts to compact forms for all other Lie groups with the given Lie algebra. Among other things, it demonstrates (by the *Weyl unitary trick*) the representation version of Weyl's complete reducibility theorem: every finite-dimensional representation $\rho\colon \mathfrak{g} \to \mathfrak{gl}(V)$ of $\mathfrak{g}$ splits as the direct sum of a finite number of irreducible representations. Indeed, one can lift this representation to a representation $\rho\colon \tilde{G} \to \mathrm{GL}(V)$ of the universal cover $\tilde{G}$, which then restricts to a representation of the compact form $\tilde{G}_c$ of $\tilde{G}$. But then by averaging some Hermitian form on V with respect to the Haar measure on $\tilde{G}_c$ one can then construct a Hermitian form with respect to which $\tilde{G}_c$ acts in a unitary fashion, at which point it is easy to take orthogonal complements and decompose V into $\tilde{G}_c$-irreducible components, which on returning to the infinitesimal action establishes a decomposition into complex vector spaces that are irreducible with respect to the action of $i\mathfrak{g}_{\mathbf{R}}$ and hence (on complexifying) $\mathfrak{g}$. A similar theorem applies for actions of simple (or semisimple) Lie groups, showing that such groups are *reductive*.

Another application of the unitary trick reveals that every simple complex Lie group G is linear, that is to say it is isomorphic to a Lie subgroup of $\mathrm{GL}_n(\mathbf{C})$ for some n (this is in contrast to real Lie groups, which can be nonlinear even when simple; the canonical example here is the *metaplectic group* $\mathrm{Mp}_n(\mathbf{R})$ that forms the double cover of the symplectic group $\mathrm{Sp}_n(\mathbf{R})$ for any $n \geq 2$). Indeed, letting G'_c be the compact form of G'_c, the *Peter-Weyl theorem* (as discussed in [**Ta2013**, §1.4]) we see that G'_c can be identified with a unitary Lie group (i.e., a real Lie subgroup of $U_n(\mathbf{C})$ for some n); in particular, its real Lie algebra can be identified with a Lie algebra $i\mathfrak{g}_{\mathbf{R}}$ of skew-Hermitian matrices. Note that $\mathfrak{g}$ can be identified with the complexification $\mathfrak{g}_{\mathbf{R}} \oplus i\mathfrak{g}_{\mathbf{R}}$. The set $\{g\exp(x) : g \in G'_c, x \in \mathfrak{g}_{\mathbf{R}}\}$ can then be seen to be a connected smooth manifold which locally is a Lie group with Lie algebra $\mathfrak{g}$, and by a continuity argument contains the group generated by a sufficiently small neighbourhood of the identity, and is therefore a Lie group with the same compact form as G, and thus descends from quotienting the universal cover $\tilde{G}$ by the same central subgroup, and so is isomorphic to G. This argument also shows that the compact form of a connected simple complex Lie group is always connected, and that every complex form of a

Lie group is associated to some linear representation of the underlying Lie algebra $\mathfrak{g}$. (For instance, the universal form is associated to the sum of the representations having the fundamental weights (the dual basis to the simple coroots) as highest weights, although we will not show this here.)

If one intersects a Cartan subalgebra $\mathfrak{h}$ with $i\mathfrak{g}_{\mathbf{R}}$ and then exponentiates and takes closures, one obtains a compact abelian connected subgroup of G_c whose Lie algebra is again $\mathfrak{h} \cap i\mathfrak{g}_{\mathbf{R}}$ (from the self-normalising property of Cartan algebras); these groups are known as (real) *maximal tori*. As all Cartan subalgebras are conjugate to each other, all maximal tori are conjugate to each other also. On a compact Lie group, the exponential map is surjective (as discussed in [**Ta2013**, §2.11]); as every element in $\mathfrak{g}$ lies in a Cartan algebra, we obtain the useful fact that every element of G_c lies in a maximal torus. The same statement lifts to other models G of the Lie group, and among other things implies that the centre $Z(G)$ of such a model is equal to the intersection of all the maximal tori in that model.

We can push the above analysis a bit further to give a more explicit description of the fundamental group of G_{ad} in terms of the root structure. We will be a bit sketchy in our presentation; details may be found for instance in [**Se2007**]. We first need a basic lemma. Let G_c be the compact form of a simple Lie group, and let T be a maximal torus in G_c. Let $N(T)$ be the normaliser of T in G_c; as Cartan algebras are self-normalising, we see that $N(T)$ has the same Lie algebra as T, and so $N(T)/T$ is a finite group, which acts on the Lie algebra $\mathfrak{t}$ of T by conjugation, and similarly acts on the dual $\mathfrak{t}^*$. It is easy to see that this action preserves the roots of $\mathfrak{t}^*$. Note that the Weyl group W of the root system, defined in the previous set of notes, also acts (faithfully) on $\mathfrak{t}^*$. It turns out that the two groups coincide:

Lemma 12.1.3 (Equivalence of Weyl groups). *We have $N(T)/T \equiv W$, with the actions on $\mathfrak{t}^*$ (or equivalently, $\mathfrak{t}$) being compatible.*

Proof. It will suffice to show that:

(a) the action of $N(T)/T$ on $\mathfrak{t}^*$ is faithful;

(b) to every element of W one can find an element of $N(T)/T$ that acts the same way on $\mathfrak{t}^*$; and

(c) for every element of $N(T)/T$ there is an element of W that acts the same way on $\mathfrak{t}^*$.

To prove (a), we establish the stronger statement that any element w of $N(T)$ that preserves a given Weyl chamber C_h of $\mathfrak{t}^*$ (for some regular $h \in \mathfrak{t}$) is necessarily in T. If w preserves the h-Weyl chamber C_h, then it permutes the h-simple roots, and thus fixes the sum $\rho = \rho_h$ of these h-simple roots. Thus, the one-parameter group $\{\exp(t\rho) : t \in \mathbf{R}\}$ lies in the connected component $Z(w)^0$ of the centraliser $Z(w) := \{g \in G_c : gw = wg\}$ of w. Of course, w also lies in $Z(w)^0$, as does any maximal torus of G_c that contains w. In particular, any maximal torus of G_c containing w is also a maximal torus in $Z(w)^0$; since all maximal tori in $Z(w)^0$ are conjugate, we conclude that all maximal tori in $Z(w)^0$ are also maximal tori in G_c; they also all contain w since w is central in $Z(w)^0$. In particular, $\{\exp(t\rho) : t \in \mathbf{R}\}$ lies in a maximal torus T' of $Z(w)^0$ (and hence in G_c) that contains w. In particular, the adjoint action of ρ fixes the Lie algebra $\mathfrak{t}'$ of T'. But ρ is regular in $\mathfrak{t}$, so its centraliser in $i\mathfrak{g}_\mathbf{R}$ is $\mathfrak{t}$. Thus $T = T'$; since $w \in T'$, we have $w \in T$ as required.

The proof of (c) is similar. Here, w need not preserve ρ, but one can select an element w' of W to maximise $\langle w'(h), w(\rho)\rangle$; arguing as in the proof of Lemma 11.7.12, we see that $(w')^{-1}w$ maps the h-Weyl chamber to itself, and the claim follows from the previous discussion.

To prove (b), it suffices to show that every reflection s_α comes from an element of $N(T)/T$. But in the rank one case (when G is isomorphic $SU_2(\mathbf{C})$) this can be done by direct computation, and the general rank case can then be obtained by looking at the embedded copy of the rank one Lie group associated to the pair of roots $\{-\alpha, \alpha\}$. $\square$

Call an element of $i\mathfrak{g}_\mathbf{R}$ *regular* if it is conjugate (under the adjoint action of G_c) to a regular element of $\mathfrak{t}$ (and hence, by the Weyl group action, to an element in the interior $\mathfrak{c}$ of the (adjoint of the) Weyl chamber); this conjugation element can be viewed as an element of G/T, which is unique by the discussion in the previous section. This gives a bijection $G_c/T \times \mathfrak{c} \to i\mathfrak{g}_\mathbf{R}^{\text{reg}}$ to the regular elements of $i\mathfrak{g}_\mathbf{R}$, which can be seen to be a homeomorphism. The nonregular elements can be computed to have codimension at least three in $i\mathfrak{g}_\mathbf{R}$ (because the centraliser of nonregular elements have at least two more dimensions than in the regular case), so $i\mathfrak{g}_\mathbf{R}^{\text{reg}}$ is simply connected; as this space retracts onto G_c/T, we conclude that G_c/T is simply connected.

From this we may now compute the fundamental group of G_c (or equivalently, of G_{ad}). By inspecting the adjoint action of T on $\mathfrak{g}$, we see that for $t \in \mathfrak{h}_\mathbf{R}$, $\exp(t)$ is trivial in T if and only if t lies in the *coweight lattice* $P := \{t \in \mathfrak{h} : \langle t, \alpha\rangle \in \mathbf{Z}\}$, so the torus T may be identified with the quotient $\mathfrak{h}_\mathbf{R}/P$. Inside P we have the *coroot lattice* Q generated by the coroots $\{h_\alpha : \alpha \in \Phi\}$; these are both full rank in $\mathfrak{h}_\mathbf{R}$ and so the quotient P/Q is finite.

Example 12.1.4. In the A_{n-1} example, $\mathfrak{h}_\mathbf{R}$ is the space $\mathbf{R}_0^n$ of vectors $(x_1, \ldots, x_n) \in \mathbf{R}^n$ with $x_1 + \ldots + x_n = 0$; the coweight lattice P is then generated by $e_i - \frac{1}{n}\sum_{j=1}^n$ for $i = 1, \ldots, n$, and the root lattice Q is spanned by $e_i - e_j$ for $1 \leq i < j \leq n$ and has index n in P.

12.1. Simple Lie groups over **C**

Call an element x of $\mathfrak{h}$ *nonintegral* if one has $\langle x, \alpha \rangle \notin \mathbf{Z}$ for all $\alpha \in \Phi$; this is a stronger condition than being regular, which corresponds to $\langle x, \alpha \rangle$ being nonzero for all α. The set of nonintegral elements of $\mathfrak{h}$ is a collection of open polytopes, and is acted upon by the group A_Q of affine transformations generated by the Weyl group and translations by elements of the coroot lattice Q. A fundamental domain of this space is the *Weyl alcove* $\mathcal{A}$, in which $\langle x, \alpha \rangle > 0$ for positive roots and $\langle x, \beta \rangle < 1$ for the maximal root β; this is a simplex in the Weyl chamber consisting entirely of nonintegral elements, such that the reflection along any of the faces of the alcove lies in $\mathcal{A}$, which shows that it is indeed a fundamental domain. (In the A_{n-1} case, the alcove consists of tuples $\theta_1, \ldots, \theta_n$ with $\theta_1 > \ldots > \theta_n > \theta_1 - 1$.)

Call an element of G_c *regular* if it is conjugate to $\exp(x)$ for some nonintegral $x \in \mathfrak{h}$; as before, the regular elements have codimension at least three in G_c, and so the fundamental group of G_c is the same as the fundamental group of the nonintegral elements of G_c. (In the case of A_{n-1}, G_c is the projective special unitary group $\mathrm{PSU}_n(\mathbf{C})$, and the equivalence class of an unitary matrix is regular if its eigenvalues are all distinct.) Observe that $\exp(ax)$ and $\exp(x)$ are conjugate whenever $a \in A_Q$; in fact, the same is true for all a in A_P, the group of affine transformations on $\mathfrak{h}$ generated by the Weyl group and translations by elements of the coweight lattice P. Because of this, we see that every element of G_c can be expressed in the form $\exp(x)^a$ where x lies in the Weyl alcove $\mathcal{A}$, g lies in G/T, and $\exp(x)^g$ is the conjugate of $\exp(x)$ by (any representative of) g. By lifting, we can then write any loop $\gamma \colon [0,1] \to G_c$ in G_c in the form

$$\gamma(t) = \exp(x(t))^{g(t)}$$

for some continuous $x \colon [0,1] \to \mathcal{A}$ and $g \colon [0,1] \to G/T$. If we fix the base point $\gamma(0) = \gamma(1) = p_0$ of γ, then we can fix the initial point $x(0) = x_0$ of x, and normalise $g(0)$ to be the identity; we then have

$$\exp(x(1))^{g(1)} = \exp(x_0),$$

which places $g(1)$ in $N(T)/T$ (since $\exp(x_0)$ and $\exp(x_1)$, being nonintegral, do not lie in any maximal torus other than T, as can be seen by inspecting its adjoint action on $\mathfrak{g}$). Thus there is an element w of W and $k \in P$ such that $x(1) = wx_0 + k$ and $g(1) = w^{-1}$; this assigns an element a of A_P to γ with the property that $ax_0 \in \mathcal{A}$; one can check that this assignment is preserved under homotopy of γ. From the simply connected nature of both G/T and $\mathcal{A}$ one can check that this assignment is injective; and by the connected nature of G/T and $\mathcal{A}$ the assignment is surjective. On the other hand, as $\mathcal{A}$ is a fundamental domain for A_Q, we see that each (right) coset of A_P in A_Q has exactly one representative a for which $ax_0 \in \mathcal{A}$, so we have obtained a bijective correspondence between $\pi_1(G_c)$ and $A_Q/A_P \equiv Q/P$.

In fact it is not difficult to show that this bijection is a group isomorphism, thus

(12.1) $$Z(\tilde{G}) \equiv \pi_1(G_{\mathrm{ad}}) \equiv \pi_1(G_c) \equiv Q/P.$$

With this formula one can now compute the fundamental group or centre (12.1) associated to any Dynkin diagram group quite easily, and it usually ends up being very small:

(i) For G_2, F_4, or E_8, the group (12.1) is trivial.

(ii) For B_n, C_n, or E_7, the group (12.1) has order two.

(iii) For E_6, the group (12.1) has order three.

(iv) For D_n, the group (12.1) has order four, and is cyclic for odd n and the Klein group for even n.

(v) As mentioned previously, for A_n, the group (12.1) is cyclic of order $n+1$.

Remark 12.1.5. The above theory for simple Lie algebras extends without difficulty to the semisimple case, with a connected Lie group defined to be semisimple if its Lie algebra is semisimple. If one restricts to the simply connected models $\tilde{G}$, then every simply connected semisimple Lie group is expressible as the direct sum of simply connected simple Lie groups. A general semisimple Lie group might not be a direct product of simple Lie groups, but will always be a *central product* (a direct product quotiented out by some subgroup of the centre).

Remark 12.1.6. The compact form G_c (and its lifts) are usually not the only real Lie groups associated to $\mathfrak{g}$, as there may be other *real forms* of $\mathfrak{g}$ than $i\mathfrak{g}_\mathbf{R}$. These can be classified by a somewhat messier version of the arguments given previously, but we will not pursue this matter here; see, e.g., [**Kn2002**]. We also caution that there are additional complications in the real case due to discrepancies between the algebraic and analytic topologies on real algebraic groups; in particular, an algebraic group over $\mathbf{R}$ may be connected in the algebraic (Zariski) sense while being disconnected in the analytic sense. In contrast, the algebraic and analytic topologies are closely related to each other in the complex case, as is most famously exemplified by Serre's GAGA paper [**Se1956**].

12.2. Chevalley groups

The theory of connected Lie groups works well in the real numbers $\mathbf{R}$ or complex numbers $\mathbf{C}$, as these fields are themselves connected in the analytic sense, although as mentioned in the previous remark, in the real case one needs to take some care to relate the analytic structure to the algebraic structure. However, the theory of Lie groups becomes more problematic

12.2. Chevalley groups

when one works with disconnected fields, such as finite fields or the p-adics. However, there is a good substitute for the notion of a Lie group in these settings (particularly when working with algebraically complete fields k), namely the notion of an *algebraic group*. Actually, in analogy to how complex Lie groups are automatically linear groups (up to isomorphism), we will be able to restrict attention to (classical) *linear algebraic groups*, that is to say Zariski-closed subgroups of a general linear group $\mathrm{GL}_n(k)$ over an algebraically closed field k. (Remarkably, it turns out that all affine algebraic groups are isomorphic to a linear algebraic group, though we will not prove this fact here.)

The following result allows one to easily generate linear algebraic groups:

Theorem 12.2.1. *Let k be algebraically closed. All topological notions are with respect to the Zariski topology, and notions of constructibility and irreducibility are in the algebraic geometry sense. If V is a connected constructible subset of $\mathrm{GL}_n(k)$ containing the identity, then the group $\langle V \rangle$ generated by V is closed (and is thus a linear algebraic group) and also irreducible.*

In particular, this theorem implies that linear algebraic groups are connected if and only if they are irreducible.

Proof. By combining V with its reflection V^{-1} we may assume that V is symmetric: $V = V^{-1}$. The product sets $V, V^2, V^3, \ldots$ are all constructible and increasing, so at some point the dimension must stabilise, thus we can find k such that V^k and V^{2k} both have dimension d. Let $A_1, \ldots, A_m$ be the d-dimensional irreducible components of V^k, and $A'_1, \ldots, A'_{m'}$ be the d-dimensional irreducible components of V^{2k}, thus every element of V^k lies in one of the sets $B_{i'} := \{g \in \mathrm{GL}_n(k) : g\overline{A_1} = \overline{A'_{i'}}\}$ for some $i' = 1, \ldots, k'$. As these sets are closed and disjoint and V^k is connected, only one of the $B_{i'}$, say B_1, is nonempty; as V^k contains the identity, we conclude that $\overline{A_1} = \overline{A'_1}$ and $V^k \subset B_1 \subset \overline{A'_1}$, thus V^k is an open dense subset of $\overline{A_1}$, which is symmetric, contains the identity, is Zariski closed, and closed under multiplication and is thus an algebraic group. This implies that V^{2k} is all of $\overline{A_1}$ (because V^k and $(V^k)^{-1}g$ intersect for all $g \in \overline{A_1}$ as they are both open dense subsets of $\overline{A_1}$) and the claim follows. $\square$

This already gives a basic link between the category of complex Lie groups and the category of algebraic groups:

Corollary 12.2.2. *Every complex simple Lie group G_ad in adjoint form is a linear algebraic group over $\mathbf{C}$.*

The same statement is in fact true (up to isomorphism) for the other forms of a complex simple Lie group (by essentially the same argument, and using the fact that the Jordan decomposition for a simple Lie algebra is universal across all representations), though we will focus here on the adjoint form for simplicity. Note though that not every real simple Lie group is algebraic; for instance, the universal cover of $SL_2(\mathbf{R})$ has an infinite discrete centre (the fundamental group of $SL_2(\mathbf{R})$ is isomorphic to $\mathbf{Z}$) and is therefore nonalgebraic. To emphasise the algebraicity of the complex simple Lie group G_{ad} (and in order to distinguish it from the more general Chevalley groups $G_{\mathrm{ad}}(k)$ which we will introduce shortly) we will now write it as $G_{\mathrm{ad}}(\mathbf{C})$.

Proof. Recall from Section 11.8 that the complex Lie algebra $\mathfrak{g}$ has a *Cartan-Weyl basis*—a complex-linear basis $(E_\alpha)_{\alpha \in \Phi}, (H_\alpha)_{\alpha \in \Pi}$ indexed by the roots Φ and the simple roots Π respectively, obeying the Cartan-Weyl relations

$$[H_\alpha, E_\beta] = A_{\alpha,\beta},$$
$$[E_\alpha, E_{-\alpha}] = H_\alpha,$$
$$[E_\alpha, E_\beta] = N_{\alpha,\beta} E_{\alpha+\beta},$$
$$[H_\alpha, H_\beta] = 0$$

where we extend H_α to all roots $\alpha \in \Phi$ by making H_α linear in the coroot of α, $A_{\alpha,\beta}$ are integers, and $N_{\alpha,\beta}$ are structure constants. Among other things, this shows that $G_{\mathrm{ad}}(\mathbf{C})$ is generated by the one-parameter unipotent subgroups $U_\alpha(\mathbf{C}) := \{\exp(t \operatorname{ad} E_\alpha) : t \in \mathbf{C}\}$ and toral subgroups $T_\alpha(\mathbf{C}) := \{\exp(t \operatorname{ad} H_\alpha) : t \in \mathbf{C}\}$ for various α. The unipotent groups U_α are algebraic because $\operatorname{ad} E_\alpha$ is nilpotent. The toral groups T_α are not quite algebraic (they aren't closed), but they are constructible, because the Cartan-Weyl relations show that $\exp(t \operatorname{ad} H_\alpha)$ is given by a diagonal matrix whose entries are monomials in $\exp(t)$, so by reparameterising in terms of $z := \exp(t) \in \mathbf{C}^\times$ we obtain the desired constructibility. The claim then follows from Theorem 12.2.1. □

Somewhat miraculously, the same construction works for any other algebraically closed fields k (and even to nonalgebraically closed fields, as discussed below), to construct an algebraic group[2] $G_{\mathrm{ad},k}$ over k that is the analogue of the adjoint form of the complex Lie group $G_{\mathrm{ad}}(\mathbf{C})$. Whereas $G_{\mathrm{ad}}(\mathbf{C})$ consisted of linear transformations from the complex vector space $\mathfrak{g}$ to itself, $G_{\mathrm{ad},k}$ consists of linear transformations on the k-vector space $\mathfrak{g}(k)$, which has the same Cartan-Weyl basis $(E_\alpha)_{\alpha \in \Phi}, (H_\alpha)_{\alpha \in \Pi}$ but now viewed as a basis over k rather than $\mathbf{C}$. The analogue $T_{\alpha,k}$ of the toral subgroups $T_\alpha(\mathbf{C}) = \{\exp(t \operatorname{ad} H_\alpha) : t \in \mathbf{C}\}$ are then the group of linear

---

[2] We use $G_{\mathrm{ad},k}$ here instead of $G_{\mathrm{ad}}(k)$, to avoid giving the impression that $G_{\mathrm{ad}}(k)$ is the set of k-points of an algebraic group G_{ad}, as this is not necessarily the case when k is not algebraically closed.

12.2. Chevalley groups

transformations $h_\alpha(z)$ on $\mathfrak{g}(k)$ that map E_β to $z^{A_{\alpha,\beta}} E_\beta$ for all roots β and annihilate all the H_γ, for some $z \in k^\times$; this is a connected constructible subgroup of $GL(\mathfrak{g}(k))$. As for the k-analogue $U_{\alpha,k}$ of the unipotent subgroups $U_\alpha(\mathbf{C}) = \{\exp(t \operatorname{ad} E_\alpha) : t \in \mathbf{C}\}$, we use crucially the fact (established in Lemma 11.8.5) that one can ensure that $N_{\alpha,\beta} = \pm(r+1)$, where r is the largest integer such that $\beta - r\alpha$ is a root. This implies in the complex setting that

$$\exp(t \operatorname{ad} E_\alpha) E_\beta = E_\beta \pm t(r+1) E_{\beta+\alpha} \pm t^2 \frac{(r+1)(r+2)}{2} E_{\beta+2\alpha} \pm \ldots$$

where the series terminates once $\beta + s\alpha$ stops being a root. The point here is that the coefficients $\pm(r+1), \pm\frac{(r+1)(r+2)}{2}$, etc. are all integers, and so one can take this as a definition for $u_\alpha(t) = \exp(t \operatorname{ad} E_\alpha)$ for $\mathfrak{g}(k)$ and any $t \in k$ regardless of what characteristic k is, and one still obtains a connected unipotent group in this way. If we then let $G_{\mathrm{ad},k}$ be the group generated by these one-parameter subgroups $T_{\alpha,k}$, $U_{\alpha,k}$, we see that this is a connected linear algebraic group defined over k, known as the (adjoint) *Chevalley group* over k associated to the given root system (or Dynkin diagram). (Other non-adjoint versions of the Chevalley group will be briefly discussed in Remark 12.2.9.)

The same construction works over fields k that are not algebraically closed, giving groups $G_{\mathrm{ad},k}$ that are also denoted $D(k)$ where D is the Dynkin diagram associated to k; for instance, $A_n(k)$ is the projective special linear group $\mathrm{PSL}_{n+1}(k)$; for the purposes of this post, we shall also refer to these groups $G_{\mathrm{ad},k}$ as Chevalley groups. However, we caution that $G_{\mathrm{ad},k}$ is *not* necessarily the k-points[3] of an algebraic group, and in this section we shall treat it purely as an abstract group instead. Despite this issue, these groups still retain a great deal of the other structure of the complex Lie group $G_{\mathrm{ad}}(\mathbf{C})$, and in particular, inherit the *Bruhat decomposition* which we now pause to recall. We first identify some key subgroups of $G_{\mathrm{ad},k}$. We first locate the *maximal torus* T_k, defined as the group generated by the one-parameter toral subgroups $T_{\alpha,k}$ for $\alpha \in \Pi$; this is an abelian subgroup of $G_{\mathrm{ad},k}$. Next, we locate the *Borel subgroup* B_k, defined as the group generated by T and the unipotent groups $U_{\alpha,k}$ for *positive* roots α; this can be seen to be a solvable subgroup of $G_{\mathrm{ad},k}$. Then, for each reflection $s_\alpha \in W$ in the Weyl group W associated to a simple root $\alpha \in \Pi$, we define the elements

$$n_\alpha(t) := u_\alpha(t) u_{-\alpha}(-t^{-1}) u_\alpha(t)$$

for $t \in k^\times$, one can check using the Cartan-Weyl relations that $n_\alpha(t)$ determines an element in a coset of T in its normaliser $N(T)$ which is independent of the choice of t. Letting N_k be the group generated by the $n_\alpha(t)$ and T_k,

---

[3] For instance, the adjoint form algebraic group associated to A_n is PGL_{n+1}, and the set of k-points of this group is $\mathrm{PGL}_{n+1}(k)$, which can be a larger group than $\mathrm{PSL}_{n+1}(k)$.

we thus see that N_k normalises T_k, and with some further application of the Cartan-Weyl relations one sees that N_k/T_k is isomorphic to W (with each $n_\alpha(t)$ projecting down to s_α); cf. Lemma 12.1.3. Indeed, if $n_w \in N_k$ is a representative of $w \in W$, one sees that the operation of conjugation $g \mapsto n_w g n_w^{-1}$ maps $U_{\alpha,k}$ to $U_{w(\alpha),k}$ for any root α.

For notational reasons we now fix an assignment n_w of a representative in N_k to each element $w \in W$, although all of the objects we will actually study will not be dependent on this choice of assignment.

The following axioms can then be verified from further use of the Cartan-Weyl relations:

(i) $G_{\mathrm{ad},k}$ is generated by B_k and N_k.

(ii) T_k is the intersection of B_k and N_k, and is normalised by N_k.

(iii) $W = N_k/T_k$ is generated by the reflections $s_\alpha, \alpha \in \Pi$, which are of order two.

(iv) No reflection s_α (or more precisely, no representative in N_k of that reflection) normalises B_k.

For each element $w \in W$ of the Weyl group, we can form the double coset $C(w) := B_k n_w B_k$; this is easily seen to be independent of the choice of representative n_w. Thus, for instance, $C(1) = B_k$. It is also clear that any two double cosets $C(w), C(w')$ are either equal or disjoint, and one has the inclusion

(12.2) $$C(ww') \subset C(w)C(w')$$

for any $w, w' \in W$, as well as the symmetry $C(w)^{-1} = C(w^{-1})$. We also have the important further inclusion relation:

Lemma 12.2.3. *For any $\alpha \in \Pi$ and $w \in W$, we have $C(w)C(s_\alpha) \subset C(w) \cup C(ws_\alpha)$.*

Proof. First suppose that $w(\alpha)$ is a positive root. Then we observe the factorisation
$$B_k = U_{\alpha,k} T_k U_{\overline{\alpha},k}$$
where $U_{\overline{\alpha},k}$ is the group generated by all the U_β for positive $\beta \neq \alpha$. From the positivity of $w(\alpha)$ one has
$$n_w U_{\alpha,k} n_w^{-1} = U_{w(\alpha),k} \subset B_k$$
and from the simplicity of α one has
$$n_{s_\alpha}(T_k U_{\overline{\alpha},k}) n_{s_\alpha}^{-1} \subset B_k$$
and thus
$$B_k \subset (n_w^{-1} B_k n_w)(n_{s_\alpha} B_k n_{s_\alpha}^{-1});$$

multiplying on the left by n_w and on the right by s_α we conclude that
$$C(w)C(s_\alpha) \subset C(ws_\alpha);$$
as the left-hand side is a nonempty union of double cosets, we in fact have equality
$$C(w)C(s_\alpha) = C(ws_\alpha).$$

Now suppose instead that $w(\alpha)$ is a negative root. Applying the previous equality with w replaced by ws_α we conclude that

(12.3) $$C(ws_\alpha)C(s_\alpha) = C(w)$$

and thus
$$C(w)C(s_\alpha) = C(ws_\alpha)C(s_\alpha)C(s_\alpha).$$

On the other hand, direct calculation with the Cartan-Weyl relations reveals that
$$C(s_\alpha)C(s_\alpha) \subset C(1) \cup C(s_\alpha)$$
and the claim then follows from (12.3). $\square$

Lemma 12.2.3 and the preceding four axioms form the axiom system, introduced by Tits, for a (B, N)-*pair*. This axiom system is convenient for abstractly achieving a number of useful facts, such as the *Bruhat decomposition*, and the simplicity of $G_{\mathrm{ad},k}$ (in most cases). We begin with the Bruhat decomposition:

Proposition 12.2.4 (Bruhat decomposition). $G_{\mathrm{ad},k}$ *is the disjoint union of $C(w)$ as w ranges over W. (Thus there is a canonical bijection between $B_k G_{\mathrm{ad},k}/B_k$ and $W = N_k/T_k$, which by slight abuse of notation can be written as $G_{\mathrm{ad},k} = B_k W B_k$.)*

Proof. We first show that the $C(w)$ cover $G_{\mathrm{ad},k}$. As the $C(w)$ cover both B_k and N_k (which together generate $G_{\mathrm{ad},k}$) and their union is symmetric, it suffices to show that $\bigcup_w C(w)$ is closed under multiplication, thus $C(w_1)C(w_2) \subset \bigcup_w C(w)$. But this is easily achieved by iterating Lemma 12.2.3 (inducting on the length of w_2, that is to say the minimal number of reflections s_α needed to generate $C(w_2)$, noting that the case $w_2 = 1$ is trivial).

Now we show that the $C(w)$ are disjoint. Since double cosets are either equal or disjoint, it suffices to show that $C(w_1) = C(w_2)$ implies $w_1 = w_2$ for all $w_1, w_2 \in W$. We induct on the length of w_2. The case when $w_2 = 1$ is trivial, so suppose that $w_2 \neq 1$ and that the claim has already been proven

for all shorter w_2. We write $w_2 = w_2' s_\alpha$ for some shorter w_2'. Then
$$\begin{aligned} w_2' = w_2 s_\alpha &\subset C(w_2) C(s_\alpha) \\ &= C(w_1) C(s_\alpha) \subset C(w_1) \cup C(w_1 s_\alpha) \\ &= C(w_2) \cup C(w_1 s_\alpha) \end{aligned}$$
and hence $C(w_2')$ is either equal to $C(w_2)$ or $C(w_1 s_\alpha)$. By induction we then either have $w_2' = w_2$ or $w_2' = w_1 s_\alpha$. The former is absurd, thus $w_2' = w_1 s_\alpha$ and thus $w_1 = w_2$ as required. $\square$

By further exploitation of the (B, N)-pair axioms and some other properties of $G_{\mathrm{ad},k}$, we can show that this group is simple in the group-theoretic sense in almost all cases (there are a few exceptions in very low characteristic). This generalises the discussion of complex Lie groups in the previous section, except now we do not need to pass through the simplicity of the associated Lie algebra (and instead work with the irreducibility of the root system).

We use an argument of Iwasawa and Tits. We first need some structural results about *parabolic subgroups* of $G_{\mathrm{ad},k}$ — subgroups that contain the Borel subgroup B (or a conjugate thereof).

Lemma 12.2.5. *Let $w \in W$ be an element of the Weyl group, with a minimal-length representation $w = s_{\alpha_1} \ldots s_{\alpha_l}$ in terms of representations. Then $n_{s_{\alpha_1}}, \ldots, n_{s_{\alpha_l}}$ lie in the group generated by B_k and $n_w B_k n_w^{-1}$.*

Proof. We may assume inductively that $l > 0$ and that the claim has been proven for smaller values of l. From minimality we know that $w^{-1}(\alpha_1)$ is a negative root, and so
$$U_{-\alpha,k} = n_w U_{-w^{-1}(\alpha),k} n_w^{-1} \subset n_w B_k n_w^{-1}$$
and $U_{\alpha,k} \subset B_k$, hence n_α, being in the group generated by $U_{\alpha,k}$ and $U_{-\alpha,k}$, is contained in the group generated by B_k and $n_w B_k n_w^{-1}$. Writing $w = s_\alpha w'$, this implies that this group contains the group generated by B_k and $n_{w'} B_k n_{w'}^{-1}$, and the claim then follows from induction. $\square$

Corollary 12.2.6 (Classification of parabolic groups). *Every parabolic group $B_k \subset P \subset G_{\mathrm{ad},k}$ containing B_k takes the form $\bigcup_{w \in W_\pi} C(w)$ for some $\pi \subset \Pi$, where W_π is the subgroup of W generated by the s_α for $\alpha \in \Pi$, and conversely each of the $\bigcup_{w \in W_\pi} C(w)$ is a parabolic subgroup of $G_{\mathrm{ad},k}$. Furthermore, all of these $2^{|\Pi|}$ parabolic groups are distinct.*

Proof. The fact that $\bigcup_{w \in W_\pi} C(w)$ is a group follows from Lemma 12.2.3. To show distinctness, it suffices by the Bruhat decomposition to show that the W_π are all distinct, but this follows from the linear independence of the simple roots. Finally, if P is a parabolic subgroup containing B_k, we can set $\pi := \{\alpha \in \Pi : n_\alpha \in P\}$, then clearly P contains $\bigcup_{w \in W_\pi} C(w)$. On the other hand, as $P = B_k P B_k$, P is the union of double cosets $C(w)$, and from Lemma 12.2.5 if P contains $C(w)$, then w is generated by reflections from π. The claim follows. $\square$

This, together with the previously noted solvability of B and the irreducibility of the root system, gives a useful criterion for simplicity:

Lemma 12.2.7 (Criterion for simplicity). *Suppose that $G_{\mathrm{ad},k}$ is a perfect group and that B_k does not contain any nontrivial normal subgroup of $G_{\mathrm{ad},k}$ (i.e., $\bigcap_{g \in G_{\mathrm{ad},k}} g B_k g^{-1} = \{1\}$). Then $G_{\mathrm{ad},k}$ is simple.*

Proof. Let H be a nontrivial normal subgroup of $G_{\mathrm{ad},k}$. Then by hypothesis H is not contained in B_k, so the group HB_k is a parabolic subgroup of $G_{\mathrm{ad},k}$ that is strictly larger than B_k, thus $HB_k = \bigcup_{w \in W_\pi} C(w)$ for some nonempty $\pi \subset \Pi$. If $\alpha \in \pi$ and $\beta \in \Pi \backslash \pi$, then H intersects $C(s_\alpha)$, and thus (by the normality of H) also intersects $n_{s_\beta} C(s_\alpha) n_{s_\beta}^{-1}$. By Lemma 12.2.3 (and (12.3)), we have

$$n_{s_\beta} C(s_\alpha) n_{s_\beta}^{-1} \subset C(s_\beta) C(s_\alpha) C(s_\beta)$$
$$= C(s_\beta) C(s_\alpha s_\beta)$$
$$\subset C(s_\beta s_\alpha s_\beta) \cup C(s_\alpha s_\beta)$$

and so at least one of $s_\beta s_\alpha s_\beta$ and $s_\alpha s_\beta$ lies in W_π. But as $s_\alpha \in W_\pi$ and $s_\beta \notin W_\pi$, we conclude that $s_\beta s_\alpha s_\beta \in W_\pi$. From this and Lemma 12.2.5 we see that any minimal representation of $s_\beta s_\alpha s_\beta$ has generators both in $\{\alpha, \beta\}$ and π, which forces $s_\beta s_\alpha s_\beta = s_\alpha$ (note that $s_\beta s_\alpha s_\beta$ cannot vanish). Thus we see that s_α commutes with s_β, contradicting the irreducibility of the root system unless $\pi = \Pi$. We thus have $HB_k = G_{\mathrm{ad},k}$. As $G_{\mathrm{ad},k}$ is perfect, this implies that $G_{\mathrm{ad},k}/H$ is also perfect; but this is a quotient of the solvable group B_k and is thus solvable also. As only the trivial group is both perfect and solvable, we conclude that $H = G_{\mathrm{ad},k}$, and the claim follows. $\square$

In the specific case of the adjoint group $G_{\mathrm{ad},k}$, the second hypothesis in Lemma 12.2.7 can be verified:

Lemma 12.2.8. *B_k does not contain any nontrivial normal subgroup of $G_{\mathrm{ad},k}$.*

As in the complex case, it turns out that nonadjoint versions of a Chevalley group have nontrivial centre that lies in every maximal torus and hence in every Borel group, so this lemma is specific to the adjoint group.

Proof. Let H be a normal subgroup of $G_{\mathrm{ad},k}$ that lies in B_k. Conjugating by the long word in W (that maps all positive roots to negative roots) we see that H actually lies in the torus T_k. In particular, for any root α, $[H, U_{\alpha,k}]$ lies in both $H \subset T_k$ and $U_{\alpha,k}$ and is thus trivial; this shows that H is central. But by the Cartan-Weyl relations we see that there are no elements of T_k that commute with all the $U_{\alpha,k}$, and the claim follows. $\square$

We remark that the above arguments can also be adapted to show that $G_{\mathrm{ad},k}$ always has trivial centre $Z(G_{\mathrm{ad},k})$ (because the above lemma and the proof of Lemma 12.2.7 then shows that $Z(G_{\mathrm{ad},k})B_k = G_{\mathrm{ad},k}$, making B_k normal in $G(k)$, which can be shown to lead to a contradiction).

From the above discussion we see that $G_{\mathrm{ad},k}$ will be simple whenever it is perfect. Establishing perfection is relatively easy in most cases, as it only requires enough explicit examples of commutators to encompass a generating subset of $G_{\mathrm{ad},k}$. It is only when the field k and the Dynkin diagram are extremely small that one has too few commutators to make a generating subset, and $G_{\mathrm{ad},k}$ fails to be perfect (and thus also fails to be simple); the specific failures turn out to be $A_1(\mathbf{F}_2)$, $A_1(\mathbf{F}_3)$, $B_2(\mathbf{F}_2)$, and $G_2(\mathbf{F}_2)$. See [**Ca1993**] for details.

Remark 12.2.9. We have focused primarily on the adjoint group $G_{\mathrm{ad},k}$ of the Chevalley groups, but much as in the complex Lie group case, to each Dynkin diagram and field k one can associate a finite number of versions of the Chevalley group, ranging from the minimal example of the adjoint form $G_{\mathrm{ad},k}$ to the maximal example $G_{u,k}$. When k is algebraically closed, these are all linear algebraic groups, and every form of the Chevalley group has an *isogeny* (the algebraic group analogue of a finite cover) to the adjoint form (arising from quotienting out by the centre) and receives an isogeny from the universal form, much as in the complex case. We still have the basic identity (12.1), but the lattices P, Q now lie over k rather than $\mathbf{R}$ or $\mathbf{C}$ (which can make the order of Q/P smaller than in the complex case if k has small positive characteristic p by quotienting out the elements of order a prime power of p, thus collapsing the number of distinct forms of the Chevalley group in some characteristics). See for instance [**St1967**] or [**GoLySo1998**] for details. As an example of the collapse phenomenon mentioned earlier, $\mathrm{SL}_2(k)$ (the universal form for $A_1(k)$) and $\mathrm{PSL}_2(k)$ (the adjoint form for $A_1(k)$) are distinct for most fields k, but coincide when k has characteristic two.

Remark 12.2.10. We caution that the way we have defined Chevalley groups here, a Chevalley group G_k over a nonalgebraically closed field is not necessarily the same as the set of k-points of the Chevalley group $G_{\overline{k}}$ of the algebraic closure $\overline{k}$, as the latter may be strictly larger. For instance, the real elements of $\mathrm{PSL}_2(\mathbf{C}) = \mathrm{PGL}_2(\mathbf{C})$ are the elements of $\mathrm{PGL}_2(\mathbf{R})$,

12.2. Chevalley groups

which is a larger group than $\mathrm{PSL}_2(\mathbf{R})$ (it also contains the projectivisation of matrices with negative determinant).

Remark 12.2.11. The Chevalley construction gives some specific families of algebraic groups over algebraically closed fields that are either simple (in the adjoint form) or almost simple (which means that the only normal groups are zero-dimensional); in the latter case they are also quasisimple as in the complex case. It is natural to ask whether there are any other (nonabelian) simple algebraic groups over an algebraically closed field. It turns out (quite remarkably) that one can perform the entirety of the classification of complex Lie algebras in the category of algebraic groups over a given algebraically closed field (regardless of its characteristic!), to arrive at the conclusion that the Chevalley groups are (up to isomorphism) the *only* nonabelian simple or almost simple connected linear algebraic groups. This is despite the lack of any reasonable analogue of the compact form G_c over arbitrary fields, and also despite the additional subtleties present in the structural theory of Lie algebras when the characteristic is positive and small. Instead, one has to avoid use of Lie algebras or compact forms, and try to build the basic ingredients of the (B, N)-pair structure mentioned above (e.g., maximal tori, Borel subgroups, roots, etc.) directly. This result, however, requires a serious amount of algebraic geometry machinery and will not be discussed here; see, e.g., [**Hu1975**] for details.

Remark 12.2.12. The Bruhat decomposition gives a parameterisation of $G_{\mathrm{ad},k}$ as

$$G_{\mathrm{ad},k} = \bigcup_{w \in W} U_k T_k n_w U_{w,k}^-$$

where U_k is the group generated by the $U_{\alpha,k}$ for all positive roots α, and $U_{w,k}^-$ is the subgroup generated by the U_α for those positive roots α for which $w(\alpha)$ is negative; every element g of $G_{\mathrm{ad},k}$ then has a unique representation of the form

$$g = u t n_w u'$$

for some $w \in W$, $u \in U_k$, $t \in T_k$, and $u' \in U_{w,k}^-$. Among other things, this allows for a computation of the order of the Chevalley group $G_{\mathrm{ad},\mathbf{F}_q}$ over a finite field of q elements,

$$|G_{\mathrm{ad},\mathbf{F}_q}| = \sum_{w \in W} q^N (q-1)^r q^{N_w},$$

where N is the number of positive roots, r is the rank (the dimension of the maximal torus), and N_w is the number of positive roots α with $w(\alpha)$ negative. If we write $q = 1 + \varepsilon$, this becomes

$$|G_{\mathrm{ad},\mathbf{F}_{1+\varepsilon}}| = \varepsilon^r (|W| + O(\varepsilon))$$

suggesting that in the limit $\varepsilon \to 0$, the Chevalley group $G_{\mathrm{ad},\mathbf{F}_1}$ over the *"field with one element"* should degenerate to something like $N_{\mathbf{F}_1} = T_{\mathbf{F}_1}.W$, an extension of the Weyl group by some sort of torus over the field with one element. Now, this calculation does not make actual rigorous sense — the currently accepted definition of a field does not allow the possibility of fields of order equal to one (or arbitrarily close to one) — but there are tantalising hints in various areas of mathematics that these sorts of formal computations can sometimes to tied to interesting rigorous mathematical statements. However, it appears that we are still a ways off from a completely satisfactory understanding of the extent to which the "field with one element" actually exists, and what its nature is.

12.3. Finite simple groups of Lie type

As discussed above, the (adjoint) Chevalley group construction $G_{\mathrm{ad},k}$, when applied to a finite field $\mathbf{F}_q$, usually gives a finite simple group $G(q) := G_{\mathrm{ad},\mathbf{F}_q}$. However, this construction does not give all of the finite simple groups that are associated to Lie groups. A basic example is the *projective special unitary group* $\mathrm{PSU}_n(\mathbf{F}_q)$ over a finite field whose order q is a perfect square: $q = \tilde{q}^2$. This field supports a Frobenius automorphism $\tau : x \mapsto x^{\tilde{q}}$ which behaves much like complex conjugation $z \mapsto \overline{z}$ does on the complex field (for instance, τ fixes the index two subfield $\mathbf{F}_{\tilde{q}}$, much as complex conjugation fixes the index two subfield $\mathbf{R}$). We can then define $\mathrm{PSU}_n(\mathbf{F}_{q^2})$ as the quotient of the matrix group

(12.4) $$\mathrm{SU}_n(\mathbf{F}_q) := \{U \in \mathrm{SL}_n(\mathbf{F}_q) : U\tau(U^T) = 1\}$$

by its centre, where $\tau(U^T)$ is the matrix formed by applying the Frobenius automorphism τ to each entry of the transpose U^T of TU. This resembles Chevalley groups such as $\mathrm{PSL}_n(k)$, but the group $\mathrm{PSU}_n(k)$ requires the additional input of the Frobenius automorphism, which is available for some fields k but not for others, and destroys the algebraic nature of the group. For instance, $\mathrm{PSU}_n(\mathbf{C})$ is not a complex algebraic group, because complex conjugation $z \mapsto \overline{z}$ is not a complex algebraic operation; it is similarly not a complex Lie group because complex conjugation is not a complex analytic operation. One can view this group as algebraic (or analytic) over an index two subgroup; for instance, $\mathrm{PSU}_n(\mathbf{C})$ is a *real* Lie group, and can also be (carefully) viewed as a real algebraic group, as long as one bears in mind that the reals are not algebraically closed. While this can certainly be a profitable way to view groups of this type (known as *Steinberg groups*), there is another perspective on such groups which extends to the most general class of finite simple groups of Lie types, which contain not only the Chevalley groups and the Steinberg groups but an additional third class, namely the *Suzuki-Ree*

12.3. Finite simple groups of Lie type

groups. To motivate this different viewpoint, observe that the definition (12.4) of the special unitary group $\mathrm{SU}_n(\mathbf{F}_q)$ can be rewritten as

$$\mathrm{SU}_n(\mathbf{F}_q) = \{U \in \mathrm{SL}_n(\mathbf{F}_q) : \sigma(U) = U\}$$

where $\sigma = \rho \circ \tau = \tau \circ \rho$, τ is the Frobenius map defined earlier (acting componentwise on each matrix entry), and τ is the transpose inverse map

(12.5) $$\rho(U) := (U^T)^{-1}.$$

Observe that τ and ρ are commuting automorphisms on $\mathrm{SL}_n(\mathbf{F}_{q^2})$ of order two, and so σ is also an automorphism of order two (i.e., it is an *involution*). Thus we see that the special unitary group is the subgroup of the Chevalley group $\mathrm{SL}_n(\mathbf{F}_{q^2})$ which is fixed by the involution σ.

This suggests that we can locate other finite simple (or at least finite quasisimple) groups of Lie type by looking at the fixed points

$$G(q)^\sigma = \{g \in G(q) : \sigma(g) = g\}$$

of automorphisms σ in a Chevalley group $G(q)$. One should look for automorphisms with a fairly small order (such as two or three), as otherwise the fixed point set might be so small as to generate a trivial group.

As the example of the special unitary group suggests, one can obtain such automorphisms σ by composing two types ρ, τ of automorphisms. On the one hand, we have the *field automorphisms* $\tau : x \mapsto x^{\tilde{q}}$, where $\tilde{q}$ is some power of the characteristic p of the field $\mathbf{F}_q$, applied to each matrix entry of Chevalley group elements. On the other hand, we have *graph automorphisms* $\rho: G(q) \to G(q)$, arising from automorphisms of the Dynkin diagram (which, as noted in Theorem 11.8.1, induces an automorphism of Lie algebras, and can also be used to induce an automorphism of Chevalley groups), which commute with field automorphisms. The transpose inverse map ρ defined in (12.5) is, strictly speaking, not of this form: it is associated to the Lie algebra involution $x \mapsto -x^T$, which maps each root $e_i - e_j$ to its negation $e_j - e_i$, so in particular, does not map simple roots to simple roots. However, if one composes ρ with the conjugation action of the *long word* in the Weyl group (an example of an inner automorphism), which in the case of SL_n is represented by an antidiagonal matrix, the associated Lie algebra involution now maps each root $e_i - e_j$ to its reflection $e_{n+1-j} - e_{n+1-i}$, and corresponds to the Dynkin diagram automorphism of A_n formed by reflection. With this conjugation by the long word, the fixed points of the resulting automorphism σ is still a special unitary group $\mathrm{SU}_n(\mathbf{F}_q)$, but the sesquilinear form that defines unitarity is not the familiar form

$$\langle (x_1, \ldots, x_n), (y_1, \ldots, y_n) \rangle := x_1 \tau(y_1) + \ldots + x_n \tau(y_n)$$

but rather an antidiagonal version
$$\langle(x_1,\ldots,x_n),(y_1,\ldots,y_n)\rangle := x_1\tau(y_n) + \ldots + x_n\tau(y_1).$$
It turns out that, up to group isomorphism, we still obtain the same projective special unitary group $\mathrm{PSU}_n(\mathbf{F}_q)$ regardless of choice of sesquilinear form, so this reversal in the definition of the form is ultimately not a difficulty.

If the graph automorphism ρ has order d, and one takes the field automorphism τ to also have order d by requiring that $q = \tilde{q}^d$, take the fixed points $G(q)^\sigma$ of the resulting order d automorphism $\sigma = \rho\tau$, we (essentially) obtain the standard form of a *Steinberg group* $^d\mathbf{D}(q)$, where $\mathbf{D}$ is the Dynkin diagram. By "essentially", we mean that we may first have to pass to a bounded index subgroup, and then quotient out by the centre, before one gets a finite simple group; this is a technical issue which we will briefly discuss later. Thus, for instance, $\mathrm{PSU}_n(\mathbf{F}_q)$ is denoted $^2A_{n-1}(q)$. (In some texts such a group would be denoted $^2A_{n-1}(\tilde{q})$ instead.) In a similar vein, the Dynkin diagrams D_n and E_6 also obviously support order two automorphisms, leading to additional Steinberg groups $^2D_n(q), ^2E_6(q)$ when $q = \tilde{q}^2$ is a perfect square. The $^2D_n(q)$ class can be interpreted as a class of projective special orthogonal groups, but the $^2E_6(q)$ family does not have a classical interpretation. A noteworthy special case is D_4, which is the unique Dynkin diagram that also supports an automorphism of order three, leading to the final class of Steinberg groups, the *triality groups* $^3D_4(q)$ when $q = \tilde{q}^3$ is a perfect cube.

In large characteristic (five and higher), the Chevalley and Steinberg groups are (up to isomorphism) turn out to be the only way to generate finite simple groups of Lie type; one can experiment with other combinations of automorphisms on Chevalley groups but they end up either giving the same groups up to isomorphism as the preceding constructions, or groups that are not simple (they do not obey the (B, N) axioms that one can use to easily test for simplicity). But in small characteristic, where the distinction between short and long roots can become blurred, there are additional Dynkin diagram automorphisms. Specifically, for the Dynkin diagrams $B_2 = C_2$ and F_4 in perfect fields of characteristic two, there is a *projective* Dynkin diagram automorphism of order two that swaps the long and short roots, which induces an automorphism ρ of the Chevalley group which is order two modulo a Frobenius map (in that ρ^2 is given by the Frobenius map $x \mapsto x^2$); see [**Ca1993**] for the construction. If one combines this automorphism with a field automorphism $\tau : x \mapsto x^{\tilde{q}}$ with q equal to $2\tilde{q}^2$, we obtain an order two automorphism σ that generates the families of *Suzuki groups* $^2B_2(2^{2n+1})$ and *Ree groups* $^2F_4(2^{2n+1})$. Similarly, the Dynkin diagram G_2 in perfect fields of characteristic three has an automorphism that swaps the short and long root, and if $q = 3\tilde{q}^2$ leads to the final class of Ree groups, $^3G_2(3^{2n+1})$.

12.3. Finite simple groups of Lie type

In contrast to the Steinberg groups, the Suzuki-Ree groups cannot be easily viewed as algebraic groups over a suitable subfield; morally, one "wants" to view $^2B_2(2^{2n+1})$ and $^2F_4(2^{2n+1})$ as being algebraic over the field of $2^{n+1/2}$ elements (and similarly view $^3G_2(3^{2n+1})$ as algebraic over the field of $3^{n+1/2}$ elements), but such fields of course do not exist[4]. One can also view the Steinberg and Suzuki-Ree groups $^d\mathbf{D}(q)$ (collectively referred to as *twisted groups of Lie type*) as being "fractal" subgroups (modulo quotienting by the centre) of the associated Chevalley group $\mathbf{D}(q)$, of relative "fractal dimension" about $1/d$, with the former group lying in "general position" with respect to the latter in some algebraic geometry sense; for instance, one could view $\mathrm{PSU}_n(\mathbf{F}_q)$ as a subgroup of $\mathrm{PSL}_n(\mathbf{F}_q)$ of approximately "half the dimension", and in general position in the sense that it does not lie in any (bounded complexity) algberaic subgroup of $\mathrm{PSL}_n(\mathbf{F}_q)$. This type of viewpoint was formalised quite profitably by Larsen and Pink in [**LaPi2011**] (and used subsequently in [**Hr2012**], [**BrGrTa2011**], [**BrGrTa2013**] and is also used in a forthcoming paper of Breuillard, Green, Guralnick, and myself).

Remark 12.3.1. We have oversimplified slightly the definition of a twisted finite simple group of Lie type: in some cases the group $G(q)^\sigma$ is not quite a simple group. As in the previous section, this can happen for very small groups (the Chevalley group examples $A_1(2), A_1(3), B_2(2), G_2(2)$ mentioned earlier, but also $^2A_2(4), {}^2B_2(2), {}^2G_2(3)$, and $^2F_4(2)$). Another issue (which already arises in the Chevalley group case if one does not use the adjoint group) is that the fixed points $G(q)^\sigma$ contain a nontrivial centre and are only a quasisimple group rather than a simple group. Usually one can quotient out by the centre (which will always be quite small) to recover the finite simple group, or work exclusively with adjoint groups which are automatically centreless. But there is one additional technicality that arises even in the adjoint group, which is that sometimes there are some extraneous fixed points of σ of $G(q)$ that one does not actually want (for instance, they do not lie in the group generated by the natural analogues of the B and N groups in this setting, thus violating the (B,N)-axioms). So one sometimes has to restrict attention to a bounded index subgroup of $G(\mathbf{F}_q)^\sigma$, such as the group $O^{p'}(G(q)^\sigma)$ generated by those "unipotent" elements whose order is a power of the characteristic p; an alternative (and equivalent, except in very small cases) approach is to work with the derived group $[G(q)^\sigma, G(q)^\sigma]$ of $G(q)^\sigma$, which turns out to kill off the extraneous elements (which are associated

---

[4] Despite superficial similarity, this issue appears unrelated to the "field with one element" discussed in Remark 12.2.12, although both phenomena do suggest that there is perhaps a useful generalisation of the concept of a field that is currently missing from modern mathematics.

to another type of automorphism was not previously discussed, namely the *diagonal* automorphisms). See [**GoLySo1998**] for a detailed treatment of these issues.

Bibliography

[Al1996] N. Alon, *On the edge-expansion of graphs*, Combin. Probab. Comput. **6** (1997), no. 2, 145–152.

[AlMi1985] N. Alon, V. D. Milman, λ_1, *isoperimetric inequalities for graphs, and superconcentrators*, J. Combin. Theory Ser. B **38** (1985), no. 1, 73–88.

[BaNiPy2008] L. Babai, N. Nikolov, L. Pyber, *Product growth and mixing in finite groups*, Proceedings of the Nineteenth Annual ACM-SIAM Symposium on Discrete Algorithms, 248–257, ACM, New York, 2008.

[BaSe1992] L. Babai, A. Seress, *On the diameter of permutation groups*, European J. Combin. **13** (1992), no. 4, 231–243.

[BaSz1994] A. Balog, E. Szemerédi, *A statistical theorem of set addition*, Combinatorica **14** (1994), no. 3, 263–268.

[BedeVa2008] B. Bekka, P. la Harpe, A. Valette, Kazhdan's property (T), New Mathematical Monographs, 11. Cambridge University Press, Cambridge, 2008.

[Bo2001] B. Bollobás, Random graphs. Second edition. Cambridge Studies in Advanced Mathematics, 73. Cambridge University Press, Cambridge, 2001.

[Bo1974] E. Bombieri, *Counting points on curves over finite fields (d'aprés S. A. Stepanov)*, Séminaire Bourbaki, 25ème année (1972/1973), Exp. No. 430, pp. 234–241. Lecture Notes in Math., Vol. 383, Springer, Berlin, 1974.

[BoSt2007] A. Booker, A. Strömbergsson, *Numerical computations with the trace formula and the Selberg eigenvalue conjecture*, J. Reine Angew. Math. **607** (2007), 113–161.

[BoHaReSe2011] R. Borcherds, M. Haiman, N. Reshetikhin, V. Serganova, *Berkeley Lectures on Lie Groups and Quantum Groups*, edited by Anton Geraschenko and Theo Johnson-Freyd. available at `math.berkeley.edu/~theojf/LieQuantumGroups.pdf`.

[BoGa2008] J. Bourgain, A. Gamburd, *Uniform expansion bounds for Cayley graphs of* $SL_2(\mathbb{F}_p)$, Ann. of Math. (2) **167** (2008), no. 2, 625–642.

[BoGaSa2010] J. Bourgain, A. Gamburd, P. Sarnak, *Affine linear sieve, expanders, and sum-product*, Invent. Math. **179** (2010), no. 3, 559–644.

[BoGlKo2006] J. Bourgain, A. Glibichuk, S. Konyagin, *Estimates for the number of sums and products and for exponential sums in fields of prime order*, J. London Math. Soc. (2) **73** (2006), no. 2, 380–398.

[BoKaTa2004] J. Bourgain, N. Katz, T. Tao, *A sum-product estimate in finite fields, and applications*, Geom. Funct. Anal. **14** (2004), no. 1, 27–57.

[BoKo2003] J. Bourgain, S. V. Konyagin, *Estimates for the number of sums and products and for exponential sums over subgroups in fields of prime order*, C. R. Math. Acad. Sci. Paris **337** (2003), no. 2, 75–80.

[BoVa2012] J. Bourgain, P. Varjú, *Expansion in $SL_d(\mathbf{Z}/q\mathbf{Z})$, q arbitrary*, Invent. Math. **188** (2012), no. 1, 151–173.

[BrGa2010] E. Breuillard, A. Gamburd, *Strong uniform expansion in* SL(2,p), Geom. Funct. Anal. **20** (2010), no. 5, 1201–1209.

[BrGrTa2011] E. Breuillard, B. Green and T. Tao, *Approximate subgroups of linear groups*, Geom. Funct. Anal. **21** (2011), no. 4, 774–819.

[BrGrTa2011b] E. Breuillard, B. Green and T. Tao, *The structure of approximate groups*, preprint. arXiv:1110.5008

[BrGrTa2011c] E. Breuillard, B. Green, T. Tao, *Suzuki groups as expanders*, Groups Geom. Dyn. **5** (2011), no. 2, 281–299.

[BrGrTa2012] E. Breuillard, B. Green, B. Guralnick, and T. Tao, *Strongly dense free subgroups of semisimple algebraic groups*, Israel J. Math. **192** (2012), no. 1, 347–379.

[BrGrTa2013] E. Breuillard, B. Green, B. Guralnick, and T. Tao, *Expansion in finite simple groups of Lie type*, preprint.

[Bu1982] P. Buser, *A note on the isoperimetric constant*, Ann. Sci. École Norm. Sup. (4) **15** (1982), no. 2, 213–230.

[Ca1993] R. Carter, Finite groups of Lie type. Conjugacy classes and complex characters. Reprint of the 1985 original. Wiley Classics Library. A Wiley-Interscience Publication. John Wiley & Sons, Ltd., Chichester, 1993.

[Ch1984] I. Chavel, Eigenvalues in Riemannian geometry. Including a chapter by Burton Randol. With an appendix by Jozef Dodziuk. Pure and Applied Mathematics, 115. Academic Press, Inc., Orlando, FL, 1984.

[Ch1970] J. Cheeger, *A lower bound for the smallest eigenvalue of the Laplacian*, Problems in analysis (Papers dedicated to Salomon Bochner, 1969), pp. 195–199. Princeton Univ. Press, Princeton, N. J., 1970.

[Ch1973] J. R. Chen, *On the representation of a larger even integer as the sum of a prime and the product of at most two primes*, Sci. Sinica **16** (1973), 157–176.

[ChGrWi1989] F. R. K. Chung, R. L. Graham, R. M. Wilson, *Quasi-random graphs*, Combinatorica **9** (1989), no. 4, 345–362.

[CoCuNoPaWi1985] J. H. Conway, R. T. Curtis, S. P. Norton, R. A. Parker, R. A. Wilson, Atlas of Finite Groups: Maximal Subgroups and Ordinary Characters for Simple Groups, Oxford, England 1985.

[Co2007] M. Collins, *On Jordan's theorem for complex linear groups*, Journal of Group Theory **10** (2007), 411–423.

[DaSaVa2003] G. Davidoff, P. Sarnak, A. Valette, Elementary number theory, group theory, and Ramanujan graphs, London Mathematical Society Student Texts, **55**. Cambridge University Press, Cambridge, 2003.

[deRoVa1993] P. de la Harpe, A. G. Robertson, A. Valette, *On the spectrum of the sum of generators for a finitely generated group*, Israel J. Math. **81** (1993), 65–96.

Bibliography

[De1977] P. Delorme, *1-cohomologie des représentations unitaires des groupes de Lie semisimple et résolubles. Produits tensoriels continus et représentations*, Bull. Soc. Math. France **105** (1977), 289–323.

[Di1901] L. E. Dickson, Linear groups: with an exposition of the Galois field theory, Leipzig: B. G. Teubner (1901).

[Di2010] O. Dinai, *Expansion properties of finite simple groups*, preprint. arXiv:1001.5069

[Do1984] J. Dodziuk, *Difference equations, isoperimetric inequality and transience of certain random walks*, Trans. Amer. Math. Soc. **284** (1984), no. 2, 787–794.

[DvKoLo2012] Z. Dvir, J. Kollár, S. Lovett, *Variety Evasive Sets*, preprint.

[ElObTa2010] J. Ellenberg, R. Oberlin, T. Tao, *The Kakeya set and maximal conjectures for algebraic varieties over finite fields*, Mathematika **56** (2010), no. 1, 1–25.

[ErSz1983] P. Erdos, E. Szemerédi, *On sums and products of integers*, Studies in Pure Mathematics, 213–218, Birkhäuser, Basel, 1983.

[Fo1995] G. Folland, A course in abstract harmonic analysis. Studies in Advanced Mathematics. CRC Press, Boca Raton, FL, 1995.

[Fr1973] G. A. Freiman, Foundations of a structural theory of set addition. Translated from the Russian. Translations of Mathematical Monographs, Vol 37. American Mathematical Society, Providence, R. I., 1973.

[Fr1973b] G. A. Freiman, *Groups and the inverse problems of additive number theory*, Number-theoretic studies in the Markov spectrum and in the structural theory of set addition, pp. 175–183. Kalinin. Gos. Univ., Moscow, 1973.

[FrIw1998] J. Friedlander, H. Iwaniec, *The polynomial $X^2 + Y^4$ captures its primes*, Ann. of Math. (2) **148** (1998), no. 3, 945–1040.

[FrIw2010] J. Friedlander, H. Iwaniec, Opera de cribro. American Mathematical Society Colloquium Publications, 57. American Mathematical Society, Providence, RI, 2010.

[FuHa] W. Fulton, J. Harris, Representation theory. A first course. Graduate Texts in Mathematics, 129. Readings in Mathematics. Springer-Verlag, New York, 1991.

[GaGa1981] O. Gabber, Z. Galil, *Explicit constructions of linear-sized superconcentrators*, Special issued dedicated to Michael Machtey. J. Comput. System Sci. **22** (1981), no. 3, 407–420.

[Ga1951] M. Gaffney, *The harmonic operator for exterior differential forms*, Proc. Nat. Acad. Sci. U. S. A. **37**, (1951). 48–50.

[Ga2002] A. Gamburd, *On the spectral gap for infinite index "congruence" subgroups of $SL_2(Z)$*, Israel J. Math. **127** (2002), 157–200.

[GiHe2010] N. Gill, H. Helfgott, *Growth in solvable subgroups of $GL_r(Z/pZ)$*, preprint. arXiv:1008.5264

[GoLySo1998] D. Gorenstein, R. Lyons, R. Solomon, The classification of the finite simple groups. Number 3. Part I. Chapter A. Almost simple K-groups. Mathematical Surveys and Monographs, 40.3. American Mathematical Society, Providence, RI, 1998.

[Go1998] W. T. Gowers, *A new proof of Szemerédi's theorem for arithmetic progressions of length four*, Geometric And Functional Analysis **8** (1998), 529–551.

[Go2008] W. T. Gowers, *Quasirandom groups*, Combin. Probab. Comput. 17 (2008), no. 3, 363–387.

[GrSo2004] A. Granville, K. Soundararajan, *The number of unsieved integers up to x*, Acta Arith. **115** (2004), no. 4, 305–328.

[GrTa2010] B. Green, T. Tao, *Linear equations in primes*, Annals of Math. **171** (2010), 1753–1850.

[GrTa2012] B. Green, T. Tao, *The Möbius function is strongly orthogonal to nilsequences*, Ann. of Math. (2) 175 (2012), no. 2, 541–566.

[GrTaZi2012] B. Green, T. Tao, T. Ziegler, *An inverse theorem for the Gowers $U^{s+1}[N]$-norm*, Ann. of Math. (2) **176** (2012), no. 2, 1231–1372.

[Gr1994] P. Griffiths, J. Harris, Principles of algebraic geometry. Reprint of the 1978 original. Wiley Classics Library. John Wiley & Sons, Inc., New York, 1994.

[Gu1980] A. Guichardet, *Cohomologie des groups topologiques et des algébres de Lie*, Cedic-F. Nathan, 1980.

[HaRi1974] H. Halberstam, H.-E. Richert, Sieve methods. London Mathematical Society Monographs, No. 4. Academic Press [A subsidiary of Harcourt Brace Jovanovich, Publishers], London-New York, 1974.

[Ha1995] J. Harris, Algebraic geometry. A first course. Corrected reprint of the 1992 original. Graduate Texts in Mathematics, 133. Springer-Verlag, New York, 1995.

[Ha1936] H. Hasse, *Zur Theorie der abstrakten elliptischen Funktionenkörper. I, II & III*, Crelle's Journal **175** (1936).

[He2001] D. R. Heath-Brown, *Primes represented by $x^3 + 2y^3$*, Acta Math. **186** (2001), no. 1, 1–84.

[He2008] H. A. Helfgott, *Growth and generation in $\mathrm{SL}_2(\mathbb{Z}/p\mathbb{Z})$*, Ann. of Math. (2) **167** (2008), no. 2, 601–623.

[He2011] H. A. Helfgott, *Growth in $SL_3(\mathbf{Z}/p\mathbf{Z})$*, J. Eur. Math. Soc. (JEMS) **13** (2011), no. 3, 761–851.

[HeSe2011] H. A. Helfgott, A. Seress, *On the diameter of permutation groups*, preprint. arXiv:1109.3550

[HoLiWi2006] S. Hoory, N. Linial, A. Wigderson, *Expander graphs and their applications*, Bull. Amer. Math. Soc. (N.S.) **43** (2006), no. 4, 439–561.

[Hr2012] E. Hrushovski, *Stable group theory and approximate subgroups*, J. Amer. Math. Soc. **25** (2012), no. 1, 189–243.

[HrWa2008] E. Hrushovski, F. Wagner, Counting and dimensions. Model theory with applications to algebra and analysis. Vol. 2, 161–176, London Math. Soc. Lecture Note Ser., 350, Cambridge Univ. Press, Cambridge, 2008.

[Hu1975] J. E. Humphreys, Linear algebraic groups. Graduate Texts in Mathematics, No. 21. Springer-Verlag, New York-Heidelberg, 1975.

[Iw1978] H. Iwaniec, *Almost-primes represented by quadratic polynomials*, Invent. Math. **47** (1978), no. 2, 171–188.

[IwKo2004] H. Iwaniec, E. Kowalski, Analytic number theory. American Mathematical Society Colloquium Publications, 53. American Mathematical Society, Providence, RI, 2004

[JiMa1987] S. Jimbo, A. Maruoka, *Expanders obtained from affine transformations*, Combinatorica **7** (1987), no. 4, 343–355.

[Jo1878] C. Jordan, *Mémoire sur les équations differentielles linéaires 'a intégrale algébrique*, J. Reine Angew. Math. **84** (1878), 89–215.

[Ka1967] D. Kazhdan, *On the connection of the dual space of a group with the structure of its closed subgroups*, Functional Analysis and Its Applications **1** (1967), 63–65.

[Ke1959] H. Kesten, *Symmetric random walks on groups*, Trans. Amer. Math. Soc. **92** (1959), 336–354.

[Ki2003] H. Kim, *Functoriality for the exterior square of GL_4 and the symmetric fourth of GL_2*. With appendix 1 by Dinakar Ramakrishnan and appendix 2 by Kim and Peter Sarnak, J. Amer. Math. Soc. **16** (2003), no. 1, 139–183.

[Kn2002] A. Knapp, Lie groups beyond an introduction. Second edition. Progress in Mathematics, 140. Birkhäuser Boston, Inc., Boston, MA, 2002.

[KoBa1967] A. N. Kolmogorov, Y.M. Barzdin, *On the realization of nets in 3-dimensional space*, Probl. Cybernet, **8** 261–268, 1967. See also Selected Works of A.N. Kolmogorov, Vol 3, pp 194–202, Kluwer Academic Publishers, 1993.

[Ko2012] E. Kowalski, *Explicit growth and expansion for SL_2*, preprint. arXiv:1201.1139.

[Ko2010] E. Kowalski, *Exponential sums over finite fields, I: elementary methods*, preprint. www.math.ethz.ch/ kowalski/exponential-sums.html

[KuSt1960] R. A. Kunze, E. M. Stein, *Uniformly bounded representations and harmonic analysis of the 2×2 real unimodular group*, Amer. J. Math. **82** (1960), 1–62.

[LaSe1974] V. Landazuri, G. Seitz, *On the minimal degrees of projective representations of the finite Chevalley groups*, J. Algebra **32** (1974), 418–443.

[LaWe1954] S. Lang, A. Weil, *Number of points of varieties in finite fields*, Amer. J. Math. **76**, (1954). 819–827.

[LaPi2011] M. Larsen, R. Pink, *Finite subgroups of algebraic groups*, J. Amer. Math. Soc. **24** (2011), 1105–1158.

[LiRo2011] L. Li, O. Roche-Newton, *An improved sum-product estimate over finite fields*, preprint. arXiv:1106.1148

[Lu2012] A. Lubotzky, *Expander graphs in pure and applied mathematics*, Bull. Amer. Math. Soc. (N.S.) **49** (2012), no. 1, 113–162.

[LuPhSa1988] A. Lubotzky, R. Phillips, P. Sarnak, *Ramanujan graphs*, Combinatorica **8** (1988), no. 3, 261–277.

[LuWe1993] A. Lubotzky, B. Weiss, *Groups and expanders*, Expanding graphs (Princeton, NJ, 1992), 95–109, DIMACS Ser. Discrete Math. Theoret. Comput. Sci., 10, Amer. Math. Soc., Providence, RI, 1993.

[Ma1973] G. A. Margulis, *Explicit constructions of expanders*, Problemy Peredaci Informacii **9** (1973), no. 4, 71–80.

[Ma1993] Y. Matiyasevich, Hilbert's 10th Problem. MIT Press Series in the Foundations of Computing. Foreword by Martin Davis and Hilary Putnam. Cambridge, MA: MIT Press, 1993.

[MaVeWe1984] C. R. Matthews, L. N. Vaserstein, B. Weisfeiler, *Congruence properties of Zariski-dense subgroups. I*, Proc. London Math. Soc. (3) **48** (1984), no. 3, 514–532.

[Ma1957] F. I. Mautner, *Geodesic flows on symmetric Riemann spaces*, Ann. of Math. (2) **65** (1957), 416–431.

[MeWi2004] R. Meshulam, A. Wigderson, *Expanders in group algebras*, Combinatorica **24** (2004), 659–680.

[Mo1966] C. C. Moore, *Ergodicity of flows on homogeneous space*, Amer. J. of Math. **88** (1966), 154–178.

[Mu1999] D. Mumford, The red book of varieties and schemes. Second, expanded edition. Includes the Michigan lectures (1974) on curves and their Jacobians. With contributions by Enrico Arbarello. Lecture Notes in Mathematics, 1358. Springer-Verlag, Berlin, 1999.

[Ne2006] M. Neuhauser, *Kazhdan constants for compact groups*, J. Aust. Math. Soc. **81** (2006), no. 1, 11–14.

[NiPy2011] N. Nikolov, L. Pyber, *Product decompositions of quasirandom groups and a Jordan type theorem*, J. Eur. Math. Soc. (JEMS) **13** (2011), no. 4, 1063–1077.

[No1987] M. Nori, *On subgroups of $GL_n(F_p)$*, Invent. Math. **88** (1987), no. 2, 257–275.

[Pi1973] M. S. Pinsker, *On the complexity of a concentrator*, 7th International Teletraffic Conference, pages 318/1-318/4, 1973.

[PySz2010] L. Pyber, E. Szabó, *Growth in finite simple groups of Lie type*, preprint, arXiv:1001.4556

[PySz2012] L. Pyber, E. Szabó, *Growth in linear groups*, preprint, arXiv:1208.2538

[Ra1930] F. P. Ramsey, *On a problem of formal logic*, Proc. London Math. Soc. Series 2 **30** (1930), 264–286.

[ReSi1975] M. Reed, B. Simon, Methods of modern mathematical physics. II. Fourier analysis, self-adjointness. Academic Press [Harcourt Brace Jovanovich, Publishers], New York-London, 1975.

[Ro1960] W. Roelcke, *Über den Laplace-Operator auf Riemannschen Mannigfaltigkeiten mit diskontinuierlichen Gruppen*, Math. Nachr. **21** (1960) 131–149.

[Ru1990] W. Rudin, Fourier analysis on groups. Reprint of the 1962 original. Wiley Classics Library. A Wiley-Interscience Publication. John Wiley & Sons, Inc., New York, 1990.

[Ru2011] M. Rudnev, *On new sum-product type estimates*, preprint. 1111.4977

[Ru2012] M. Rudnev, *An improved sum-product inequality in fields of prime order*, Int. Math. Res. Not. IMRN 2012, no. 16, 3693–3705.

[SGVa2012] A. Salehi-Golsefidy, P. Varjú, *Expansion in perfect groups*, Geom. Funct. Anal. **22** (2012), no. 6, 1832–1891.

[Sa2013] T. Sanders, *The structure theory of set addition revisited*, Bull. Amer. Math. Soc. **50** (2013), 93–127.

[Sa2004] P. Sarnak, *What is... an expander?*, Notices Amer. Math. Soc. **51** (2004), no. 7, 762–763.

[SaXu1991] P. Sarnak, X. Xue, *Bounds for multiplicities of automorphic representations*, Duke Math. J. **64** (1991), no. 1, 207–227.

[Sc1980] J. Schwartz, *Fast probabilistic algorithms for verification of polynomial identities*, Journal of the ACM **27** (1980), 701–717.

[Se1965] A. Selberg, *On the estimation of Fourier coefficients of modular forms*, in Whiteman, Albert Leon, Theory of Numbers, Proceedings of Symposia in Pure Mathematics, VIII, Providence, R.I.: American Mathematical Society, pp. 1–15, 1965.

[Se2007] M. Sepanski, Compact Lie groups. Graduate Texts in Mathematics, 235. Springer, New York, 2007.

[Se1956] J.-P. Serre, *Géométrie algébrique et géométrie analytique*, Université de Grenoble. Annales de l'Institut Fourier **6** (1956), 1–42.

[Se1966] J.-P. Serre, *Algébres de Lie semi-simples complexes*, W. A. Benjamin, Inc., New York-Amsterdam 1966.

[Sh1997] I. R. Shafarevich, Basic Algebraic Geometry. Translated from the Russian by K. A. Hirsch. Revised printing of Grundlehren der mathematischen Wissenschaften, Vol. 213, 1974. Springer Study Edition. Springer-Verlag, Berlin-New York, 1977. xv+439 pp.

[Sh2000] Y. Shalom, *Rigidity of commensurators and irreducible lattices*, Invent. Math. **141** (2000), no. 1, 1–54.

[So2009] J. Solymosi, *Bounding multiplicative energy by the sumset*, Adv. Math. **222** (2009), no. 2, 402–408.

[St1967] R. Steinberg, Lectures on Chevalley groups, 1967. Prepared by John Faulkner and Robert Wilson. Available at www.math.ucla.edu/~rst/YaleNotes.pdf.

[St1932] M. H. Stone, *On one-parameter unitary groups in Hilbert Space*, Annals of Mathematics **33** (3), 643–648.

[Sz1975] E. Szemerédi, *On sets of integers containing no k elements in arithmetic progression*, Acta Arithmetica **27** (1975), 199–245.

[Sz1978] E. Szemerédi, *Regular partitions of graphs*, Problèmes combinatoires et théorie des graphes (Colloq. Internat. CNRS, Univ. Orsay, Orsay, 1976), pp. 399–401, Colloq. Internat. CNRS, 260, CNRS, Paris, 1978.

[Ta2008] T. Tao, Structure and Randomness: pages from year one of a mathematical blog, American Mathematical Society, 2008.

[Ta2009] T. Tao, Poincaré's Legacies: pages from year two of a mathematical blog, Vol. I, American Mathematical Society, Providence RI, 2009.

[Ta2009b] T. Tao, Poincaré's Legacies: pages from year two of a mathematical blog, Vol. II, American Mathematical Society, Providence RI, 2009.

[Ta2009c] T. Tao, *The sum-product phenomenon in arbitrary rings*, Contrib. Discrete Math. **4** (2009), no. 2, 59–82.

[Ta2010] T. Tao, An epsilon of room, Vol I., Graduate Studies in Mathematics, 117, American Mathematical Society, 2010.

[Ta2010b] T. Tao, An epsilon of room, Vol II., American Mathematical Society, 2010.

[Ta2011] T. Tao, An introduction to measure theory, American Mathematical Society, Providence RI, 2011.

[Ta2012] T. Tao, Higher order Fourier analysis, American Mathematical Society, Providence RI, 2012.

[Ta2013] T. Tao, Hilbert's fifth problem and related topics, Graduate Studies in Mathematics, 153. American Mathematical Society, Providence, RI, 2014.

[Ta2013b] T. Tao, Compactness and contradiction, American Mathematical Society, Providence, RI, 2013.

[Ta2013c] T. Tao, Spending symmetry, in preparation.

[TaVu2006] T. Tao, V. Vu, Additive combinatorics. Cambridge Studies in Advanced Mathematics, 105. Cambridge University Press, Cambridge, 2006.

[Te1985] A. Terras, Harmonic analysis on symmetric spaces and applications. I. Springer-Verlag, New York, 1985.

[Ti1966] J. Tits, *Sur les constantes de structure et le théorème d'existence des algèbres de Lie semi-simples*, Inst. Hautes Études Sci. Publ. Math. No. **31** (1966) 21–58.

[Ti1972] J. Tits, *Free subgroups in linear groups*, Journal of Algebra, **20** (1972), 250–270.

[Va2012] P. Varjú, *Expansion in $SL_d(\mathcal{O}_K/I)$, I squarefree*, J. Eur. Math. Soc. (JEMS) **14** (2012), no. 1, 273–305.

[Wa1991] S. Wassermann, *C^*-algebras associated with groups with Kazhdan's property T*, Ann. Math. **134** (1991), 423–431.

[We1949] A. Weil, *Numbers of solutions of equations in finite fields*, Bull. Amer. Math. Soc. **55** (1949), 497–508.

[Wo1999] N. C. Wormald, *Models of random regular graphs*, Surveys in combinatorics, 1999 (Canterbury), 239–298, London Math. Soc. Lecture Note Ser., 267, Cambridge Univ. Press, Cambridge, 1999.

[Zi1979] R. Zippel, *Probabilistic algorithms for sparse polynomials*, Proceedings of Symbolic and Algebraic Computation, EUROSAM '79, An International Symposium on Symbolic and Algebraic Computation, Marseille, France, June 1979.

Index

(B, N)-pair, 283

abelianisation, 230
adjacency matrix, 5
adjoint, 208
adjoint form, 271
adjoint representation, 231
affine algebraic variety, 187
almost prime, 144
approximate group, 86
asymptotic notation, xiii
autocorrelation function, 50

Balog-Szemerédi theorem, 91
Balog-Szemerédi-Gowers lemma
    approximate group form, 93
    product set form, 92
Bertini's theorem, 198
beta sieve, 154
bipartite graph, 88
Bochner's theorem, 49
Bonferroni inequalities, 19
Borel subgroup, 111, 281
bounded functional calculus, 214
Bourgain-Gamburd expansion machine, 87
Bruhat decomposition, 283
Brun's theorem, 155

Cartan decomposition, 247
Cartan matrix, 261
Cartan semisimplicity criterion, 232
Cartan solvability criterion, 240
Cartan subalgebra, 242

Cartan's theorem, 270
Casimir operator, 264
Cayley graph, 23, 167
Cayley transform, 69
centre of a Lie algebra, 231
characteristic subalgebra, 239
Cheeger constant, 10
Chevalley basis, 263
Chevalley group, 281
Chevalley normalisation, 262
chromatic number, 14
class equation, 117
classical Lie algebra, 260
closure of an operator, 207
coboundary, 183
cochain, 177
cocycle, 35, 179
combinatorial sieve, 150
compact form, 272
complemented subspace, 241
complete graph, 4, 8
complexity of an algebraic variety, 187
composition factors, 157
composition series, 157
concentration of measure, 15
concrete Lie algebra, 228
conjugacy class, 114
conjugacy of Cartans, 244
Connectivity of an expander graph, 14
converse to Balog-Szemerédi-Gowers, 95
convolution, 62
coroot, 249
coweight lattice, 276

Coxeter diagram, 253
cusp, 70
cyclic vector, 217

depolarisation, 214
derivation, 229
derived algebra, 230
derived series, 230
diagram, 169
Diameter of an expander graph, 14
dimension, 126
dimension of an algebraic variety, 187
direct product, 229
direct sum, 229
direct sum of representations, 29
discrete Cheeger inequality, 13
discrete Cheeger inequality, weak, 10
Dynkin diagram, 255

edge expansion, 9
Engel's theorem, 236
escape from subspaces, 121
essentially self-adjoint operator, 209
exceptional Lie algebra, 260
expander family, 6
expander mixing lemma, 13
extended Dynkin diagram, 257

flattening lemma, 83, 98
Fokker-Planck equation, 81
Frobenius endomorphism, 190
Frobenius lemma, 59
functional calculus, 203
fundamental domain, 70

generalised eigenspace, 233
girth, 14
Goursat's lemma, 159
graph, 4
graph Laplacian, 6
graph metric, 14

Hardy's inequality, 73
Hasse-Weil bound, 190
heat flow, 218
Hellinger-Toeplitz theorem, 209
Herglotz representation theorem, 213
horocycle flow, 54
hyperbolic plane, 68

indecomposable representation, 241
independent set, 14
indicator function, xiii

induced representation, 39
invariant subspace, 217
invariant vector, 30
involved torus, 124
irreducible representaiton, 57
irreducible representation, 241
irreducible root system, 250
isogeny, 286

Jacobi identity, 227
Jacobi operator, 212
Jordan's theorem, 67
Jordan-Chevalley decomposition, 234
Jordan-Holder theorem, 157

Katz-Tao lemma, 107
Kazhdan constant, 30
Killing form, 231

Landau conjectures, 143, 155
Lang-Weil bound, 189
Lang-Weil with parameters, 192
large sieve, 156
Larsen-Pink inequality, 113, 115, 122
Lefschetz principle, 140
Legendre identity, 148
Legendre sieve, 148
Leibniz rule, 229
Levi decomposition, 232
Lie algebra, 227
Lie algebra ideal, 229
Lie's second theorem, 269
Lie's theorem, 237
Lie's third theorem, 269
linear, 45
Littlewood-Paley projection, 223
lower central series, 230
Lubotzky's 1-2-3 problem, 136

Möbius function, 148
Mautner phenomenon, 53
maximal torus, 111, 281
Mersenne prime, 143
mixing inequality, 63
modular curve, 193
Moore ergodic theorem, 54
multiplicative energy, 92

nilpotent Lie algebra, 230
nilpotent operator, 234
nonconcentration estimate, 87
normaliser, 242
notation, xii

Index

one-sided expander, 6

parabolic subgroup, 284
perfect group, 59
perfect Lie algebra, 230
pivot argument, 106
Poincaré disk, 69
Poincaré half-plane, 68
Poincaré inequality, 9
polar decomposition, 272
polarisation identity, 51
polycyclic Lie algebra, 231
primorial, 148
principal congruence subgroup, 74
principal modular curve, 70
principal series representation, 60
product theorem, 87, 101, 120

quasirandom group, 57
quasiregular representation, 29
quasisimple group, 271

radical of a Lie algebra, 232
Ramanujan graph, 7
rank of a Lie algebra, 244
Rayleigh quotient, 72
Ree groups, 290
regular element, 277
regular element of a Lie algebra, 242
regular graph, 4
regular representation, 29
regular semisimple element, 111
regular unipotent element, 111
representation (of a Lie algebra), 228
resolvent, 210
Resolvent identity, 210
root system, 250
root vector, 248
Ruzsa covering lemma, 95, 107
Ruzsa triangle inequality, 93, 107

Schrödinger propagator, 220
Schreier graph, 24
Schwarz-Zippel type bound, 188
Selberg sieve, 156
Selberg's 3/16 theorem, 75
Selberg's conjecture, 74
Selberg's expander construction, 68
self-adjoint operator, 209
semidirect product, 229
semisimple Jordan decomposition, 241
semisimple Lie algebra, 232
semisimple operator, 234

short exact sequence, 228
simple Lie algebra, 231
simple Lie group, 271
solvable Lie algebra, 230
spectral gap, 71
split extension, 229
Steinberg group, 290
Steinberg representation, 60
Stone's theorem, 220
strong approximation property, 159
subrepresentation, 30
sum-product theorem, 105
sumset estimates, 107
Suzuki groups, 290

Tits alternative, 138
transitivity of induction, 40
trivial representation, 28
twin prime, 143
twisted group of Lie type, 291
two-sided expander, 6

unipotent group, 111
unitary representation, 28
universal cover, 269

variety, 126
virtually quasirandom group, 66

wedding cake decomposition, 12
weight vector, 235
weighted Balog-Szemerédi-Gowers lemma, 86
weighted Balog-Szemerédi-Gowers theorem, 91
Weyl alcove, 277
Weyl chamber, 258
Weyl group, 257
Weyl's complete reducibility theorem, 232, 263
Whitehead's lemma, 264